CLASSIC TRAINS

AND

RAILROAD ENGINEERING

VOLUME 2

ANTIQUE LOCOMOTIVE OPERATIONS PART 2

Edited by Mark Bussler

Classic Trains and Railroad Engineering: Volume 2
Antique Locomotive Operations Part 2

Restored and Edited by Mark Bussler
Cover Design by Mark Bussler

Volume 2 of 8

www.CGRpublishing.com

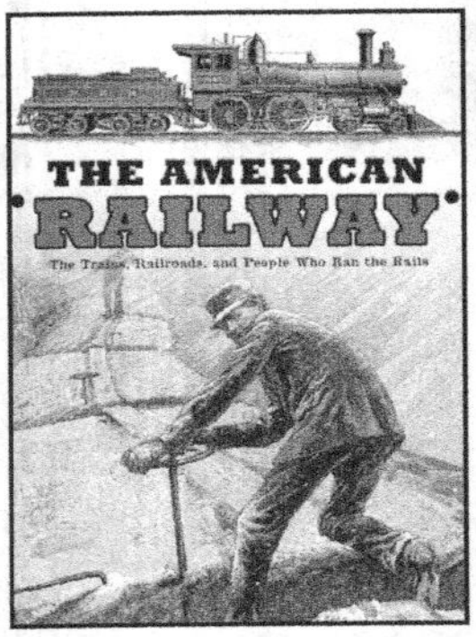

The American Railway: The Trains, Railroads, and People Who Ran the Rails

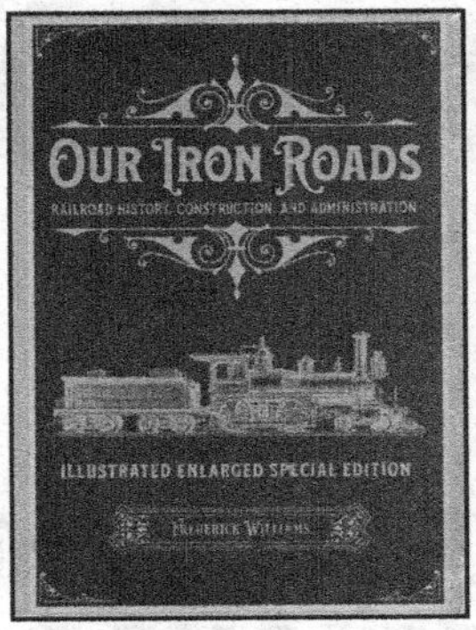

Our Iron Roads: Railroad History, Construction, and Administration

Sinking of the Titanic: The Greatest Disaster at Sea

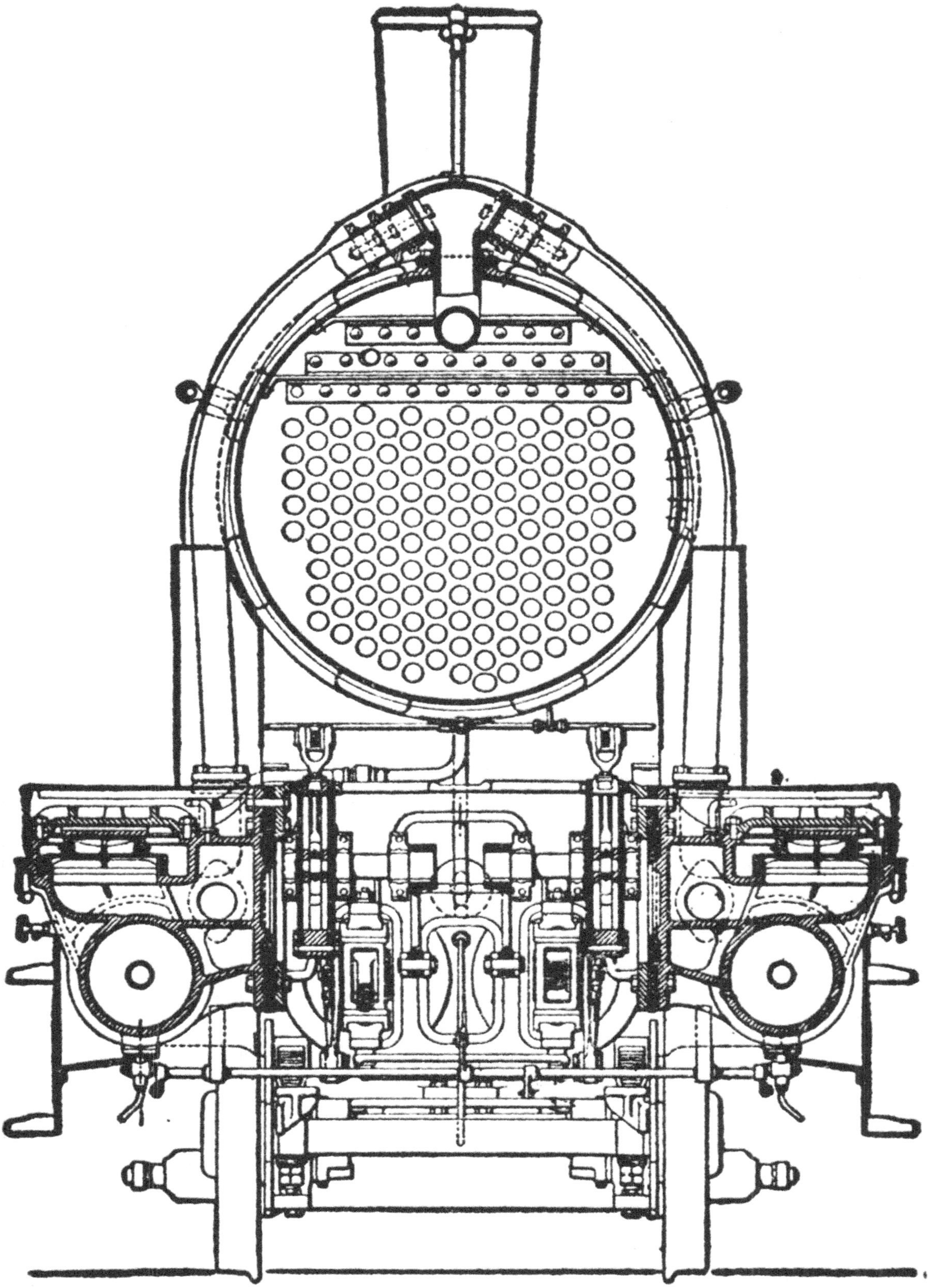

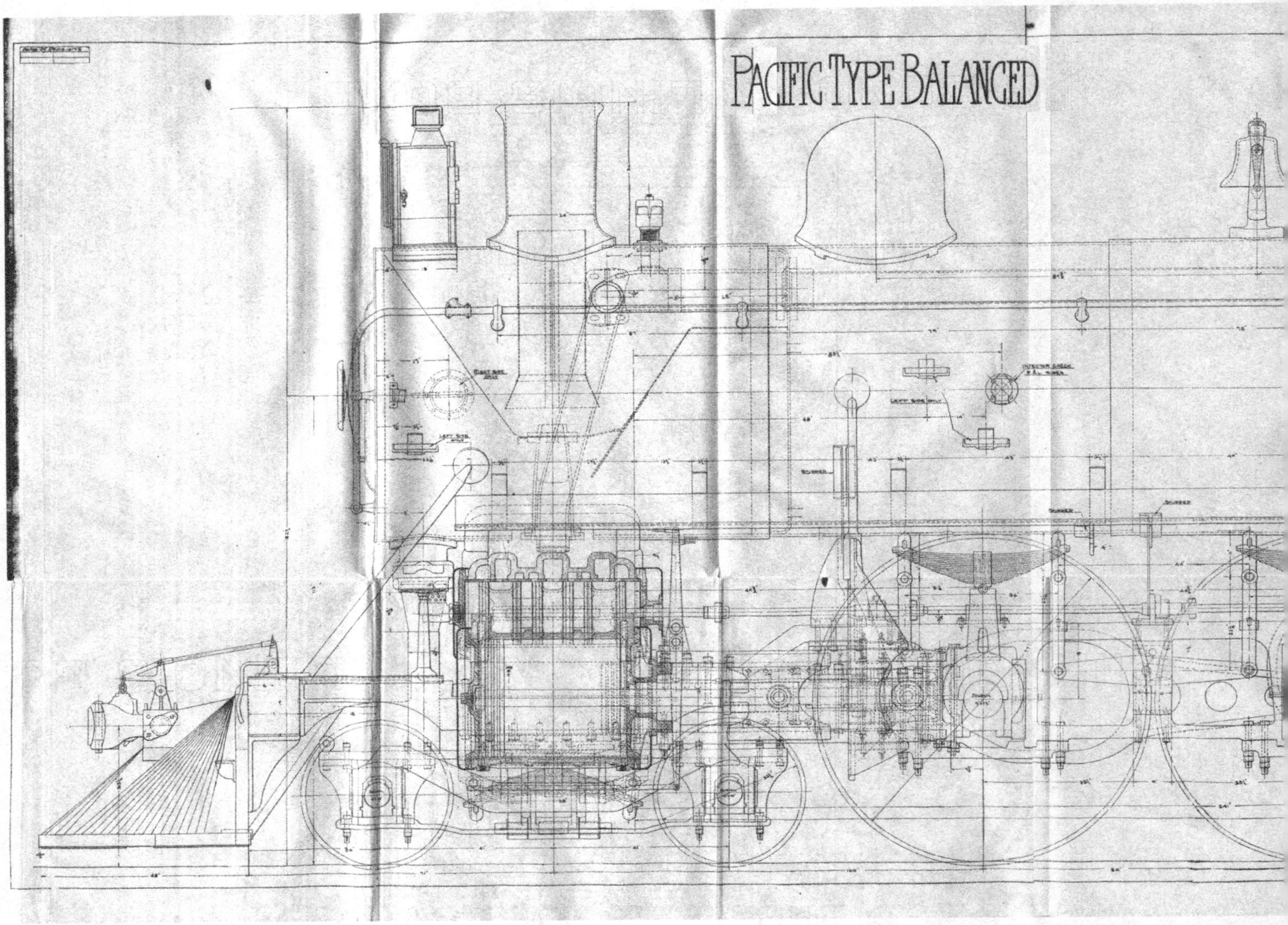
PACIFIC TYPE BALANCED

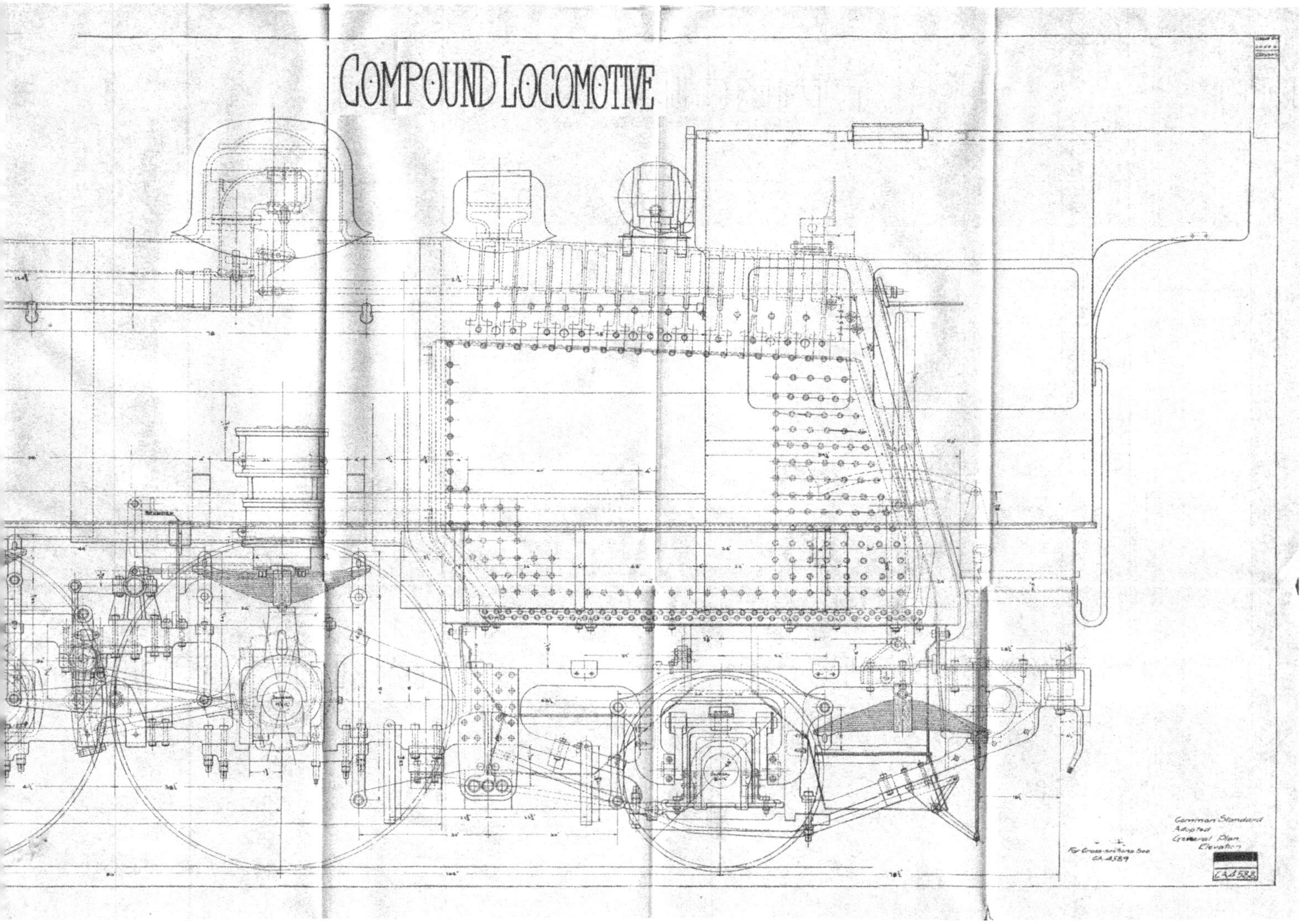
COMPOUND LOCOMOTIVE
For Cross-sections See
CA 4589
Common Standard
Adopted
General Plan
Elevation
CA 4588

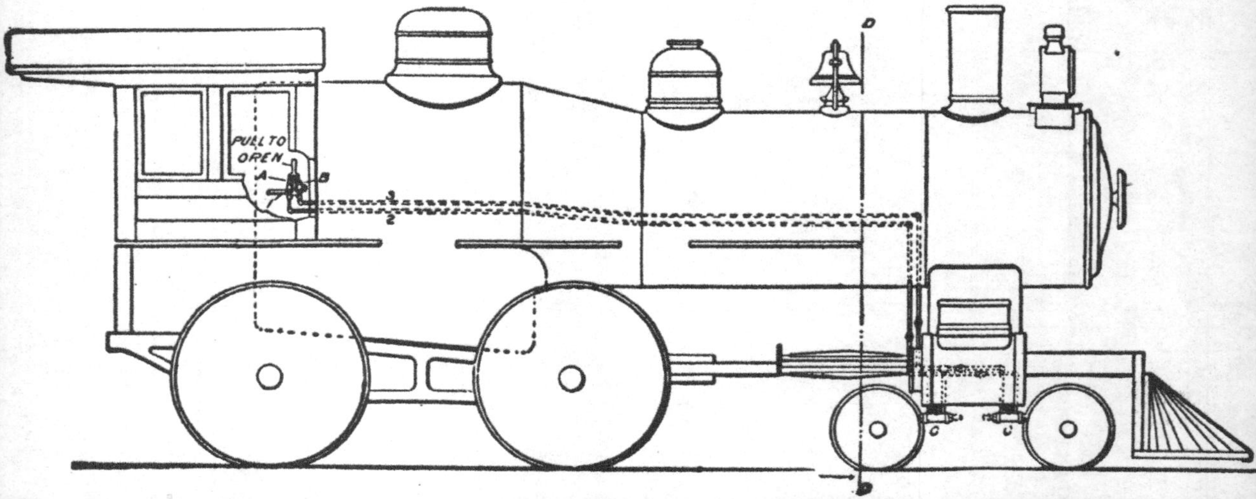
PULL TO
OPEN
A
B
1
2
D
C
C
D

TABLE OF CONTENTS

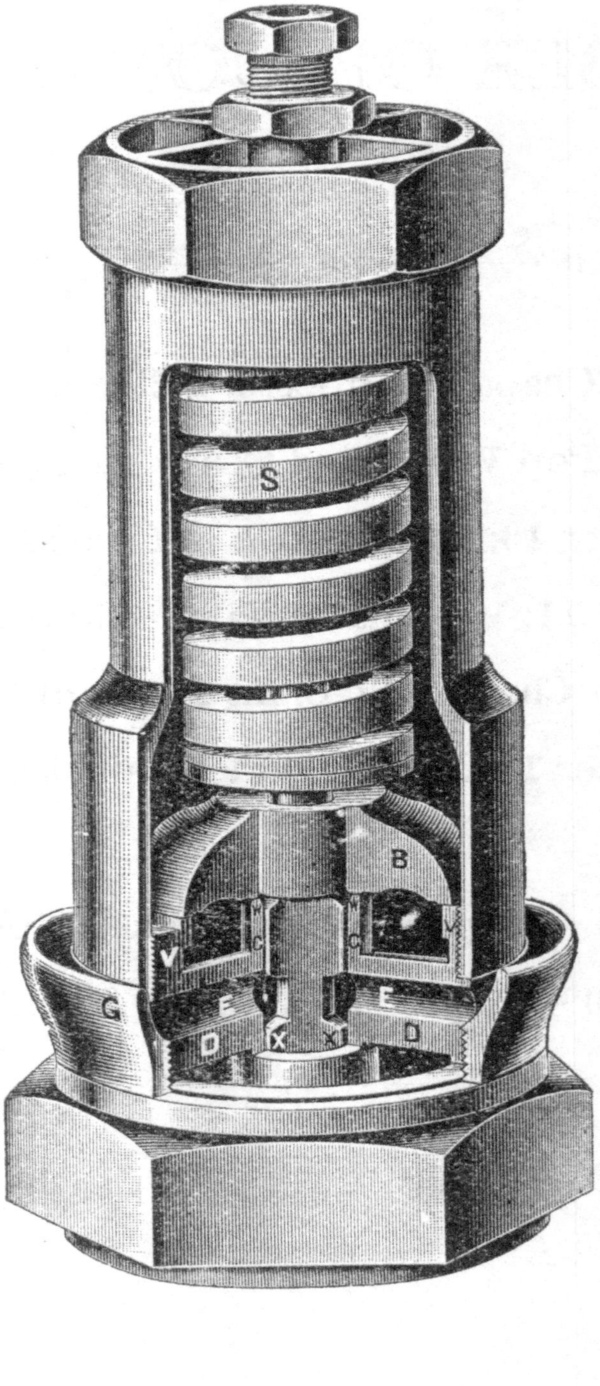
S
B
V
G
E
E
D
X
X
D

CHAPTER IX

INJECTORS, STEAM GAUGES, POP VALVES AND OTHER FITTINGS

Injectors. The proper method of feeding water to a boiler while in operation under a high pressure, is a problem that demands the constant and earnest attention of the engineer, not only as a matter of personal safety, but the efficiency of the boiler depends in a large measure upon the manner in which the feed water enters the boiler. Theoretically the supply should just equal the demand at all times; that is to say, there should be a constant ingoing of water into the boiler during all the time that the fire is active, and the volume of water entering the boiler should exactly equal the volume of water that is being evaporated within the boiler. But these conditions are hardly possible in practice. Especially is this so in locomotive practice, where the service differs so greatly from marine or stationary service. The judicious use of the injector on a locomotive is a subject that engineers and firemen should study to familiarize themselves with.

The importance of this matter is shown in the following extract from the report of a committee of the Traveling Engineers' Association: "It would hardly cut any figure how careful an engineer might be in the handling of his train, with the skill he uses in regulating speed or in the adjustment of the throttle and the reverse lever, if the water was not put into the boiler at the right time and the right place. In our experience we have known almost remarkable results to be brought about in an engine's fuel performance by explaining this matter to engineers who perhaps had not given it the thought that the subject deserves." It will be noticed that the committee emphasizes the importance of "putting the water into the boiler at the right time and the right place," if economy in fuel is to be attained, and this certainly is a worthy object for every locomotive engineer to have in view at all times.

Now as to the "right time" for putting water into a

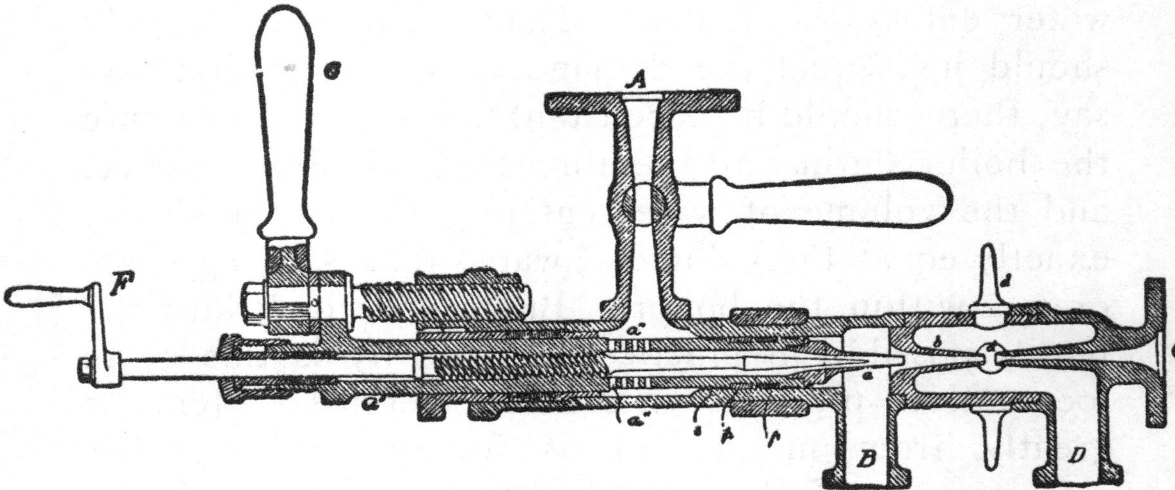

Figure 155

Original Form of the Giffard Injector

locomotive boiler: A good time to use the injector to advantage is while standing at a station. To this end, it is the practice of some engineers when approaching a stopping point, to allow the water level to drop below the normal point, thus utilizing the heat already stored in the water that is in the boiler to enable them to get into the station. When the throttle is closed for making the stop the injector may be started, and much of the heat that would otherwise be wasted at the pop valve will be utilized in forcing a new supply of water into the boiler. Another "right" time and place to use the injector is just after passing the summit of a long hill, when the throttle can be eased off. This will prevent the pop from rising so freely on the down grade, and thus another source of economy will be taken advantage of. There are many other "right" times and places for using the injector, that an observant and careful engineer will, by a little thinking, be enabled to figure out for himself. Much, of course, depends upon the kind of an injector a man has on his engine. If it has a wide range of capacities, and can be throttled so as to feed a very small jet without breaking, it may be used almost continuously, especially if the track is straight, and there are not very many heavy grades. The modern injector, as it is furnished at the present time by the leading manufacturers, approaches very nearly to being a perfect boiler feeder for locomotives Ever since the time of the invention of the injector in 1858 by that eminent French engineer Henri Giffard, and its introduction into this country in 1860 by Wm. Sellers & Co., of Philadelphia, it has been constantly improved upon, and developed by various inventors and manufacturers, and it is to-day, without doubt, the most simple, the most economical, and the best device for feeding water into locomotive boilers. As a short study of the philosophy of the action of the injector is not only useful, but should be interesting to engineers and firemen, a space will be devoted to this subject. The leading types of injectors and inspirators will also be described and illustrated.

How an Injector Works.* How can an injector lift

*Stricland L. Kneass, C. E., from Sellers' Hand-Book of Injectors.

and force large volumes of water into the boiler, against the same or even higher pressure than that of the steam?"

"An injector works because the steam imparts sufficient velocity to the water to overcome the pressure of the boiler."

This is a statement of fact; to explain the action, we will take up the important parts of the question separately.

Why should an injector work? Let us assume that the boiler pressure is 180 pounds—that is to say, every square inch of the sheets, top and bottom, receives an

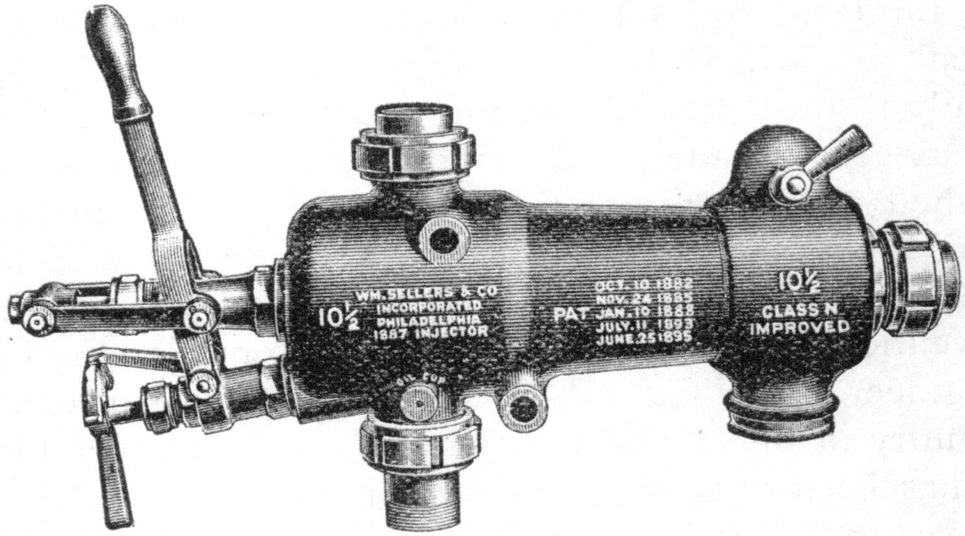

FIGURE 156

THE SELF-ACTING INJECTOR, CLASS N IMPROVED P. R. R. STANDARD

internal pressure of 180 pounds. If the thermometer is placed inside, it is found that both the water and the steam are at the same temperature, 379 deg. But the steam contains more heat than the water, because after water is heated, more coal must be burned to break up the drops of water to change them into steam; this heat is stored in the steam and represents work done by the burning of the coal. Steam not only exerts a pressure of 180 pounds per square inch, but also can expand eight to twenty-six times its original volume, depending upon whether it exhausts into the air or into a partial vacuum; water under the same pressure would be discharged in a solid jet and without expansion. Either steam or water can be used in the cylinder of an engine or to drive the vanes of a steam or water turbine, but one pound of steam is capable of much more work than one pound-weight of water, on account of the heat which has been used to change it into steam. This is easily seen by comparing the velocities of discharge from a steam nozzle and a water nozzle under 180 pounds pressure; steam would expand while issuing, reaching at the end of the nozzle a velocity of about 3600 feet per second, while the water, having no expansion, would have a velocity of only 164 feet per second, about $\frac{1}{22}$ of that of the steam. The same weight of steam discharging per second would therefore have vastly more power for doing work than the water jet.

If a steam or water jet comes in contact with a body in front of it, the tendency is to drive the body forward. The force which tends to move the body is called "momentum," and is equal to the weight of water or steam discharged by the jet in one second, multiplied by its velocity per second. If 1 pound of both the water and the steam are discharged per second, the "momentum" of the steam jet is 3600; because 1 multiplied by 3600 = 3600; the momentum of the water jet is 164. If the water jet discharged about twenty-two pounds per second, its momentum would be the same as that of the steam, because 22 multiplied by 164 is nearly 3600. The two jets are discharged under the same pressure, but the steam has twenty-two times as much "momentum" or force as the water jet; it could, therefore, easily enter a boiler at 180 pounds pressure if we could reduce it to the size of the hole of the water nozzle.

How ought an injector to work? Here a practical difficulty is reached. A steam jet 6 in. from the nozzle is much larger than at the opening, and it would appear almost impossible to make it enter a smaller tube. Even at the narrowest part of the nozzle it is more than sixteen times larger in diameter than a

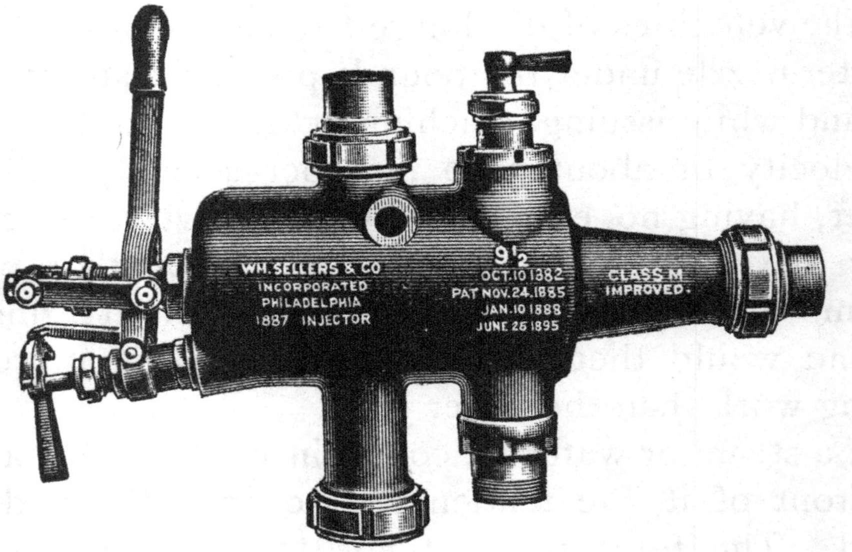

Figure 157

The Self-Acting Injector, Class M Improved

water jet discharging the same weight per second; therefore, if the steam is changed to water without reducing its velocity, it would pass through a hole one-sixteenth the diameter of the "steam nozzle" at a velocity of 3600 feet per second. The simplest and best way to reduce its size is to condense it, and to use water for this purpose, especially as water is needed in the boiler. To condense the steam and utilize its velocity, the water must be brought into close contact with it, without interfering with the direct line of discharge; a funnel or "combining tube" suitably placed will compel water to enter evenly all around the steam jet. The mouth of this funnel must not be too large, or too much water will enter and swamp the jet; if too small, insufficient water will enter to condense the steam. The effect of condensing the steam is to reduce the diameter of the jet; therefore the funnel or combining tube must be a smooth, converging taper, to lead the combined jet of water and condensed steam into the smaller hole of the delivery tube. The effect of the impact of the steam is to give the water its momentum, so that a solid stream shall issue from the lower end of the tube. Each little drop of entering water is driven ahead faster and faster by the vast number of little atoms of steam moving hundreds of times as rapidly, until the steam and water thoroughly combine into one swiftly-moving jet of water and condensed steam, which contracts sufficiently in diameter to enter the smaller delivery tube.

Why does the jet enter the boiler? The combined jet now passes from the end of the combining tube into the delivery tube; why does it enter the boiler?

If a pipe shaped like a fire-hose nozzle or a "delivery tube" is connected to a tank or boiler carrying 180 pounds, the water will issue in a solid jet with a velocity of about 164 feet per second; or, if we could force water into the tube at a speed of 164 feet per second at the same part of the tube, this water would enter and fill up the boiler or tank against 180 pounds pressure. Therefore to enter the boiler the combined jet of water and steam issuing from the combining tube must have a velocity of at least 164 feet per second.

Now, what is the velocity of the combined jet at the lower end of the combining tube? If the steam nozzle discharges one pound per second at 3600 feet velocity, the momentum of the steam is 1 multiplied by 3600, or 3600. If the vacuum caused by the condensation of the steam lifts and draws into the combining tube ten pounds of water per second at a velocity of forty feet, its momentum is 400; and that of the combined jet is 3600 added to 400, or 4000. The weight of the combined jet is eleven pounds, and at the time of entering the delivery tube its velocity ought to be equal to 4000 divided by 11, or 366 feet per second; but as the water and the steam do not meet in precisely the line of discharge there is a loss of momentum, and the velocity in the delivery tube is only 198 feet per second. But the jet only needs a velocity of 164 feet to enter the boiler or tank carrying 180 pounds pressure, therefore the actual jet in the delivery tube is able to overcome a pressure of 206 pounds per square inch, or twenty-six pounds above that of the steam, because the velocity of a jet of water under a head or pressure of 206 pounds would be 198 feet per second. This excess is more than sufficient to overcome the friction of the delivery piping and the resistance of the main check valve. Therefore:

"The action of the injector is due to the high velocity with which a jet of steam strikes the water entering the combining tube, imparting to it its momentum and forming with it during condensation a continuous jet of smaller diameter, having sufficient velocity to overcome the pressure of the boiler."

The Sellers Improved Self-acting Injector. *Description.* This injector is simply constructed and contains few operating parts. The lever is used in starting only, and the water valve for regulation of the delivery. It is self-adjusting, with fixed nozzle, and restarts automatically. All the valve seats that may need refacing can be removed; the body is not subject to wear and will last a lifetime.

The action is as follows: Steam from the boiler is admitted to the lifting nozzle by drawing the starting lever (33) about one inch, without withdrawing the plug on the end of the spindle (7) from the central part of the steam nozzle (3). Steam then passes through the small diagonal-drilled holes and discharges by the outside nozzle, through the upper part of the

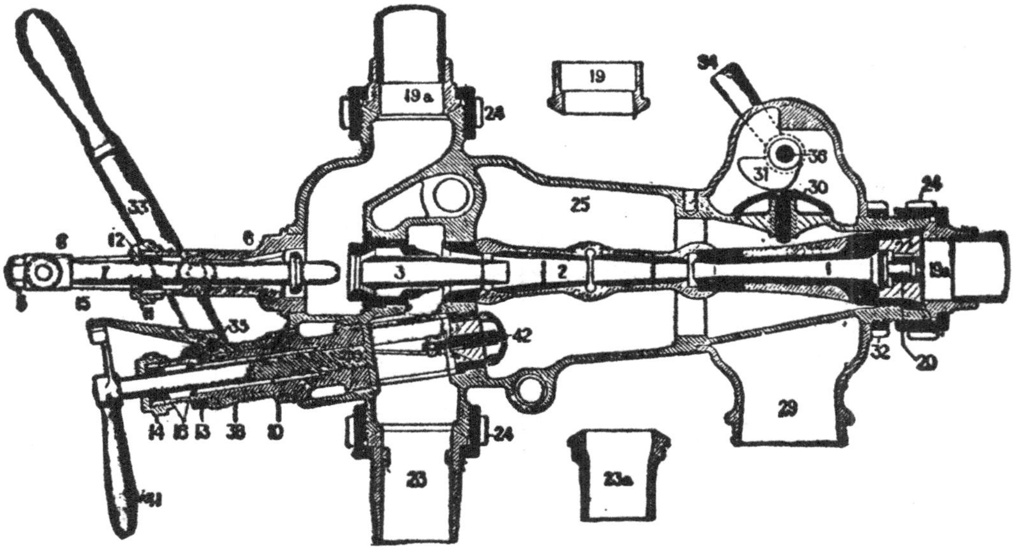

FIGURE 158

THE SELF-ACTING INJECTOR, CLASS N IMPROVED

P. R. R. STANDARD SELLERS STANDARD FORM

combining tube (2) and into the overflow chamber, lifts the overflow valve (30), and issues from the waste pipe (29). When water is lifted the starting lever (33) is drawn back, opening the forcing steam nozzle (3), and the full supply of steam discharges into the combining tube, forcing the water through the delivery tube into the boiler pipe.

At high steam pressure there is a tendency in all injectors having an overflow to produce a vacuum in the chamber (25). In the Improved Self-Acting Injector this is utilized to draw an additional supply of water into the combining tube by opening the inlet valve (42); the water is forced by the jet into the boiler, increasing the capacity about 20 per cent.

The water-regulating valve (40) is used only to adjust the capacity to suit the needs of the boiler. The range is unusually large.

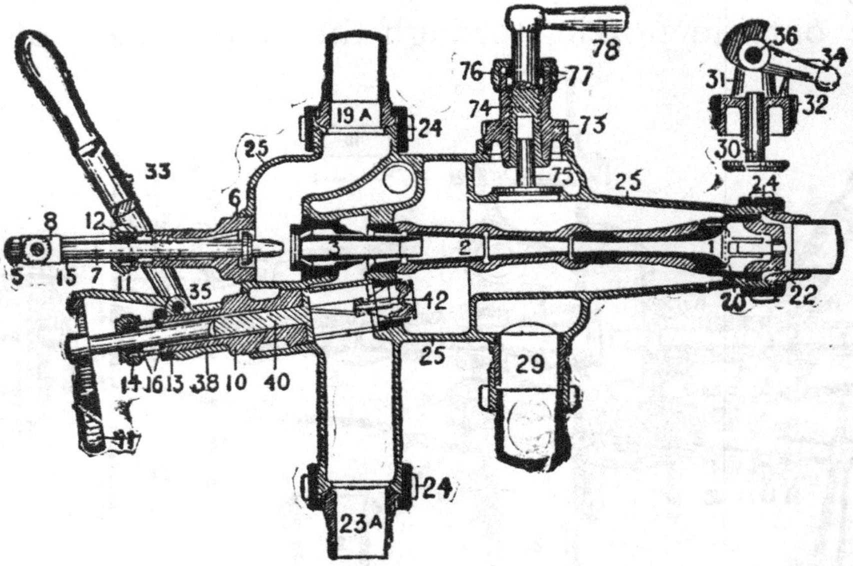

Figure 159

Self-Acting Injector, Class M Improved

SPECIAL FORM, INTERCHANGEABLE WITH MONITOR, OHIO, ETC.

The cam lever (34) is turned toward the steam pipe to prevent the opening of the overflow valve when it is desired to use the injector as a heater or to clean the strainer. The joint between the body (25) and the waste-pipe (29) is not subject to other pressure than that due to the discharging steam and water during starting; the metal faces should be kept clean and the retaining nut (32) screwed up tight.

To tighten up the gland of the steam spindle, push in the starting lever (33) to end of stroke, remove the little nut (5) and draw back the lever (33). This frees the crosshead (8) and links (15), which can be swung out of the way, and the follower (12) tightened on the packing to make the gland steam-tight.

The Improved Self-Acting Injector is specially adapted to railroad service, as its efficient, positive action and wide range of capacities at 200 pounds steam render its application to high-pressure locomotive boilers very advantageous. It will work from the

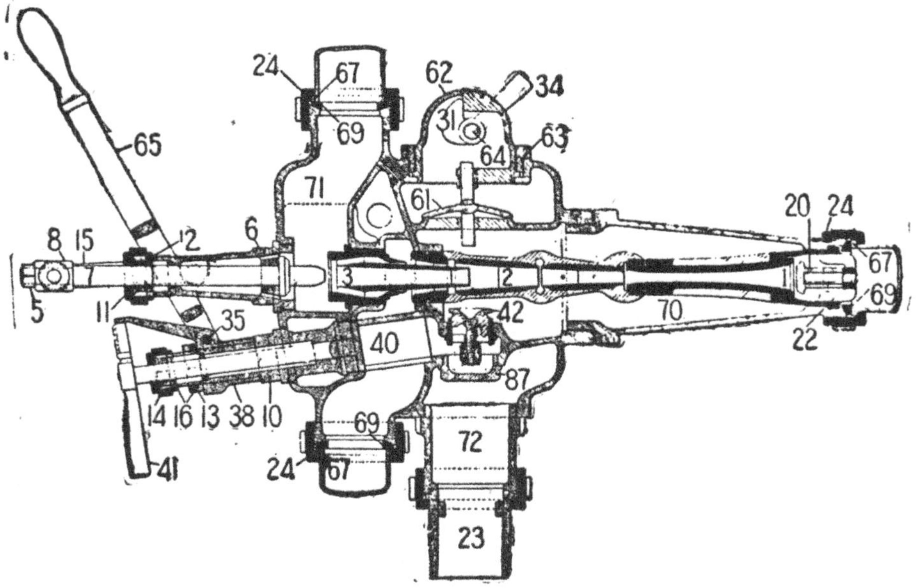

FIGURE 160

SELF-ACTING INJECTOR, CLASS P, SPECIAL $10\frac{1}{2}$ AND $11\frac{1}{2}$ ONLY

highest steam pressures used on locomotives down to 35 pounds steam without adjustment and without wasting at the overflow, and by regulating the water-supply valve on the injector it can be operated at 15 pounds. As it restarts instantly under all conditions of service, it can always be depended upon to force all the water into the boiler, so that the engineer can give his whole attention to his other duties.

Sizes of Injectors for Locomotives.* In determining the size of injector required for locomotives, the size

*From "Practice and Theory of the Injector," Wiley & Sons, New York.

of the cylinder is usually taken as the standard, although the diameter of the boiler and the kind of service for which the locomotive is intended has a modifying influence.

TABLE 17

Diam. of Cyl., inches	Size of Injector	Diam. of Cyl., inches	Size of Injector	Diam. of Cyl., inches	Size of Injector	Diam. of Cyl., inches	Size of Injector
9	$4\frac{3}{10}$	13	$5\frac{4}{10}$	17	$7\frac{1}{2}$	21	$9\frac{1}{2}$†
10	$4\frac{3}{10}$	14	$6\frac{1}{2}$	18	$8\frac{1}{2}$	22	$10\frac{1}{2}$
11	$5\frac{4}{10}$	15	$6\frac{1}{2}$	19	$8\frac{1}{2}$†	23	$10\frac{1}{2}$
12	$5\frac{4}{10}$	16	$7\frac{1}{2}$	20	$9\frac{1}{2}$	24	$11\frac{1}{2}$
						25	$11\frac{1}{2}$
						26	$12\frac{6}{10}$

† Use next size larger with specially large boiler.

TABLE 18

IMPROVED SELF-ACTING INJECTOR

MAXIMUM AND MINIMUM CAPACITIES, ALL CLASSES

Gallons per Hour—5 Feet Lift. ($7\frac{1}{2}$ Gallons = 1 Cubic Foot.)

Size	60 Lbs. Steam		120 Lbs. Steam		180 Lbs. Steam		200 Lbs. Steam	
	Max.	Min.	Max.	Min.	Max.	Min.	Max.	Min.
‡$4\frac{3}{10}$	427	158	562	208	517	345	500	350
$5\frac{4}{10}$	667	247	907	340	1027	395	1035	455
$6\frac{1}{2}$	967	358	1320	489	1492	568	1516	667
$7\frac{1}{2}$	1290	477	1755	650	1987	757	2010	885
$8\frac{1}{2}$	1657	613	2257	835	2550	970	2587	1138
$9\frac{1}{2}$	2070	766	2820	1044	3150	1197	3187	1402
$10\frac{1}{2}$	2535	938	3450	1280	3900	1482	3952	1740
$11\frac{1}{2}$	3037	1124	4132	1530	4672	1775	4725	2079
$12\frac{6}{10}$	3650	1351	4968	1847	5616	2134	5700	2450

‡ Class N. Imp. not made $4\frac{3}{10}$ size; only supplied in Classes L, M, and N.

Things to be Remembered. With Locomotives Carrying High Steam Pressure (180 to 225 Pounds). Set the injector just above the top water level of the tank. At 8 feet lift, 200 pounds, the capacity is about 10 per cent less than the list.

Cold water is best for the injector. Hot water reduces the life and efficiency. At 120 deg. the capacity is about one-third below list given in the table. The range of capacities is reduced and no injector lifts as promptly.

Use large suction pipe and tank valve connections. If the diameter is increased one size, the gain in capacity is from 5 to 10 per cent.

Use large strainer with small holes. Small strainers require frequent cleaning. If the holes are large, cinders and coal pas through and wear the tubes. If the strainer is too small, the injector does not give full capacity. Be sure that the gasket between hose and suction pipe is not squeezed so as to close opening.

FIGURE 161

LOCOMOTIVE FEED-WATER STRAINER FOR RIGHT- OR LEFT-HAND SIDE OF ENGINE

Suction pipe must be absolutely tight. Any leak of air reduces the capacity and makes the overflow valve jump.

Delivery pipe and main check valve must be of ample area. If an injector gives high back pressure, it is using too much steam. If the delivery opening is too small, the power of the injector is wasted in increased friction in the pipes.

Take good care of the injector. Keep all glands steam-tight, and watch carefully for leaks in the suction pipe. Do not force the steam valve hard against its seat; close the valve gently. Start the injector in the same way; at very high pressure the delivery pipe is liable to burst if the lever starting valve is jerked open. Keep the injector clean and report at once if not working properly. Do not run with the water-regulating valve wide open all the time.

List of Parts, Self-Acting Injector, Class N Improved

1. Delivery Tube.
2. Combining Tube.
3. Steam Nozzles.
5. Spindle Nut.
6. Steam Stuffing Box.
7. Spindle.
8. Cross-Head.
10. Water Stuffing Box.
11. Follower.
12. Packing Ring.
13. Lock Nut.
14. Follower for No. 10.
15. Links.
16. Packing Ring.
19. Plain } Rings for Copper Pipe.
19a. Reduc. } Rings for Copper Pipe.
20. Check Valve.
22. Guide for No. 20.
23. Plain } Unions for Iron Pipes.
23a. Reduc. } Unions for Iron Pipes.
24. Coupling Nuts.
25. Injector Body.
27. Wrench.
29. Waste Pipe.
30. Waste Valve.
31. Waste Valve Cam.
32. Jam Nut for No. 29.
33. Starting Lever.
34. Cam Lever.
35. Pin, Nos. 38 and 33.
36. Cam Shaft.
37. Washer on 36.
38. Collar and Index.
39. Funnel.
40. Plug Water Valve.
41. Regulating Handle.
42. Inlet Valve.
57. Closed Overflow Connection.

List of Parts, Self-Acting Injector, Class M Improved

1. Delivery Tube.
2. Combining Tube.
3. Steam Nozzles.
5. Spindle Nut.
6. Steam Stuffing Box.
7. Spindle.
8. Cross-Head.
10. Water Stuffing Box.
11. Follower.
12. Packing Ring.
13. Lock Nut.
14. Follower for No. 10.
15. Links.
16. Packing Ring.
19. Plain } Rings for Copper Pipe.
19a. Reduc. } Rings for Copper Pipe.
20. Check Valve.
22. Guide for No. 20.
23. Plain } Unions for Iron Pipes.
23a. Reduc. } Unions for Iron Pipes.
24. Coupling Nut.
25. Injector Body.
27. Wrench.
29. Waste Pipe.
30. Waste Valve.
31. Guide for No. 30.
32. Jam Nut for 31.

33. Starting Lever.
34. Cam Lever.
35. Pin, Nos. 38 and 33.
36. Pin through 31 and 34.
38. Collar and Index.
39. Funnel.
40. Plug Water Valve.
41. Regulating Handle.
42. Inlet Valve.
57. Closed Overflow Connection.
73. Guide for Overflow Valve 75.
74. Heater Stem.
75. Overflow Valve.
76. Follower.
77. Pack Ring in 73.
78. Heater Lever.

The Sellers' Self-Acting Injector. Class P—Sizes 10½ and 11½. This is a special form of body, designed to be applied to the back-head of the locomotive boiler, with the starting lever and water regulating valve placed directly over the brake valve and within convenient reach when the engineer is seated. The coupling nuts and sizes of the pipe are Pennsylvania Railroad (Sellers') Standard, but the branches are located so as to avoid the fire-door and boiler attachments.

Hints to be Read Before Connecting the Injector.

1. Blow out all pipes carefully with steam before attaching the injector, tapping the pipe with a hammer in order to loosen all the scale.

2. When drip pipe is attached close to overflow of injector, it must be same size as given in table.

3. Always use a dry pipe attachment to insure perfectly dry steam.

4. The diameter of the strainer should be large enough to give an ample supply of water even when some of the holes are choked.

5. Keep all valves steam-tight; all leaks tend to increase rapidly, owing to the velocity with which steam passes through the smallest opening.

6. Keep the steam pipe and chamber free from dirt and chips from the threads on the pipes, and the steam nozzles perfectly clean. The steam nozzle is the life

of an injector, and should be maintained in best condition. If the injector is new and the lifting nozzle should fill up, remove from body as described.

7. When grinding the steam valve, place a rubber washer over the holes leading to the lifting nozzle to prevent the sand from working into the lifting jet; this washer should, of course, be provided with a hole large enough to admit the plug on the end of the spindle; then screw the steam stuffing box rather tightly against its shoulder to insure its proper alignment. Keep the steam valve perfectly tight.

8. To remove lime and scale, immerse the tubes or the whole injector in a bath composed of ten parts of water to one part muriatic acid. Remove as soon as scale is dissolved.

Emergency Methods of handling the improved self-acting injector (Locomotive Firemen's Magazine) The improved self-acting injector in good working order is the most satisfactory boiler feeder that can be used, and fully deserves the confidence placed in it by careful enginemen. But there are times when even the best injector refuses to work. It may not be the fault of the injector itself; in fact, it seldom is to blame. Sometimes the trouble is due to careless handling, to the leaky condition of the steam valves, joints of the suction pipe and hose coupling; to cinders and dirt in the tank; and under such conditions many an injector is struggling, which not only reduces the efficiency and length of service, but finally prevents it from delivering water to the boiler.

It is of course difficult to do much in the way of repair to an injector when out on the road; even the pipe-coupling wrench is apt to be missing, and few tool boxes have wrenches for the removal of the tubes.

A special feature of the self-acting injector is that the combining and delivery tubes can be removed with ordinary tools, but when it is necessary to take an injector apart the leakage from the steam and main check valves makes the work very disagreeable; but when an injector does not work it is something more than aggravating, it is often serious, especially if the left-hand injector has not been used for some time and also refuses to start. Then is the time for quick thinking and quick acting.

Suppose that an injector suddenly stops working. Probably the tubes, hose, suction pipe or strainer are stopped up. The last two can probably be cleared out by closing the cam over the overflow valve and drawing the starting lever quickly; if the hose lining has become loose it will let the steam flow back and close up again as soon as the injector is started, disabling this injector until a short nipple or coiled wire can be forced up the hose or a new hose obtained. If the next station cannot be reached before water is needed the left-hand injector must be made to work, unless the train is stopped and the injector and pipes thoroughly examined. Treat the left-hand injector exactly as the right. Open the tank valve and draw the injector starting lever; if the water is lifted, but will not enter the boiler, set the lazy cock at half capacity and tap the main check valve on cap with hammer to loosen it in its guide; at half capacity, because at that point the injector gives a higher back pressure than with the lazy cock wide open. This will probably be effective.

To Remove Tubes. The sectional views show very clearly how the tubes are held in the body. Uncouple the feed pipe from the injector and swing it out of the way; place a monkey wrench on the guide (22) for the line check (20) and unscrew; in some of the older patterns of injector it may be necessary to insert an old file or flat piece of iron, or perhaps two pieces in opposite openings; at any rate, it can be removed quite easily unless the seats are heavily lined up. This draws out the combining (2) and delivery (3) tubes, which can be separated and carefully examined inside; here is where the trouble will usually be found, and

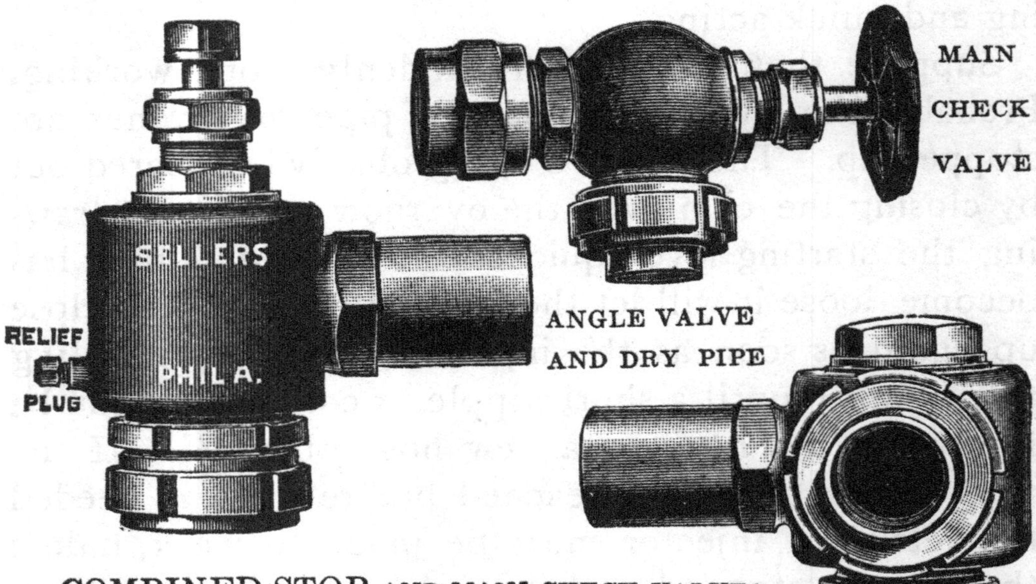

COMBINED STOP AND MAIN CHECK VALVE; BY CLOSING THE STOP VALVE, THE CHECK VALVE AND SEAT CAN BE REMOVED WITHOUT BLOWING OFF BOILER. SCREW OR FLANGED.

FIGURE 162

the impediment must be taken out without bruising the surface or bending the tubes. When the parts are replaced, test before recoupling the feed pipe. No steam should issue from suction branch.

Frequently the cause of stoppage is the absence of a strainer in the tank or suction pipe, or due to the fact that the holes in the straining plate are too large. An admirable arrangement of fixed strainer is shown in Fig. 161 placed between the hose and the suction pipe.

Suppose that steam nozzles (piece 3) require cleaning. Stoppage of the lifting tube is usually gradual and is shown by a slow falling off in the working of the injector. These tubes are more difficult to remove unless a wrench to fit the hexagon is at hand. Sometimes a large iron chip or heavy piece of scale is carried into the nozzle (piece 3) by the steam, due to carelessness when cleaning the boiler, but this is of infrequent occurrence. If this happens it is better to make running repairs to the other injector and leave it for the men in special charge of injector repairs.

At times the main check valve does not seat and all efforts to close it prove unavailing; if the line check valve has been omitted during repair, the water from the boiler rushes back into the injector. With injectors having no lazy cock, the only method of preventing the burning of the crown sheet is to draw the fires; but with the self-acting close the overflow valve by means of the cam, then quickly shut the lazy cock. The check pipe and injector body will then carry full boiler pressure until the roundhouse is reached, when the fire can be drawn and the pressure blown off.

Leakage of air into the suction pipe is usually the cause for unsatisfactory working of the injector. Enginemen should be especially careful about this and always tighten the joints so that no air can enter; even a slight drip from any of the joints under the pressure of the water in the tank indicates a large enough opening to admit sufficient air to affect the working of the injector, especially when the water level in the tank is low. Another point is the tightness with which the cover of the manhole of the tank fits on its seat; if air does not enter freely upon the top of the water the capacity of the injector will be reduced, the effect being more marked at high steam pressures and long lift than under ordinary conditions.

Lime and salts contained in the supply water coat the surfaces of the tubes; the accumulation occurs slowly, destroying the restarting feature, the promptness of lifting, and reduces the capacity; this should be at once reported to the proper authorities.

Inlet Valve (42). When the improved injector is feeding, the overflow chamber—the part of the body between the water branch and the waste pipe—is filled with cold water; if it does not feel cold to the hand the inlet valve (42) is not open and working properly. This method of surrounding the tubes with cold water tends to prevent the formation of scale, and this pattern of injector gives longer service in districts where the supply water contains lime than those that do not contain this feature. Crude oil introduced into the steam or water pipe softens the scale and is often helpful. Bosses on both the water and steam branches may be tapped for self-feeding oil cups, but all the joints should be tight.

Maintain the injector in good working order.

The Metropolitan "1898" Locomotive Injector, Figs. 163 and 164, is a double-tube injector, composed of a lifting set of tubes which lifts the water and delivers it to the forcing set of tubes under pressure, which in turn forces the water into the boiler.

The lifting set of tubes act as a governor to the forcing tubes, delivering the proper amount of water required for the condensation of the steam, thus enabling the injector to work without any adjustment under a great range of steam pressure, handle very hot water and admit of the capacity being regulated for light or heavy service under all conditions.

Figure 163
Metropolitan "1898" Locomotive Injector

This injector will start with 30 to 35 lbs. steam pressure, and without any adjustment of any kind will work at all steam pressures up to 300 lbs. In fact, at all steam pressures and under all conditions its operation is the same. When working, all the water must be forced into the boiler. It is impossible for part or all the water to waste at the overflow should the steam pressure vary.

The injector is easily handled. The lever works very freely and can be handled without care, for there is no sensitiveness whatever in starting, as is the case with most injectors; consequently any one can operate it.

Regulation of capacity is an important, in fact indispensable feature of the perfect locomotive injector. With Metropolitan "1898" locomotive injectors the capacity can be regulated for light or heavy service under all steam pressures and with hot as well as with cold feed water. While most injectors will admit of the capacity being regulated with low steam pressures and cold feed water, this injector is the first that admits of a successful regulation with steam pressures up to and above 250 lbs. and with the feed water heated.

Model H Metropolitan "1898" locomotive injectors will interchange and fit the Monitor coupling connections. Model H injectors, sizes 5 and 6, have same size body and interchange. Sizes 8 and 9 have same size body and interchange. Sizes 11 and 12 have same size body and interchange.

The Model H and Model I types of these injectors differ solely in the pipe connections and the form of the main casing or shell. All the parts for each are the same for corresponding sizes.

Model I Metropolitan "1898" locomotive injectors will interchange and fit the Sellers coupling connections. Model I injectors, sizes 6 and 7, have the same size body and interchange.

The Metropolitan locomotive injector is manufactured by the Hayden and Derby Manufacturing Company of New York, who furnish the following directions for connecting and operating it

Pipe Connections. The injector should be located inside the cab, so that it can be conveniently handled by the engineer. It should be located with the overflow nozzle about 4 in. above the top of the tank. It is necessary that the steam pipe and the openings in the main steam valve should be as large or larger than the inside diameter of the sizes of copper pipe given in the list below, so that the injector will receive a full supply of dry steam. The openings in the goose neck and tank valve should not be smaller than the size of suction pipe called for in the list below.

TABLE 19

Size	Capacity per Hour		Pipe Connections							
	Steam Pressures		Steam		Suction		Delivery		Overflow	
	160 Pounds	210 Pounds	Iron	Copper	Iron	Copper	Iron	Copper	Iron	Copper
5	1180 gals.	1210 gals.	1½	1¾	1½	1¾	1½	1¾	1¼	1½
6	1605 gals.	1647 gals.	1½	1¾	1½	1¾	1½	1¾	1¼	1½
7	2095 gals.	2151 gals.	1½	1¾	1½	1¾	1½	1	1¼	1½
8	2651 gals.	2723 gals.	2	2	2	2¼	2	2	1½	1¾
9	2954 gals.	3034 gals.	2	2	2	2¼	2	2	1½	1¾
10	3961 gals.	4068 gals.	2	2¼	2½	2¾	2	2¼	2	2¼
11	4700 gals.	4810 gals.	2½	2¾	3	3¼	2½	2¾	2	2¼
12	5700 gals.	5950 gals.	2½	2¾	3	3¼	2½	2¾	2	2¼

Operation. To start the Metropolitan "1898" locomotive injector, the lever, part 292, Fig. 164, is drawn back, lifting the auxiliary steam valve, part 213, from its seat. This allows steam to flow through the lifting steam jet, part 224, into the lifting combining tube, part 225, thereby creating a vacuum in the suction chamber, causing the water to flow through the lifting combining tube, part 225, condensing the steam, then out through the overflow valve, part 215, and through the final overflow valve, part 234, through the overflow pipe to the atmosphere. A further movement of

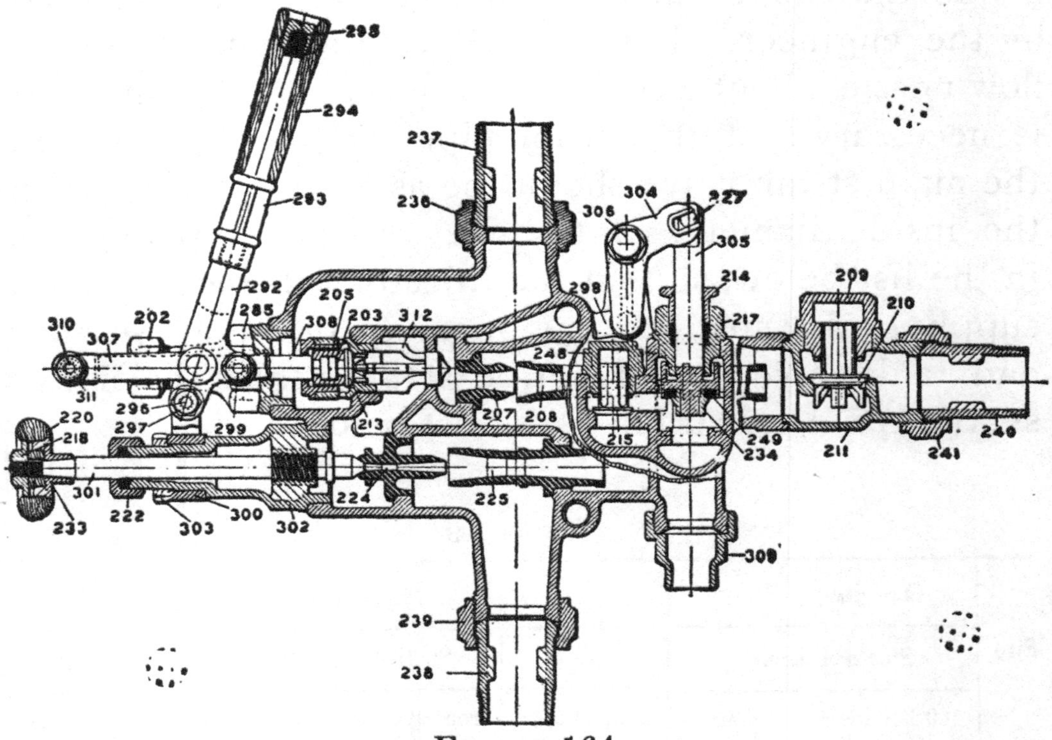

FIGURE 164

METROPOLITAN "1898" INJECTOR, SECTIONAL VIEW

the lever, part 292, opens the steam valve, part 206, which admits steam to the forcing steam jet, part 207, which is condensed by the water which is in the intermediate chamber and in the forcing combining tube, part 208, creating a pressure in the delivery chamber of the injector, which is sufficient to close the overflow valve, part 215; and a further movement of the lever, part 292, closes the final overthrow valve thereby turning the water from the overflow into the boiler, thus opening the check valve, part 210. When the injector is working, the overflow valve is closed and held to its seat by pressure equal to the boiler pressure.

The capacity of the Metropolitan "1898" locomotive injector is regulated by increasing or decreasing the amount of steam to the lifting steam jet, part 224, by

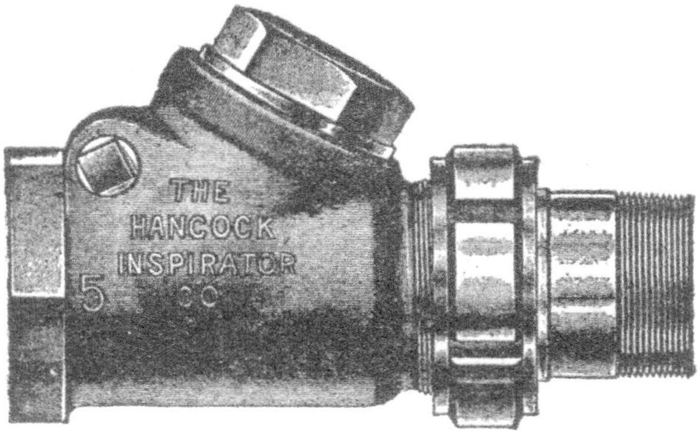

FIGURE 165

SWING INTERMEDIATE CHECK VALVES, EXTERIOR VIEW

means of the regulating valve, part 301. When this valve is wide open, the lifting steam jet, part 224, receives a full amount of steam, which enables the lifting apparatus of the injector to lift the greatest quantity of water and deliver it to the forcing apparatus. When this regulating valve, part 301, is partially closed, it will partially close the opening into the lifting steam jet, part 224, decreasing the flow of steam, which will decrease the amount of water lifted by the lifting apparatus. This arrangement has been found to be far better than the old method of throttling the water supply. It enables the injector to run steadier when working at its minimum capacity and also enables the capacity to be reduced more.

To use the injector as a heater, lift the side links, part 286, by means of the small handle on same, and pull the links back until the pin, part 287, drops into the notch. This operation causes the final overflow to be closed and a small amount of steam can be admitted, enough to heat the injector. When it is desired to operate the injector after using it as a heater, the lever is simply pushed in, which will place the injector in position to be operated.

If the injector breaks or will not start promptly, see if there is a leak in the suction connection. If the openings into the tank are too small, or the hose strainer clogged, or the hose kinked, or the hose lining is collapsed, the injector will not get a sufficient supply of water. If the injector will lift the water but will not deliver it into the boiler, see that the intermediate or line check valve, or the main boiler check valve are in proper working order, also examine the suction pipe for leaks. A leak in the suction pipe, while it may not prevent the injector from lifting, will prevent the water being forced into the boiler. If the main steam pipe or the main steam valve are not of sufficient size, or if there is a leak in the dry pipe, the injector will not receive a supply of steam sufficient to force the water into the boiler. If the overflow pipe is smaller than the overflow nozzle, there will be a back pressure, which will prevent the injector from lifting the water promptly. The overflow nozzle and overflow pipe should be kept free from lime or scale. This is very important.

Repairing. When the tubes become worn they should be renewed. The forcing tubes are removed by removing the check valve casing, part 211, by breaking the flanged joint. The lifting tubes are removed by removing the regulating center piece, part 302. Should the steam valves leak they should be re-ground. Overflow valve, part 215, must seat tightly. If this valve leaks, it will cause the hot water from the delivery chamber of the injector to be forced into the intermediate chamber and drawn into the combining tube, part 208, causing the injector to break. This is very important.

The final overflow valve has a soft disk, part 249. This disk is made soft so that in case the valve should close on to any hard substance, it will not injure the valve seat. These disks can be removed very easily and are very inexpensive.

Swing Intermediate or Line Check Valves (Hancock Pattern, Fig. 165). These check valves can be applied to any locomotive injector.

There are no wings or guides to become incrusted with scale or deposit while the valve is open, which would prevent its closing, and the liability therefore of damage or delay caused by the valve failing to close is obviated.

The Monitor Injector. (Figs. 166-167, made by the Nathan Manufacturing Co., New York.) The proper position of this injector is in the cab above the level of the water in tender, convenient to engineer. Should the Monitor have to be placed outside, it must be provided with connecting rods extending into the cab. Steam should be taken from dome or highest part of boiler to insure best effects.

Its Range. It does not waste water at overflow by

ordinary variation of working steam pressure, but steadily performs its duty, whether the water-valve is wide open or throttled down until almost shut.

Steadiness. It works steadily, whether the engine is running fast or slow; while reversing, applying brakes, and during ordinary stoppages. It is also capable of

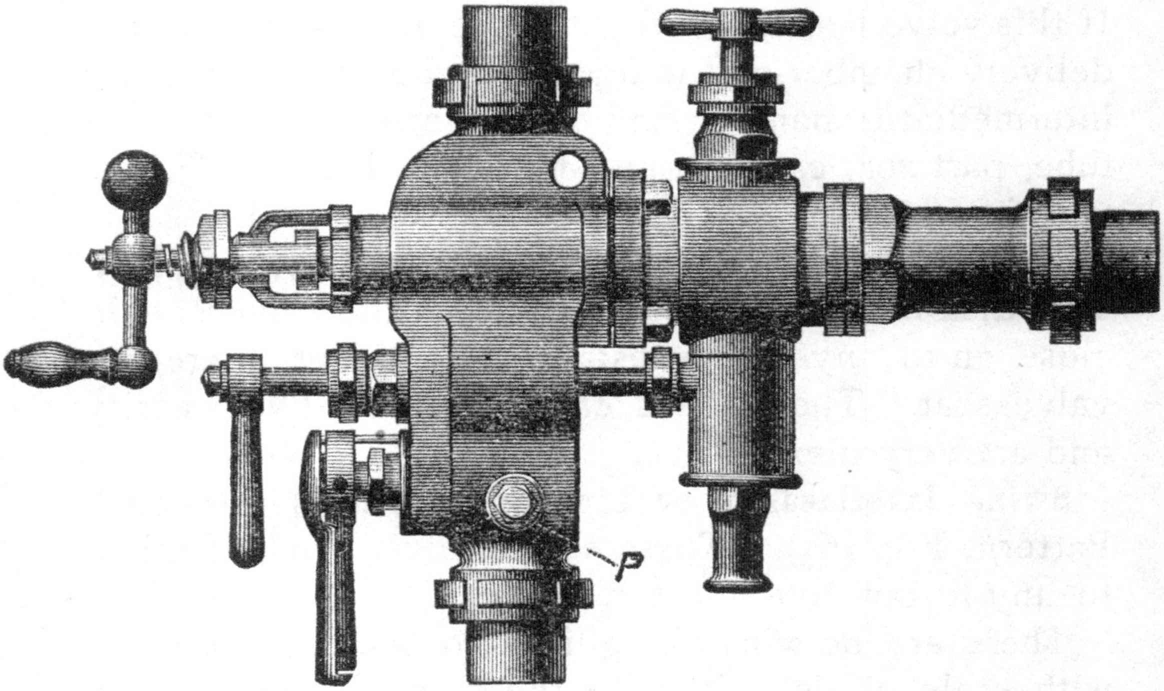

FIGURE 166

THE "MONITOR," EXTERIOR VIEW

running heavy as well as light trains, the quantity of water needed being easily regulated by the water-valve attached.

Reliability. It is provided with an independent lifting jet, which enables the injector to start promptly at all times. This is a peculiar feature and very important, because it allows the injector to start as promptly after doing its duty as a heater cock, as at first.

Flanged Monitors. Since 1885 the body of the Monitor has been divided into two parts, which are firmly held together by a double flange securely bolted. This very convenient arrangement enables the interior parts to be taken out readily, for cleaning or renewal, when necessary, without injury to the injector.

Recent Improvements. The steam valve spindle is provided with an improved patent yoke stuffing box, which makes it possible to place the threaded part of

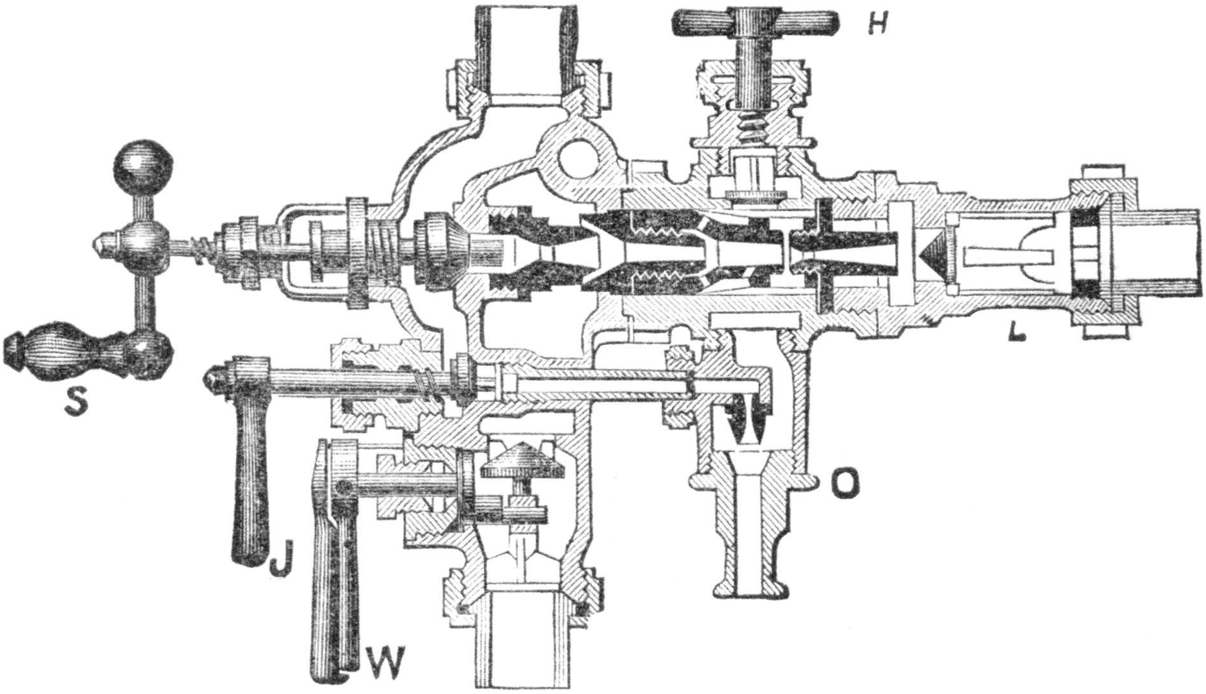

FIGURE 167
THE "MONITOR," INTERIOR VIEW

the spindle outside the steam chamber, diminishing its wear. The packing can be adjusted and tightened by means of a large central nut, still preserving the simplicity and convenience of an ordinary stuffing box.

The water valve has been provided with a double handle with index pin, which engages with notches, cut into the stuffing box cap, thereby keeping the water valve steady in any position against any jar or vibration of the engine.

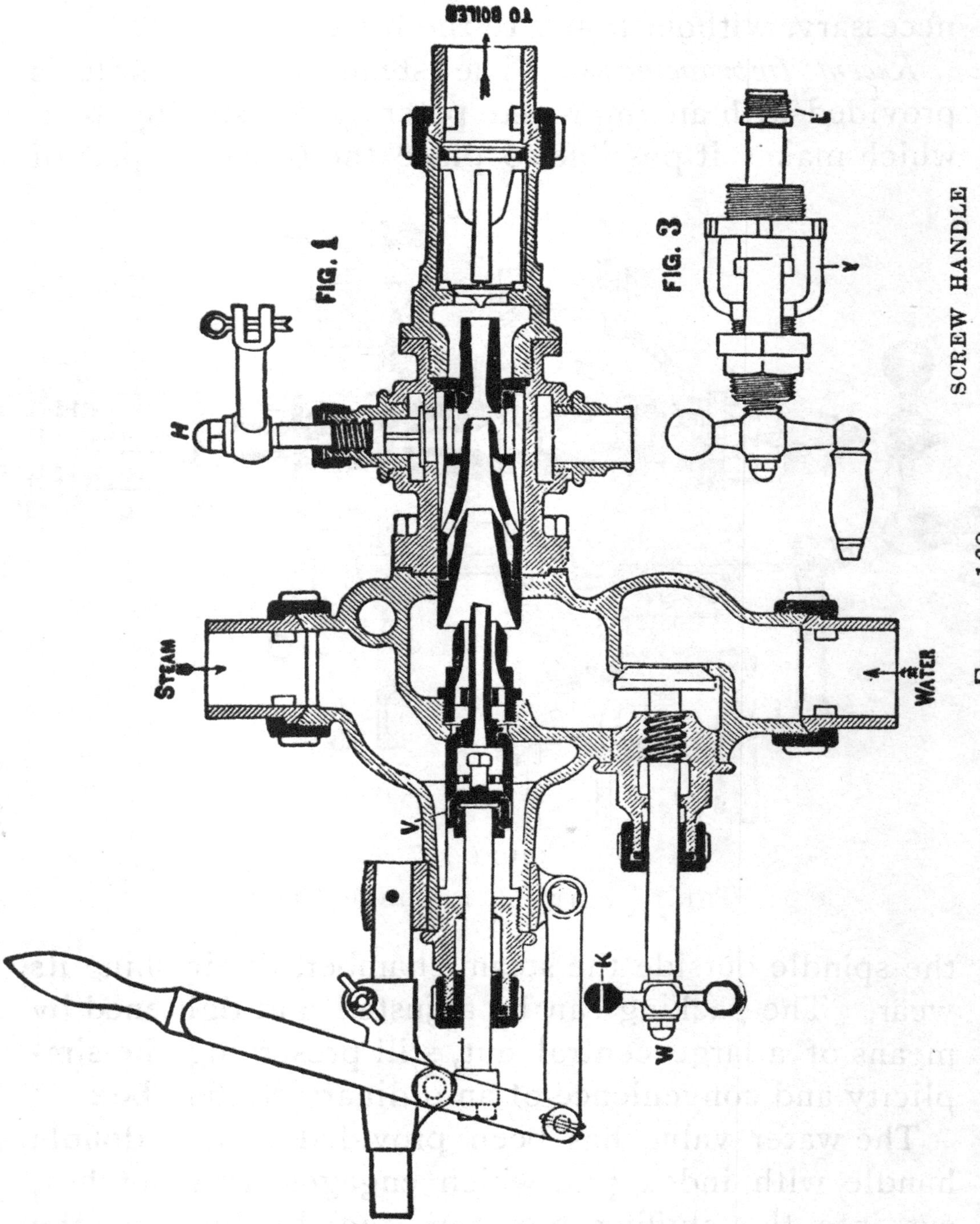

FIGURE 168
"'88 MONITOR," INTERIOR VIEW

Description of '88 Monitor, Fig. 168. This injector is a modification of the well-known locomotive injector of that name, and is designed to supply the demand for a lever-handled injector, and embody in a new combination all the best qualities of the former instrument. The most prominent feature of the '88 Monitor is the facility with which it can be started and stopped by the new lever-handle attachment, or the single screw spindle motion, whichever may be preferred. The quantity of water which the new injector is capable of throwing, will command attention, and the range of its capacity, running as it does from 100 per cent at maximum to less than 50 per cent at minimum, makes it equally applicable to the moving of heavy or light trains, as the case may happen.

It will lift the feed-water 5 ft. at 30 lbs. pressure, and at standard working pressure, to a height not likely to arise in ordinary locomotive practice.

Its pipe connections are the same as the other Monitors and interchangeable therewith, so that in the use of the new instrument, the old fittings, if they are good, need not be disturbed.

The construction of the starting arrangements is such that the screw attachment can be readily substituted for the lever-handle, should the former method be preferred.

Directions for Application. Place the injector above water level in tender. Take steam from dome, or highest part of boiler, through dry pipe. This will insure the best effects.

Instructions to Operate the Injector. *With Lever Motion.* To start: Pull out the lever a short distance to lift the water; when water runs from the overflow, steadily draw back the lever until overflow ceases.

Do not increase the steam supply after overflow has ceased.

Regulate for quantity with water-valve W

To stop: Push in the lever.

With Screw Motion. To start: Open the steam valve one quarter of a turn to lift the water. When water runs from the overflow, open steam valve until overflow ceases. Do not increase the steam supply after overflow has ceased.

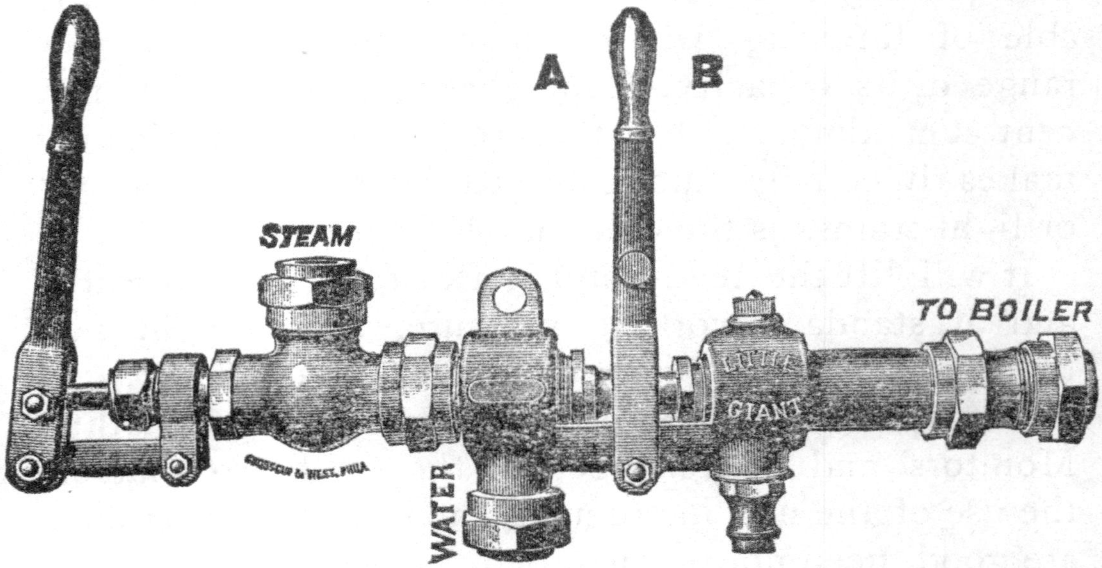

FIGURE 169

Regulate for quantity with water-valve W.

To stop: Close steam valve.

Note 1. To grade injector: Throttle water by valve W; if this is not sufficient, reduce the steam by pushing in lever handle about half-way, and in case of the screw motion, by screwing in the steam spindle about half-way.

2. To use as a heater: Close valve H and pull out lever all the way, and in case of screw motion open valve full. At all other times valve H must be kept open.

3. The heater cock can be worked from the cab by means of arm A, adapted for the attachment of an extension rod. Arm A is held on the heater cock spindle by friction, and by loosening cap C it can be set at any angle to suit the most convenient position for the extension rod.

4. The hole in the top knob K of water handle W indicates the position of the water valve. One turn of the handle fully opens, or entirely closes, the water passage.

In either case, the knob with the hole in should be in an upright position. Intermediate positions of the knob K indicate corresponding openings in the water passage.

The Little Giant Locomotive Injectors, Fig. 169, made by the Rue Manufacturing Co., Philadelphia, have been on the market for many years. This injector is simple in construction, and is not liable to get out of order.

These injectors are fitted with a movable combining tube, operated by a lever which allows them to be adjusted to work correctly at different pressures of steam, and under the many conditions which injectors are required to work.

TABLE 20—SIZE AND CAPACITY

Size of Injector	Copper Pipe Outside		Iron Pipe Inside		Gallons of Water per hour
	Steam	Water & Delivery	Steam	Water & Delivery	
4	$\frac{7}{8}$	$\frac{7}{8}$	$\frac{3}{4}$	$\frac{3}{4}$	600
5	$1\frac{1}{8}$	$1\frac{1}{8}$	1	1	950
6	$1\frac{3}{8}$	$1\frac{3}{8}$	$1\frac{1}{4}$	$1\frac{1}{4}$	1275
7	$1\frac{3}{8}$	$1\frac{3}{8}$	$1\frac{1}{4}$	$1\frac{1}{4}$	1800
8	$1\frac{3}{4}$	$1\frac{3}{4}$	$1\frac{1}{2}$	$1\frac{1}{2}$	2250
9	$1\frac{3}{4}$	$1\frac{3}{4}$	$1\frac{1}{2}$	$1\frac{1}{2}$	2800
10	$1\frac{3}{4}$	$2\frac{1}{4}$	$1\frac{1}{2}$	2	3500

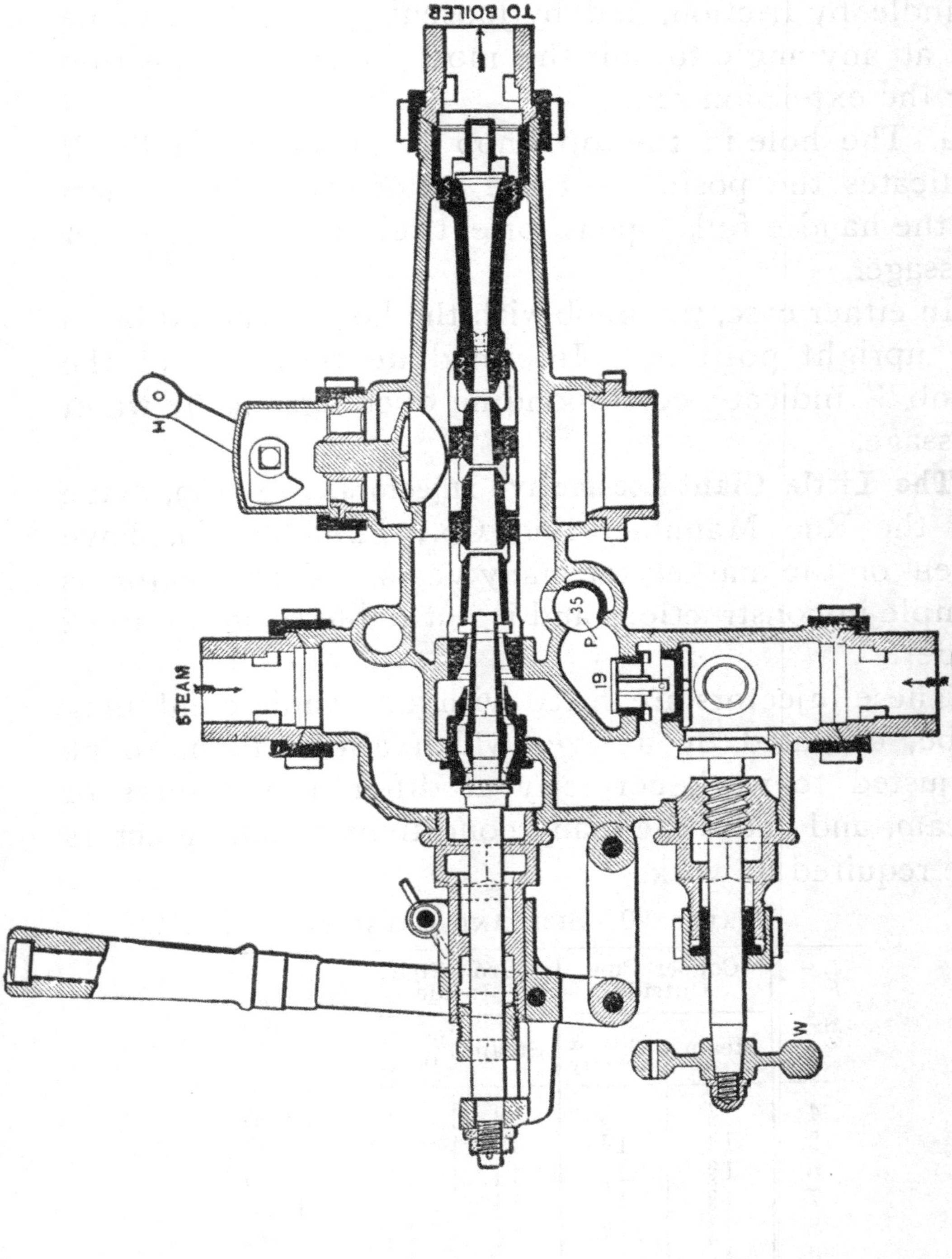

FIGURE 170
THE SIMPLEX INJECTOR

To Operate. Have the combining tube in position to allow a sufficient quantity of water to condense the steam when the starting valve is full open, then open the starting valve slightly; when water shows at overflow, open full. Regulate the water by moving the combining tube. To use as a heater, close overflow by moving the combining tube up against the discharge, then open starting valve enough to admit the quantity of steam required.

The Simplex Locomotive Injector, Fig. 170, has been designed to meet the severe requirements of modern locomotive practice, especially where it is desired that the instrument be self-adjusting and re-starting.

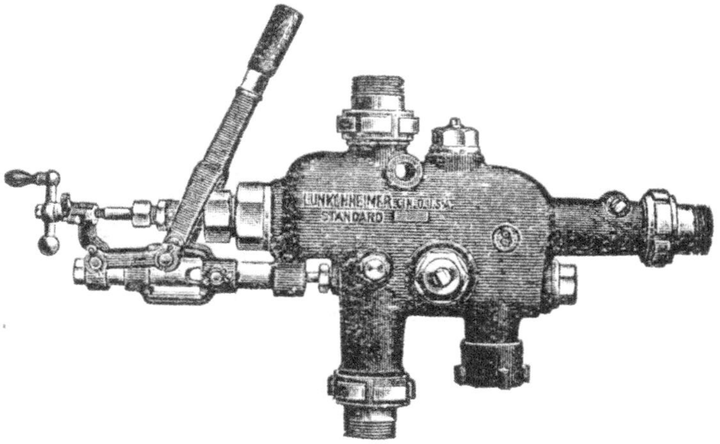

FIGURE 171

LUNKENHEIMER '99 MODEL STANDARD INJECTOR

Attention is called to the largely increased delivering capacities at the high steam pressure of locomotive engines of to-day.

This injector is of the "re-starting" type, and if the water supply should happen to be temporarily interrupted, the instrument will start again without any manipulation, just as soon as the water supply is again within reach. The instrument is also "self-regulating," and requires no water valve regulation above 50 lbs. steam pressure, to prevent spilling at the overflow. Its lifting qualities are of the highest order, and

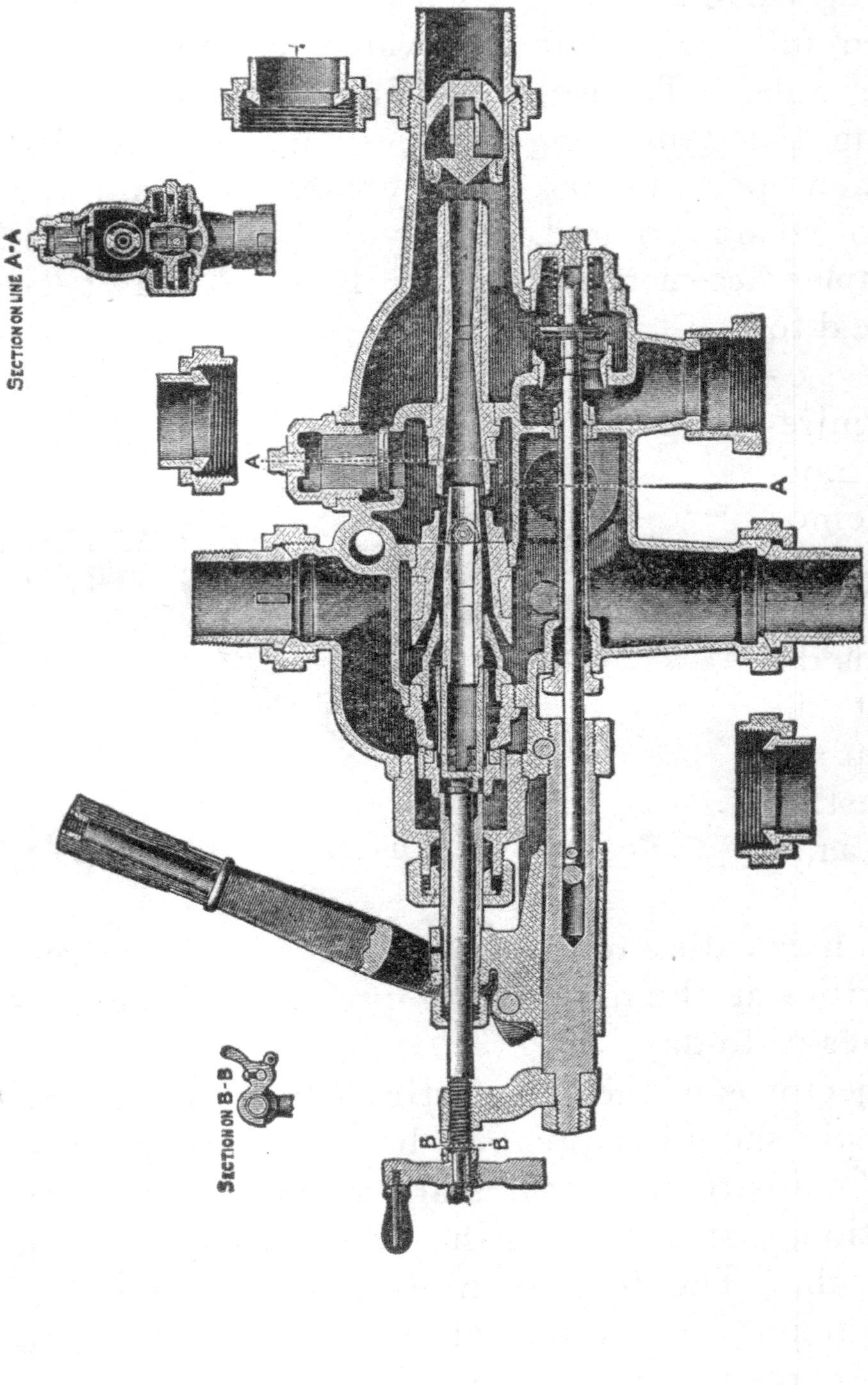

FIGURE 172
SECTIONAL VIEW
LUNKENHEIMER '99 MODEL STANDARD INJECTOR

it may be relied upon to start promptly after doing duty as a heater.

The throttling capacity is fully 50 per cent of the maximum capacity under ordinary variations of lift and of feed water temperature.

If it is desired that the instrument be placed outside the engine cab, and operated by means of extension

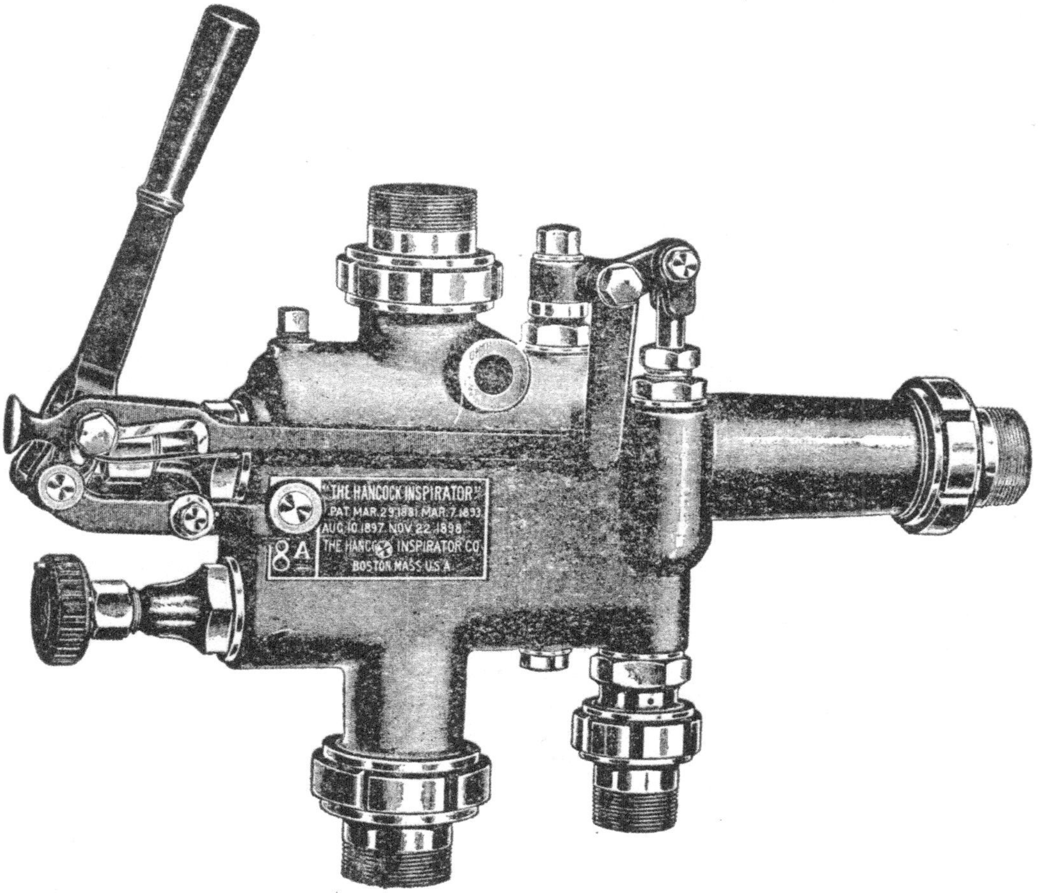

FIGURE 173

THE HANCOCK INSPIRATOR, SECTIONAL VIEW

rods, a quick motion screw attachment can be readily substituted in the place of the lever handle.

The construction of the injector is such that all its interior nozzles and other component parts are easily accessible for examination and repairs.

Method of Operating. To start: Pull out the lever.

To stop: Push in the lever.

Regulate for quantity by means of the water valve.

To use as heater for the feed water: Close heater cock and draw out the lever.

In starting on high lifts and in lifting hot water, pull the lever out slowly.

In case a pipe is attached to the overflow, its inside diameter must under no circumstances be less than the inside diameter of the overflow nozzle.

If the water inlet valve (part 19 of details) should leak and prevent the prompt lifting of the feed water, it will only be necessary to turn around key 35, so that the letter "S" (not shown on cut) on the square spindle-end will be "up." This will close passage "P," and permit the continued use of the instrument until valve 19 can be repaired.

The Simplex injector is made by the Nathan Mfg. Co. of New York, who furnish the following table of capacities.

TABLE 21

Size	Capacity in Gallons p'r Hour		Inside Diameter of Iron Pipes, in Inches			Outside Diameter of Copper Pipe, in Inches		
	125 lbs.	200 lbs.	Steam	Suction	Delivery	Steam	Suction	Delivery
5	990	1140	1¼	1¼	1¼	1½	1½	1½
6	1260	1440	1¼	1¼	1¼	1½	1½	1½
7	1830	1950	1½	1½	1½	1¾	1¾	1¾
8	2280	2580	1½	2	2	1¾	2	2
9	2880	3240	1½	2	2	1¾	2¼	2
10	3450	3800	2	2	2	2¼	2¼	2¼

Lunkenheimer '99 Model Standard Injector. Figs. 171-172. This injector embodies in its construction all desirable features which tend to make an injector high grade and efficient. Under high steam pressures it is necessary to have a machine which can be operated as efficiently as under low pressures, and one which admits of sufficient range of work to cover all conditions of service.

The construction is simple, manipulation easy, and the results attained show a high degree of efficiency. It can be started promptly, under most conditions, at all pressures from 30 to 250 lbs., and can be handled without fear of uncertainty of action, as it is not sensitive in this respect. It will work without adjustment of steam or water from 40 to 250 lbs. and higher, and the capacity can be reduced over 50 per cent at all points. This feature makes it especially suitable for severe service, such as is found on railroads, steamboats and high-pressure power plants, and in other places where the load varies and it is necessary to have an injector in which the capacity can be reduced within wide limits.

The design of the machine is excellent, the parts are well proportioned, the operating mechanism which controls the steam and overflow valves is not complicated and is all contained within the body of the injector, and there are no outside connecting rods, usually found in machines of this class.

TABLE 22. CAPACITIES OF THE LUNKENHEIMER STANDARD INJECTOR

Size No.	Pipe Con. Steam Suction Delivery	Pipe Connections Overflow	Maximum Capacities at Various Steam Pressures Feed Water, 76° F. Lift, 5 feet.				
			125lbs.	150lbs.	175lbs.	200lbs.	225lbs.
8½	¾″	½″	650	665	682	700	715
9½	1″	¾″	835	855	876	900	925
10½	1¼″	1″	1110	1140	1170	1200	1230
11½	1¼″	1″	1485	1520	1560	1600	1640
12½	1½″	1¼″	1865	1910	1950	2000	2050
13½	2″	1½″	2320	2375	2440	2500	2560
14½	2″	1½″	2780	2850	2925	3000	3080
15	2″	1½″	3430	3518	3610	3700	3795
15½	2″	1½″	3820	3910	4000	4100	4200
16½	2½″	2″	4640	4760	4875	5000	5130

The Hancock Locomotive Inspirator. The Hancock inspirator consists of one apparatus for lifting and one for forcing. The original stationary type embodied this feature by incorporating two chambers side by side, connected at the top for steam, and at the bottom for water, so that it was apparent to the casual observer that each carried its own set of apparatus. The Hancock inspirator works successfully under the most severe conditions. With high or low steam pressure, on all lifts up to 25 feet, when taking feed water under a head, with hot feed water as well as cold, for all steam pressures and for all conditions, its operation is the same and it requires no adjustment for varying steam pressures.

The lifting apparatus consisted of a steam nozzle and a combining tube. The throat of the combining tube being so much larger than the smallest opening in the steam nozzle enables it to increase or diminish the amount of water as the pressure of steam increases or decreases. As pressure of steam increases, the pressure in the forcing chamber or delivery chamber of the lifter is increased, enabling the water to enter the forcer combining tube against the increased tension of the steam from the forcing nozzle, thus enabling it to work from low pressures to high without any adjustment of either steam or water supply. The forcing or combining tube being made without any openings between its mouth and discharge end, permits of the steam and water combining up to a very high temperature, the overflow being closed positively.

The above described apparatus as made and used was provided with a separate valve for opening and closing the intermediate overflow in starting, and valve for opening and closing the steam to the forcer and the final overflow valve. While each was very simple in construction and the operating of the inspirator was easily understood, still for locomotive purposes it was considered that the functions performed by the valves above mentioned should be brought under the control of one operating lever.

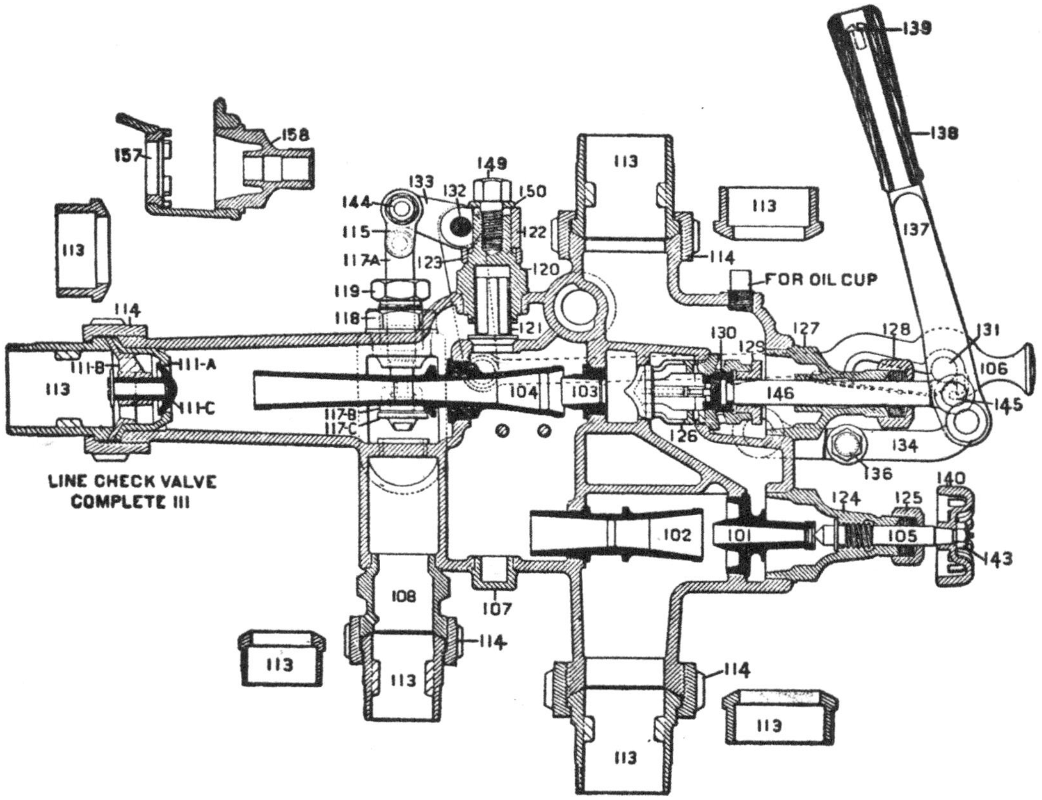

FIGURE 174

THE HANCOCK INSPIRATOR, SECTIONAL VIEW

To accomplish this, the Hancock locomotive inspirator, Fig. 173, operated by a single lever, was evolved, and is made in different types to suit different connections.

The present Hancock locomotive inspirator will work successfully with pressures of steam from 35 lbs. to 350 lbs. without any adjustment of either steam or water, and the proportions are such as to increase its quantity of water from 35 lbs. to 200 lbs., this being about the average pressures carried on locomotives, and while its maximum capacity is at 200 lbs. the percentage of decrease from 200 lbs. down to 160 lbs. is not enough to interfere with the requirements of the locomotive. It will lift water promptly on the highest lifts encountered in locomotive practice, even if the suction becomes heated or filled with hot water.

It will take feed water at a temperature of 125° reliably with a steam pressure of 200 lbs. Tests have been made where the inspirator has taken water on a lift of two feet at 132° with 200 lbs. of steam and at 140° with 140 lbs. of steam.

Regulation from Maximum to Minimum. The regulation of this machine from maximum to minimum is accomplished by simply reducing the amount of steam supplied to the lifting apparatus.

As has been before mentioned, the combining tube has no openings between its mouth and delivery end, and it admits of a positively closed overflow; hence all water passing through the combining tube must go to the boiler and cannot escape at the overflow. This condition is possible on account of the two sets of tubes, the lifting tube acting as a regulator and governor for the forcer, hence requiring no adjustment from the lowest to the highest steam pressures within its entire range.

Internal Arrangement. The intermediate overflow valve operates automatically, its only function being to give direct relief to the lifter steam nozzle when lifting or priming, and comes to its seat when the forcer steam is applied and is held there by the pressure exerted by the forcer.

An observation of the internal parts of this instrument, Fig. 174, will show at once the simplicity of its construction and the ease with which it can be repaired and parts renewed.

Material. In the manufacture of the Hancock locomotive inspirator the formulae used in the composition of the material for the several parts have been selected especially for the service each part has to perform. The tubes and valves are made of composition that does not contain zinc. Other parts are constructed of material that will produce the least wear with its companion piece. As with the high temperatures incident upon high pressure of steam, care has been taken in the selection of the composition of parts that would work in harmony without abrasion. The bodies are made of a composition containing about 10 per cent of tin.

The Hancock Inspirator Co. furnish the following table of capacities and sizes of pipe connections for type A:

TABLE 23. CAPACITIES AND SIZES OF PIPE CONNECTIONS

Size	Capacity per Hour		Pipe Connections							
	Steam Pressures		Steam		Suction		Delivery		Overflow	
	160 Pounds	210 Pounds	Iron	Copper	Iron	Copper	Iron	Copper	Iron	Copper
5	1180 gals.	1210 gals.	1¼	1½	1¼	1½	1¼	1½	1¼	1½
6	1605 gals.	1647 gals.	1¼	1½	1¼	1½	1¼	1½	1¼	1½
7	2095 gals.	2151 gals.	1½	1¾	1½	1¾	1½	1¾	1½	1¾
8	2651 gals.	2723 gals.	2	2	2	2¼	2	2	1½	1¾
9	2954 gals.	3034 gals	2	2	2	2¼	2	2	1½	1¾
9½	3274 gals.	3362 gals.	2	2	2	2¼	2	2	1½	1¾
10	3961 gals.	4068 gals.	2	2¼	2½	2¾	2	2¼	1½	1¾
11	4215 gals.	4450 gals.	2	2¼	2½	2¾	2	2¼	1½	1¾

Type A Hancock inspirator will interchange and fit the Monitor coupling connections.

Hancock inspirators, type A, sizes 5 and 6, have the

same size body and interchange. Sizes 8, 9 and $9\frac{1}{2}$ have the same size body and interchange.

Type B Hancock inspirator will interchange and fit the Sellers coupling connections.

Hancock inspirators, type B, sizes 5, 6 and 7, have the same size body and interchange. Sizes 8, 9 and $9\frac{1}{2}$ have the same size body and interchange.

The Hancock inspirators, types A, B and D, differ only in the form of the bodies and connections. All internal parts are the same for corresponding sizes. All external parts are the same for corresponding sizes, except part No. 106 (connecting rod).

Directions for Connecting and Operating. To obtain the best results, locate the inspirator with the overflow nozzle about 4 in. above the water in the tank. Take the steam through a dry pipe from the dome. Connections from the inspirator to the dome must not be smaller than the inside diameter of the size of copper pipe given in Table 23. The openings in the suction or feed pipe connections from the inspirator to the tank must not be smaller than the inside diameter of the sizes of iron pipe given in Table 23.

Overflow Pipe. An ordinary source of annoyance very often occurs from the overflow nozzle or overflow pipe becoming filled up, contracting the openings so that the inspirator will not lift or prime promptly. Sometimes it occurs that when the overflow nozzle is all free and clear the overflow pipe is apt to escape the attention of the person doing the repairs or overhauling the inspirator, and are reported not working satisfactorily. It is almost impossible to ascertain this without removing the pipe. These pipes should always be looked over and kept free.

Intermediate or Line Check Valve. The intermediate

or line check valve in the delivery pipe should receive attention. The line check valve in the delivery end of the inspirator, in case of impure water, should be looked after frequently. When the inspirator is provided with a swing check valve, care should be taken to keep this valve clear from deposits resulting from impurities in the water.

Suction Pipes. Where iron suction pipes are used, especially if the pipe is in two pieces and connected by unions, they should be carefully watched to see that they are absolutely tight and well supported, as a very slight leakage of air will materially reduce the capacity of the inspirator, and if too large a quantity of air is admitted it will cause the inspirator to break.

In General. It is very important that there should be ample steam and water supply to all types of the Hancock inspirator. It sometimes occurs that the inspirator will not work satisfactorily with the regulating valve wide open or at its maximum, but will work when this valve is partially closed or when at its minimum. This indicates clearly an insufficient steam supply. It may be due to the contracted openings in the valve next to the boiler, combination box, or too small dry pipe leading to the combination box, and should be remedied. An insufficient supply of water caused by too small size or restricted opening in the tank valve, too small opening in the goose neck leading to the tank, too small area in the strainer, a kinked or partially collapsed hose, or leaks in the suction pipe, would cause the inspirator to break.

Operation. To start the inspirator, draw the lever, part No. 137, Fig. 174, back to lift the water, then draw it back to the stop. When the lever, part No. 137, is drawn back slightly, steam is admitted to the lifter steam valve, part No. 130, through the forcer steam valve, part No. 126, to the lifter steam nozzle, part No. 101. The flow of the steam into the lifter tube, part No. 102, creates a vacuum, and causes the water to flow through the lifter tube, part No. 102, condensing the steam, and out through the intermediate overflow valve, part No. 121, and through the final overflow

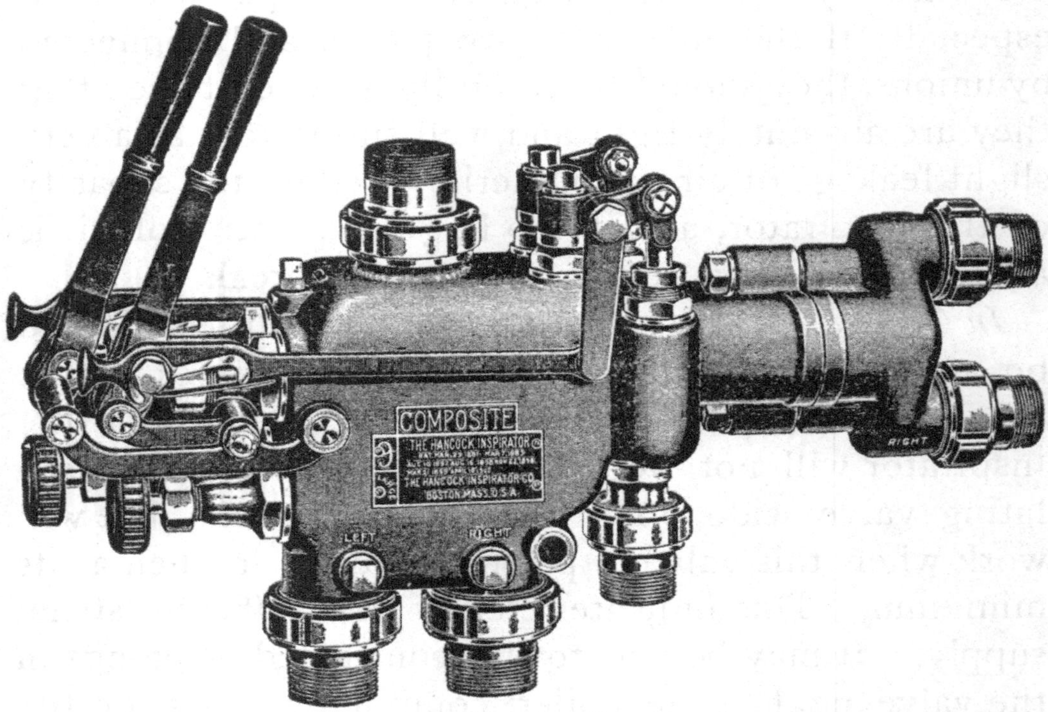

FIGURE 175

HANCOCK INSPIRATOR, TYPE COMPOSITE

valve, part No. 117, in the delivery chamber. A further movement of the lever, part No. 137, opens the forcer steam valve, part No. 126, admitting steam to the forcer steam nozzle, part No. 103, and to the forcer combining tube, part No. 104, creating a pressure in the delivery chamber sufficient to close the intermediate overflow valve, part No. 121, and open the intermediate or line check valve, part No. 111. The final overflow valve, part No. 117, will be closed and the inspirator in full operation when the lever is drawn back to the stop. When the pin in the wheel of the regulating valve is at the top, the inspirator will deliver its maximum quantity of water; to reduce the feed, turn the regulating wheel to the right.

Regulating. To use the patent heater attachment, lift the connecting rod, part No. 106, until disengaged from the stud in the lever, part No. 131, then draw back the connecting rod to close the overflow valve, part No. 117. Draw the lever back to the point used in lifting. This will usually give all the steam that is required for a heater. If the amount going back is too large, regulate it by the regulating wheel to give just the amount required, as with the lever in the position described all the steam blowing back would pass through the lifter nozzle. Thereby the closing of the main steam valve at the boiler becomes unnecessary.

FIGURE 176 a
THE HANCOCK MAIN BOILER CHECK VALVE

Type Composite. The Hancock composite inspirator, Fig. 175, consists of two separate and individual inspirators within one body or casing, which can be operated separately or simultaneously, as desired. Where it may be desired to locate both injectors on one side of the locomotive, convenient to either the engineer or fireman who has charge of pumping the engine, or on the boiler butt, available to both, the advantages of the composite are apparent. Owing to the limited room in the cab, it is generally difficult to locate both instruments so that they can both be operated by the engineer and be equally convenient.

It occupies but little more space than a single inspirator or injector, and owing to its compactness it has been found that it can be located in positions where in the past it has not been possible to locate two separate instruments.

It places both instruments directly under control of the engineer, and both are equally convenient to operate, the result being that both instruments are operated and kept in good order.

Each instrument has an independent suction pipe, delivery pipe and line check valve, thus enabling each to be operated independent of the other.

In attaching the composite inspirator (either to back head or side of boiler), one steam valve, one steam pipe, one overflow pipe and one opening into the boiler are dispensed with, thus effecting a very considerable saving of material and labor which would be required with two separate instruments.

The operation of the Hancock composite inspirator is the same as the Hancock inspirator, types A, B and D. To operate either instrument, draw the lever back until the water is lifted, then draw it back as far as it will go. To put both instruments in operation, start one and then the other.

It is desirable to use a double check valve in connection with the composite inspirator.

The Hancock Boiler Washer. Fig. 177. A perfectly simple and durable apparatus for washing boilers.

Will either lift water from 15 to 20 feet or take it under a head.

At a steam pressure of 100 lbs. the temperature of the delivery water will be about 120° Fahrenheit, or as hot as it can be conveniently handled.

The hose or delivery end of the boiler washer is left blank, to be threaded to fit hose couplings in use, or will be threaded as desired.

TABLE 24. CAPACITIES AND SIZES OF PIPE CONNECTIONS

Size	Capacity per Hour, Steam Pressure 60 Pounds	Pipe Connections		Discharge Nozzle for end of Delivery Hose
		Steam	Suc. and Delivery	
Small.	2200 gals.	¾ inch.	1½ inch.	¾ inch.
Medium . . .	3900 "	1 "	2 "	1 "
Large	6000 "	1¼ "	2½ "	1¼ "

Pipe Connections. For steam, suction and delivery connections, see above table.

Place a globe valve in the steam pipe for a starting valve, and another valve in suction pipe for a water valve.

FIGURE 176 b
SECTIONAL VIEW OF THE HANCOCK MAIN BOILER CHECK VALVE

If the boiler washer is to lift water, there should be an overflow or outlet pipe not smaller than one inch in size connected to the delivery pipe. Place a valve in this overflow pipe.

Operation. If the boiler washer takes the water

under a head, open the water valve in the suction pipe and then open the valve in the steam pipe.

Vary the temperature of the delivery water by regulating the steam and water supplies with either the starting or water valve or both.

If the boiler washer lifts the water, open both the valve in the overflow pipe and the water valve in the suction pipe, and give steam with the starting valve.

When water appears at the overflow, close the valve in the overflow pipe, and vary the temperature of the

FIGURE 177

THE HANCOCK BOILER WASHER

delivery water by regulating the steam and water supplies with either the starting or water valve or both.

STEAM GAUGES

The theory and action of the Bourdon spring gauge has already been discussed to some extent, and will not be enlarged upon except to give some illustrations of the latest improved types of pressure gauges with which modern locomotives are fitted.

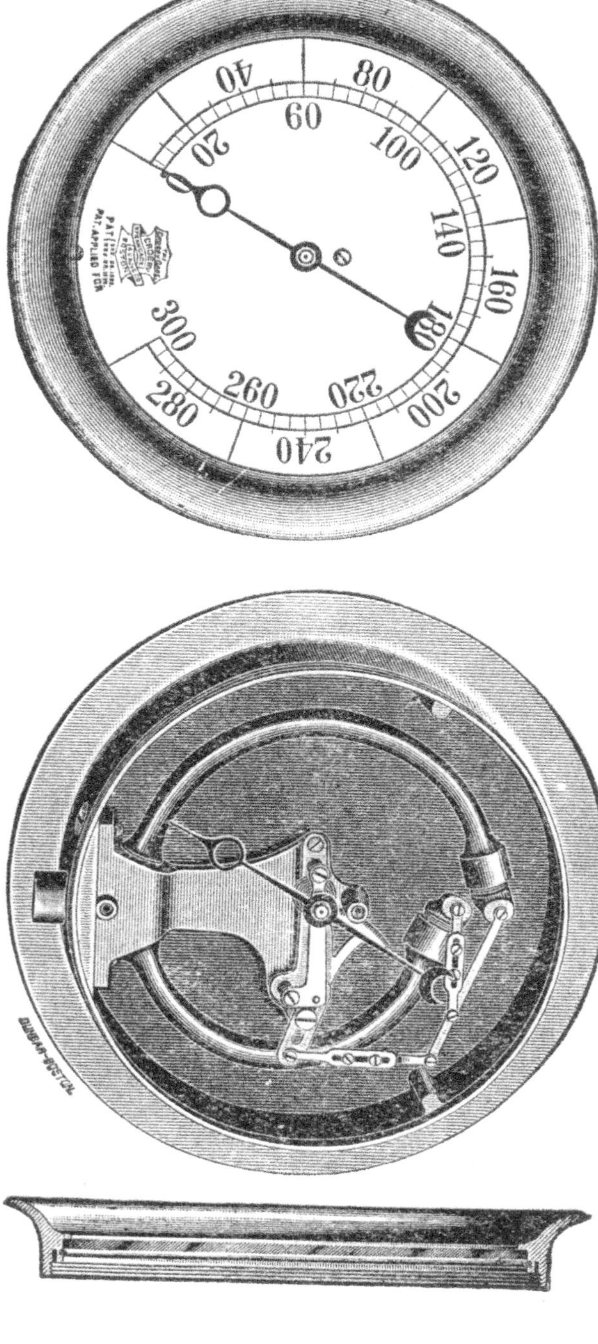

SECTIONAL VIEW

FIGURE 178

CROSBY LOCOMOTIVE PRESSURE GAUGE

Crosby Improved Locomotive Pressure Gauge. Attention is called to the cut, Fig. 178, of the Crosby improved locomotive pressure gauge, showing the Bourdon tube springs and their mechanism. It will be observed that the tube springs are attached to the socket and to the tips, to which the lever mechanism is connected in a new way.

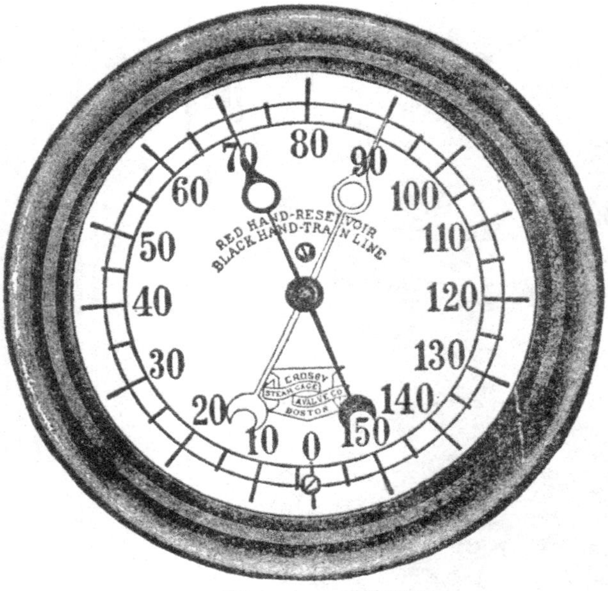

FIGURE 179
CROSBY DUPLEX GAUGE

The method practiced is to have these attachments of the tube springs made by means of solder or other metal of low-fusing point, and, while this may be safe in all low-pressure gauges, or where in their location in use they are not subjected to great heat, yet it is hazardous where high pressures of steam are to be measured by them, especially where there is liability of the admission of such steam into the tube springs. In such case the soldering material may soften, and under the high pressure be forced out, causing a leak and the destruction of the gauge.

In the Crosby improved locomotive pressure gauge, the tube springs are connected at each end with their respective parts by screw threads, without the use of any soldering material whatever, thus insuring tight joints under all conditions of heat and pressure.

In addition to the improved Bourdon tube springs so employed, careful attention has been given to the lever mechanism which transmits the free movements of these Bourdon tube springs to the index. They have been designed and constructed not only to convey the full movement of the tube springs, but so that they may be renewed without difficulty in case of repairs or reconstruction.

American Locomotive Gauge, with non-corrosive movement, Figs. 181 and 182. It is constructed of metals of superior quality, to withstand the constant vibration to which it is subjected. The spring is made of very heavy seamless drawn tube of superior quality. The connections are made of hard phosphor bronze. The movement is made with a wide faced sector which will outwear three of the ordinary sectors. The pinion and sector shafts are made of hard phosphor bronze, and the hair spring is made of bronze, making the gauge non-corrosive, rigid, and adding materially to the life of the gauge.

POP VALVES

Safety Valves. One of the prime causes of boiler explosion is the gradual and insidious increase of the pressure of steam beyond the endurance of the boiler; but to every boiler there is a limit of pressure within which it is substantially safe. This point should be ascertained by hydraulic test annually, and no excess of pressure beyond this limit should be allowed at any time. The only sure preventive is a safety valve which is all its name implies. The diameter of a safety valve is not a test of its efficiency. A valve is effective in direct proportion to its lift, other things

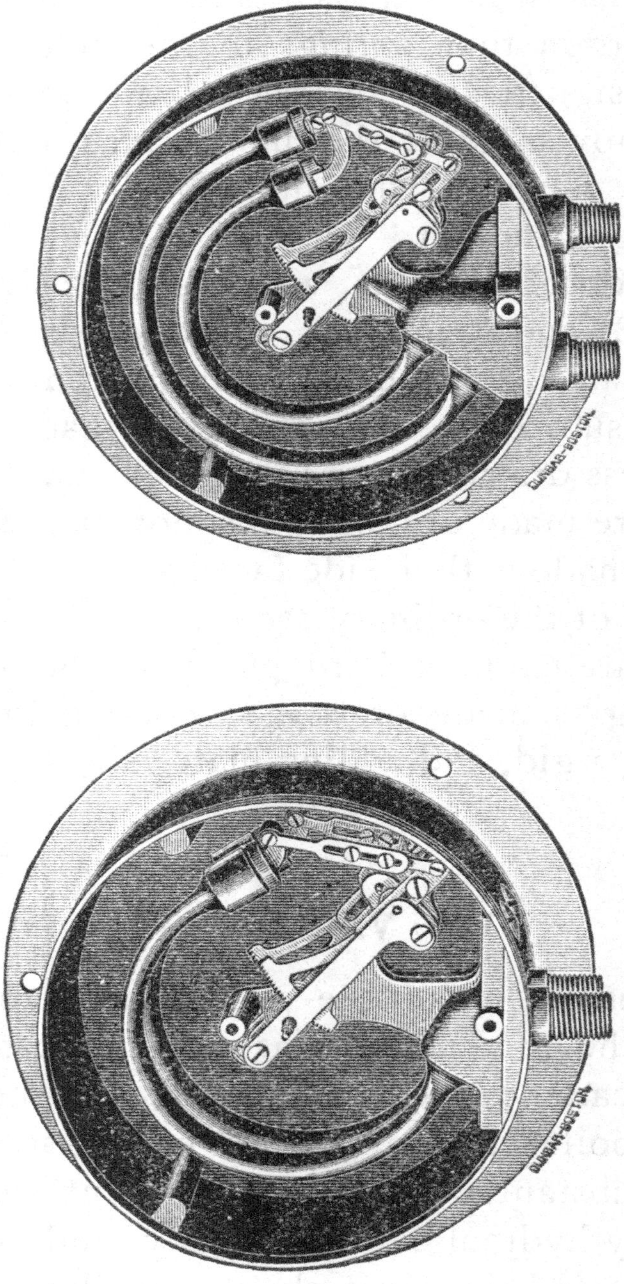

FIGURE 180
CROSBY DUPLEX GAUGE, SECTIONAL VIEW

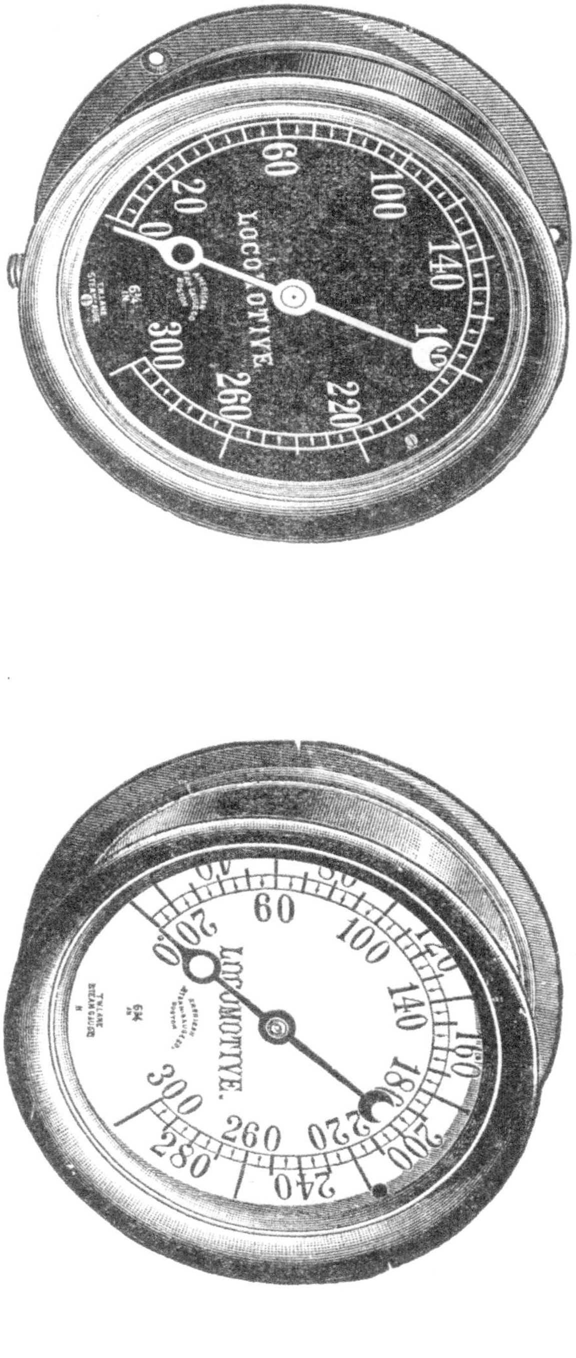

BLACK DIAL

WHITE DIAL

FIGURE 181

THE AMERICAN LOCOMOTIVE PRESSURE GAUGE

being equal. The higher the pressure of steam, the less will a common safety valve lift; at most, its lift is very slight, and with the increasing pressure of steam, the lift will not increase sufficiently to relieve the boiler under all circumstances; but the pressure can and may increase until an explosion occurs, while the valve is in operation. The common safety valve has much to answer for. Owing to the great friction of its parts, it will not open until the pressure is above what it is set at; it will continue to blow after the pressure of the steam has fallen far below the point of opening; it wastes large quantities of valuable steam in operation. There are other grave faults, but these stated are sufficient to condemn it. Instead of standing guard over the boiler, a sentry has to be set over it, and should he by accident, ignorance or negligence, not properly attend to his duties, the boiler is without any safeguard whatever. Hence the importance of any device which shall reduce the danger to a minimum. A safety valve which is automatic, certain in its action, prompt in opening and closing at the required point of pressure, and which can be fully relied upon to relieve the boiler under all circumstances, is what is necessary.

FIGURE 182
SECTIONAL VIEW
AMERICAN PRESSURE GAUGE

The Crosby Locomotive Pop Safety Valve. Fig. 186. *Description of the Valve.* The valve proper **B B** rests

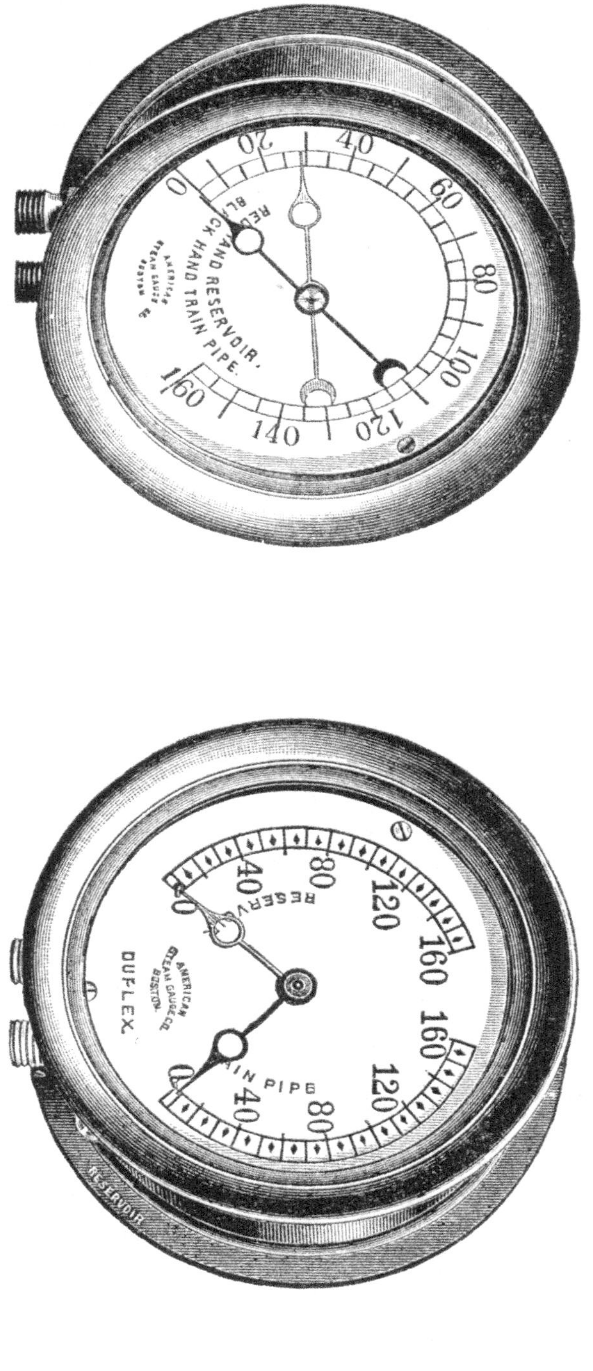

FIGURE 183

AMERICAN DUPLEX AIR-BRAKE GAUGE, WESTINGHOUSE STYLE

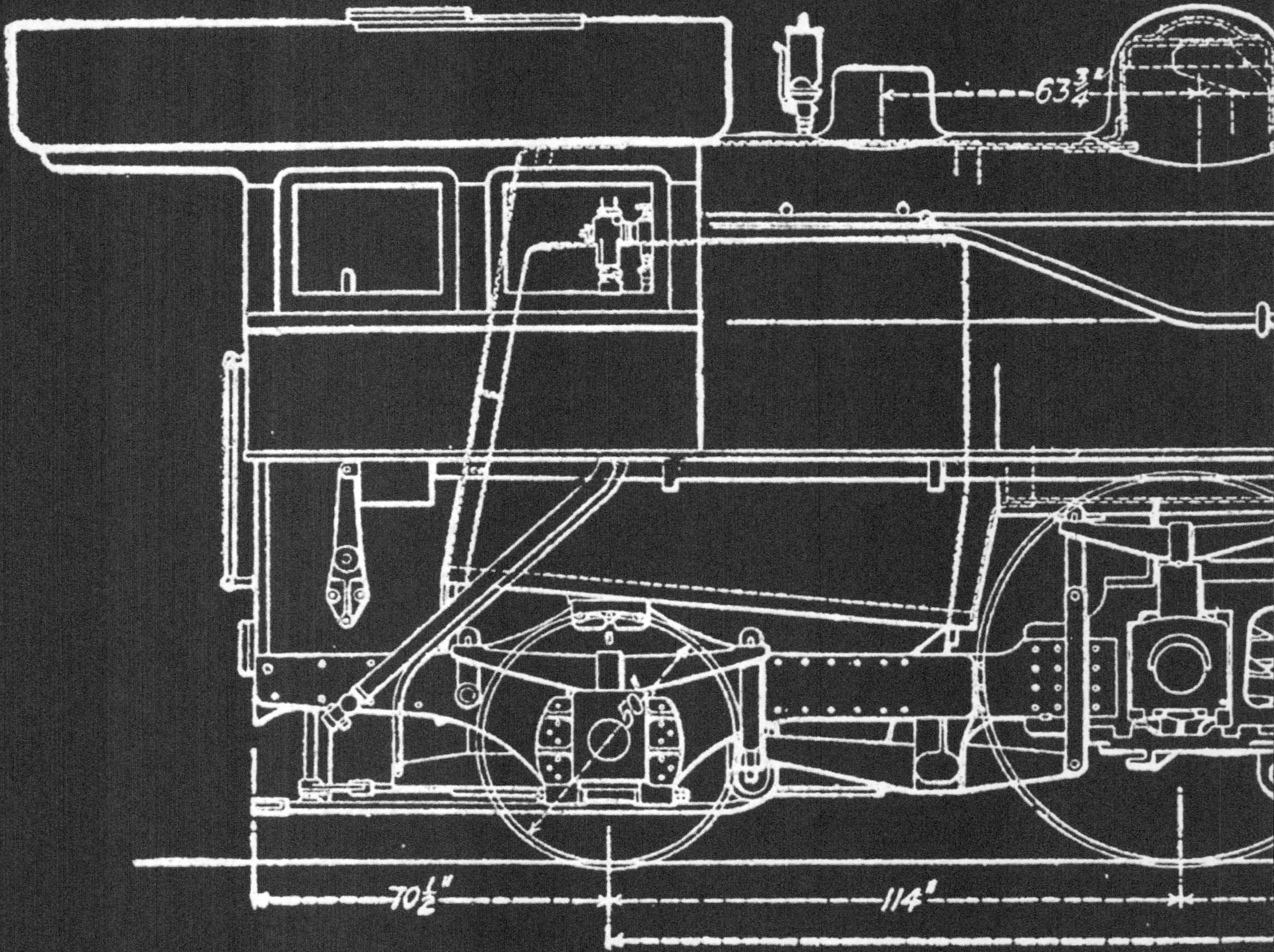
63 3/4"
50
70 1/2"
114"

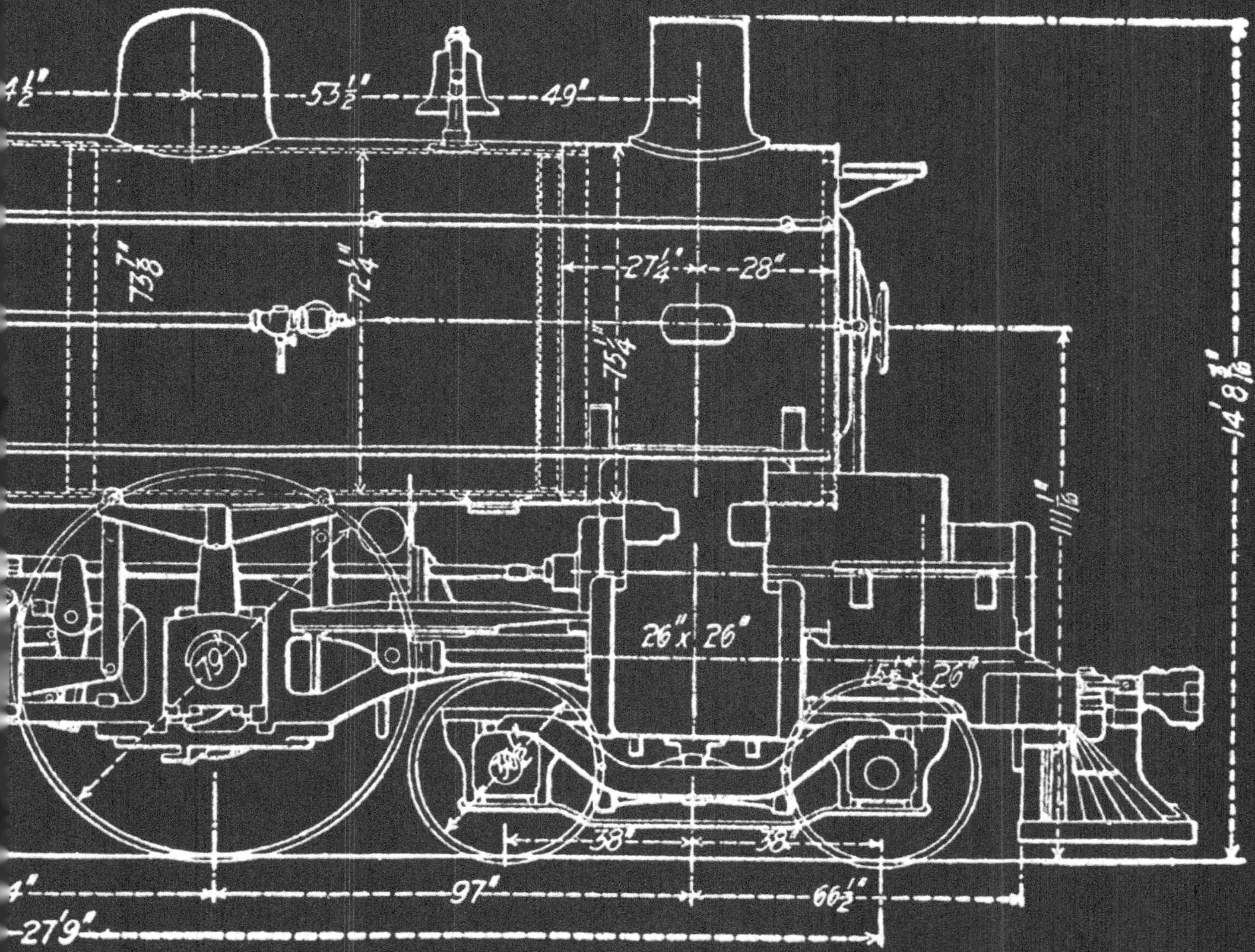

4½"
53½"
49"
7⅜"
72¼"
27¼"
28"
75¼"
14'8 3/16"
111 1/16"
26" x 26"
15½" x 26"
79"
38"
38"
4"
97"
66½"
27'9"

upon two flat annular seats V V and W W on the same plane, and is held down against the pressure of steam by the steel spiral spring S. The tension of this spring is obtained by screwing down the threaded bolt L at the top of the cylinder K. The area contained between the seats W and V is what the steam pressure acts upon ordinarily to overcome the resistance of the spring. The area contained within the smaller seat W W is not acted upon until the valve opens.

FIGURE 184
AMERICAN DUPLEX AIR-BRAKE GAUGE, WESTINGHOUSE STYLE

The larger seat V V is formed on the upper edge of the shell or body of the valve A. The smaller seat W W is formed on the upper edge of a cylindrical chamber or well C C, which is situated in the center of the shell or body of the valve, and is held in its place by arms D D, radiating horizontally, and connecting it with the body or shell of the valve. These arms have passages E E for the escape of the steam or other fluid from the well into the air when the valve is open. This well is deepened so as to allow the wings X X of the valve proper to project down into it far enough to act as guides, and the flange G is for the purpose of modifying the size of the passages E E and for turning upward the steam issuing therefrom.

Action of the Valve when Working under Steam. When the pressure under the valve is within about one pound of the maximum pressure required, the valve opens slightly, and the steam escapes through the outer seat into the cylinder and thence into the air; the steam also enters through the inner seat into the well, and thence through the passages in the arms to the air. When the pressure in the boiler attains the maximum point, the valve rises higher and steam

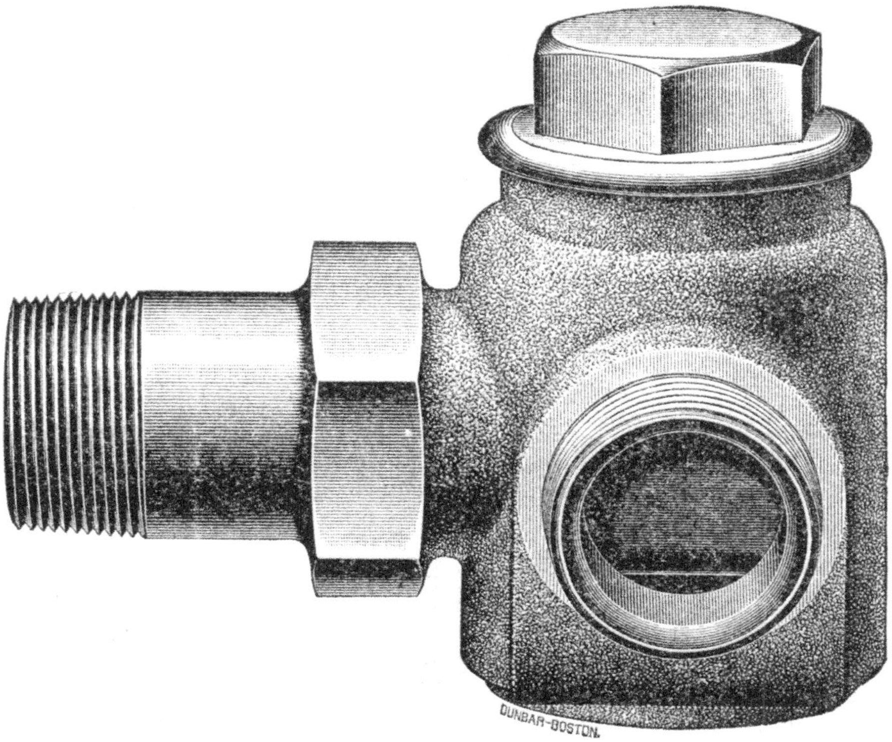

FIGURE 185

CROSBY SPRING-SEAT LOCOMOTIVE CHECK VALVE

is admitted into the well faster than it can escape through the passages in the arms, and its pressure rapidly accumulates under the inner seat; this pressure, thus acting upon an additional area, overcomes the increasing resistance of the spring, and forces the valve wide open, thereby quickly relieving the boiler. When the pressure within the boiler is lessened, the flow of steam into the well is also lessened, and the pressure therein diminishing, the valve gradually settles down; this action continues until the area of the opening into the well is less than the area of the apertures in the arms, and the valve promptly closes.

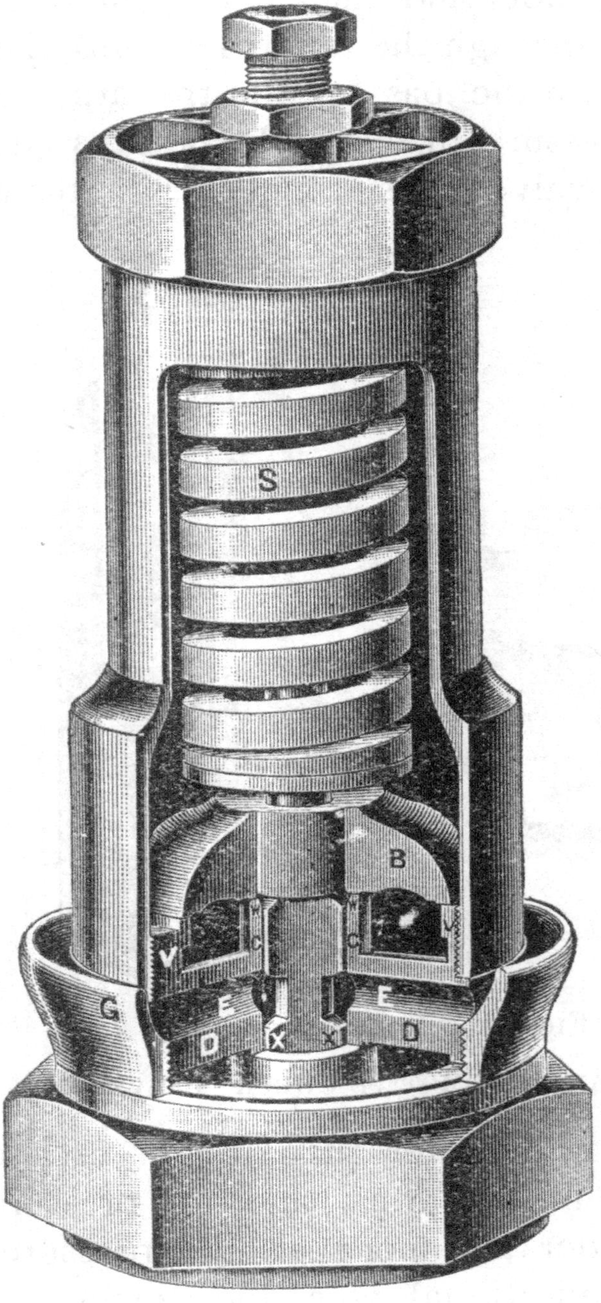

FIGURE 186
CROSBY LOCOMOTIVE POP SAFETY VALVE

The point of opening can be readily changed while under steam by screwing the threaded bolt at the top of the cylinder up for diminishing, or down for increasing, the pressure.

The seats of this valve are flat, and do not cut or wear out and leak so readily as beveled seats. The valve is made of the best gun metal.

Directions. *Setting.* Screw the head-bolt which compresses the spring up for diminishing, or down for increasing, the pressure, until the valve opens at the pressure desired, as indicated by the steam gauge;

secure the head-bolt in this position by means of the lock-nut; for regulating the loss of escaping steam, turn the screw ring G up for increasing, or down for decreasing it.

Caution. Care should be taken that no red lead,

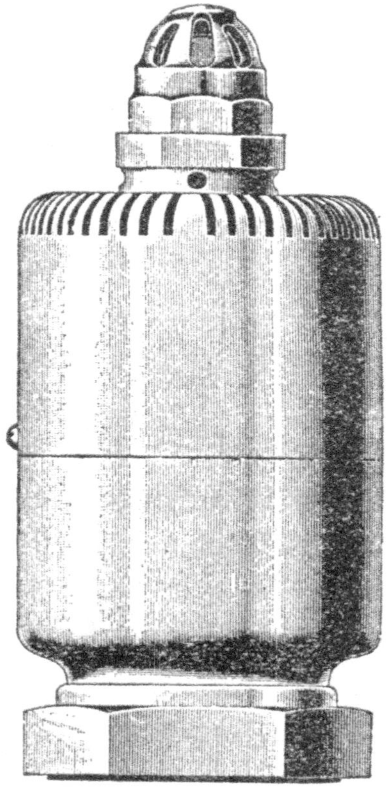

EXTERNAL VIEW

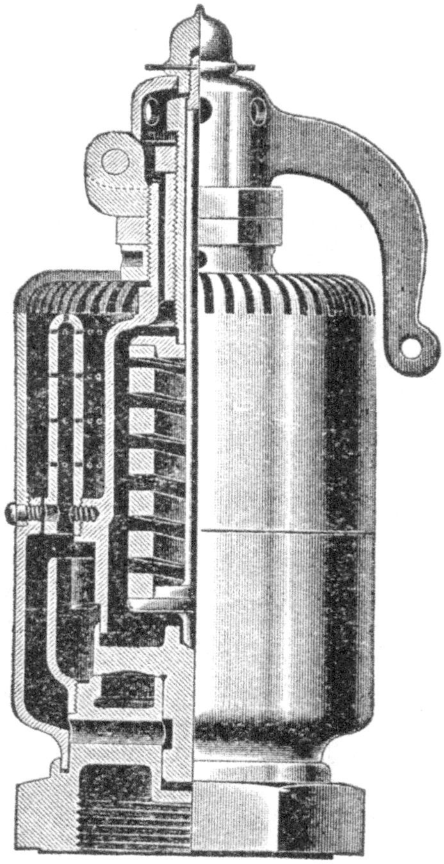

SECTIONAL VIEW WITH LEVER

FIGURE 187

CROSBY MUFFLED LOCOMOTIVE POP SAFETY VALVE

chips, or any hard substance be left in the pipes or couplings when connecting the valve with the boiler. Never make a direct connection by screwing a taper thread into the valve, but make the joint with the valve by the shoulder.

Repairing. This valve, having flat seats on the same plane, is very easily made tight if it leaks, by following these directions, viz.: With an ordinary lathe slightly turn off the two concentric seats of the valve and valve shell or base respectively, being careful that this is done in the same plane and perpendicular to the axis of the valve. The valve will then fit tightly on

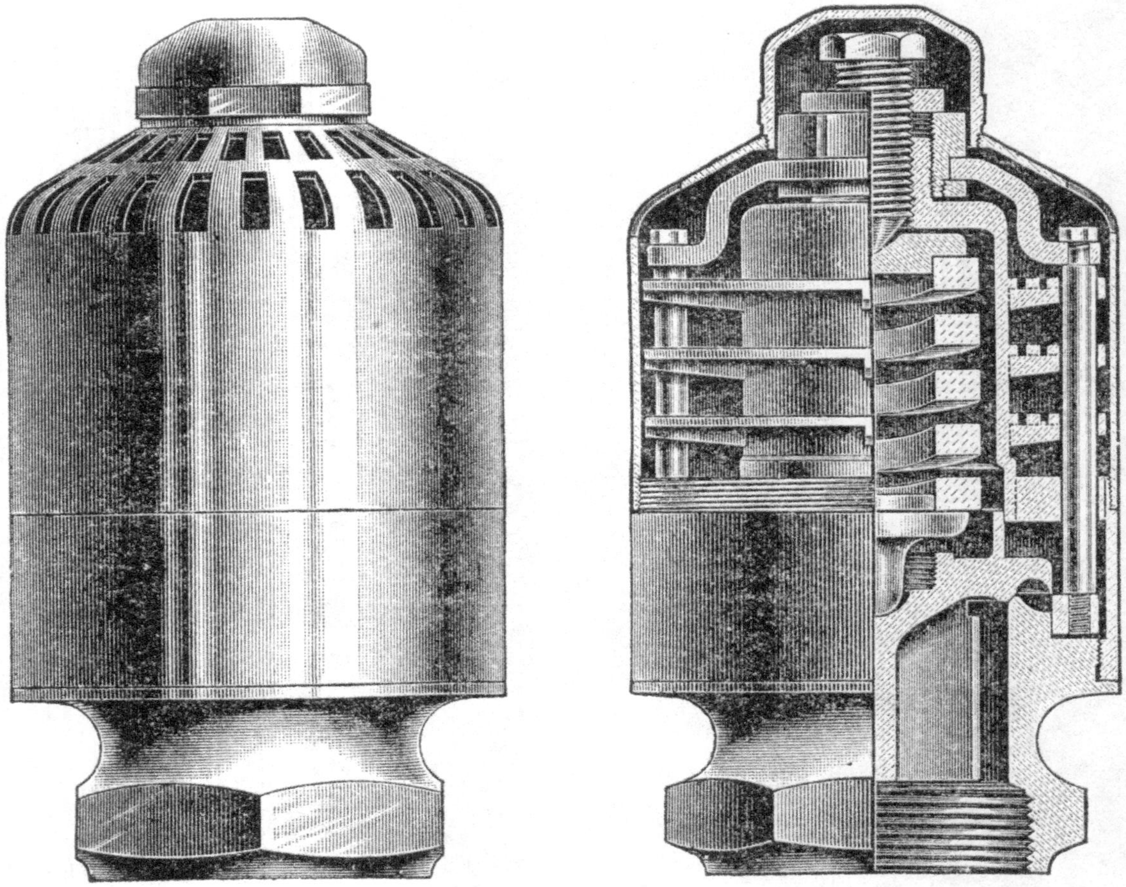

FIGURE 188

AMERICAN LOCOMOTIVE SAFETY VALVE

the valve shell. If no lathe is at hand, then grind the valve proper on a perfectly flat surface of iron or steel, until its two bearings are exactly on a plane and with good smooth surfaces; then take the shell and grind its seats in precisely the same manner; rinse both parts in water and put together, and the valve will be found to be tight; to ascertain when the bearings are on the same plane, use a good steel straight edge. Do not grind the valve to its seats on the shell by grinding them together, but grind each part separately as above stated.

The Crosby pop safety valve and muffler combined, possesses outside means for adjusting or regulating it under changes of pressure, without disturbing its connections or parts.

American Improved Locomotive Muffled Pop Safety Valve. This valve has many new features, one of which is the manner in which it is adjusted; both the blowing-off pressure and the blow-down are adjusted from the top of the valve without removing the muffler casing—simply remove the small top cap. The blowing-off pressure is adjusted by means of a compression screw, the same as is used in all locomotive valves. The blow-down is adjusted by means of a hexagon nut just below the compression screw. This nut is connected by means of a yoke and standards to the relief ring, which is raised or lowered to adjust the blow-back where the case may require, the nut acting as a swivel. The same lock-nut locks both the compression screw and the blow-down nut, as shown in the sectional cut, Fig. 188.

FIGURE 189
SPRING FOR AMERICAN LOCOMOTIVE POP VALVE

The noise of the escaping steam is lessened by the muffling device, as shown in cut.

Springs. These springs are made of the highest grade of steel, carefully tempered, and are ground square on the end. Each spring is tested to double its capacity.

Crane's Patent Improved Pop Safety Valves. Fig. 190. Their construction embodies a self-adjusting feature automatically regulating the "pop" of the valve; in other words, maintains the least waste of steam between the opening and closing points, an improvement which will be readily recognized, as there is no necessity of readjusting to regulate the "pop" on reasonable changes in the set pressure.

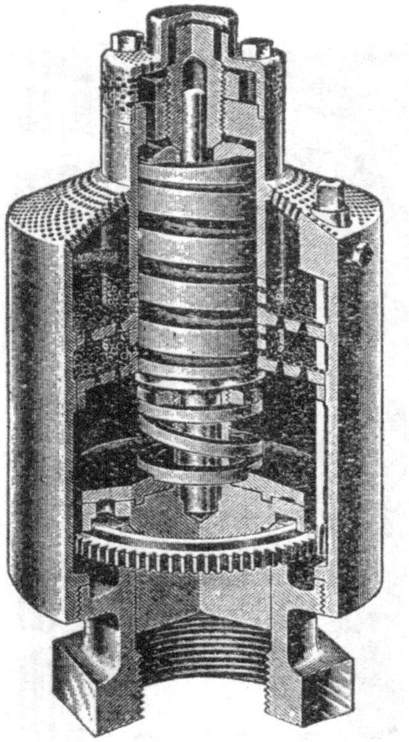

FIGURE 190
CRANE POP VALVE, SECTIONAL VIEW

This is more clearly explained, as follows: In all pop safety valves it is necessary to have a "pop," or huddling chamber into which the steam expands when main valve opens, thereby creating an additional lifting force proportionate to this increased area and greater than the force of spring, thus holding the valve open until pressure is relieved. Means must also be provided to relieve this "pop" chamber of pressure in order to allow the valve to close promptly and easily. This is accomplished by a self-adjusting auxiliary valve and spring, which are entirely independent of the main valve and spring; and to further explain their operation, the steam in "pop" chamber finds a passage through holes or ports into an annular space provided in the auxiliary valve or disc, and by reason of the light auxiliary spring, this pressure lifts the auxiliary valve and allows the steam in "pop" chamber to gradually escape, thus permitting a greater range in setting pressures with the least waste of steam and at the same time supplying a cushion or balancing medium, thereby preventing any chattering or hammering and affording the easiest possible action in closing.

To change pressure, unscrew the top bolts and remove the cap, slacken lock nut; to increase pressure, turn screw plug down (to the right); to decrease pressure, turn screw plug up (to the left), then tighten lock nut.

Should the valve waste too much steam between the opening and closing point, turn the outside pop regulator to the left until the desired waste is obtained. Should it work too close, turn regulator to the right for greater waste.

FIGURE 191
CRANE MUFFLED POP VALVE

Crane's Patent Locomotive Muffler Pop Safety Valve. Fig. 191. This valve is made with the outside "pop" regulator conveniently arranged at the top of the valve, for quick and easy adjustment, and while the boiler is under pressure.

TABLE 25

Size Valves, inches..........	2½	3	3½
Dome Connection, inches...	2	2½	3

The Kunkle Lock-Up Pop Safety Valve. Fig. 194. The pressure screw of this valve is made of hard tempered brass and can be removed at any time without being troubled with rust. The whole of the valve is made of brass throughout, with the exception of the pressure spring, which is made of the best steel and is thoroughly excluded from steam or dampness, making every part of the valve adjustable and free from rust. The spring is set between two tapered points, as will be seen by reference to open view. Another point in its favor is the uniform bearing of the valve upon its seat by reason of its long ribbed chamber, which guides the valve so accurately as not to allow it to cap on one side, thereby preventing the steam from cutting away the seat.

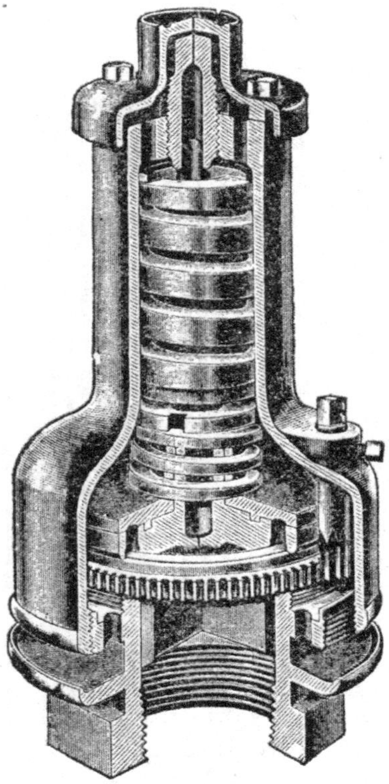

FIGURE 192
CRANE MUFFLED POP VALVE
SECTIONAL VIEW

The valve with its regulator can be adjusted to go off suddenly without a loud pop or without loss of boiler pressure.

The Kunkle Lock-Up Pop Safety Valve and Muffler Combined. Figs. 194 and 195. It relieves itself of all over-pressure, without raising above what it is set at; will close down upon its seat without losing any of the desired or fixed pressure, and by means of its lock-up device, so ingeniously constructed, all unauthorized persons are prevented from tampering with it. It is provided with a pressure screw with the tapering point resting in the center of the top plate on top of the spring, and is locked with a key, and a jam-nut holding it firmly to is place, which makes it perfectly safe from being tampered with.

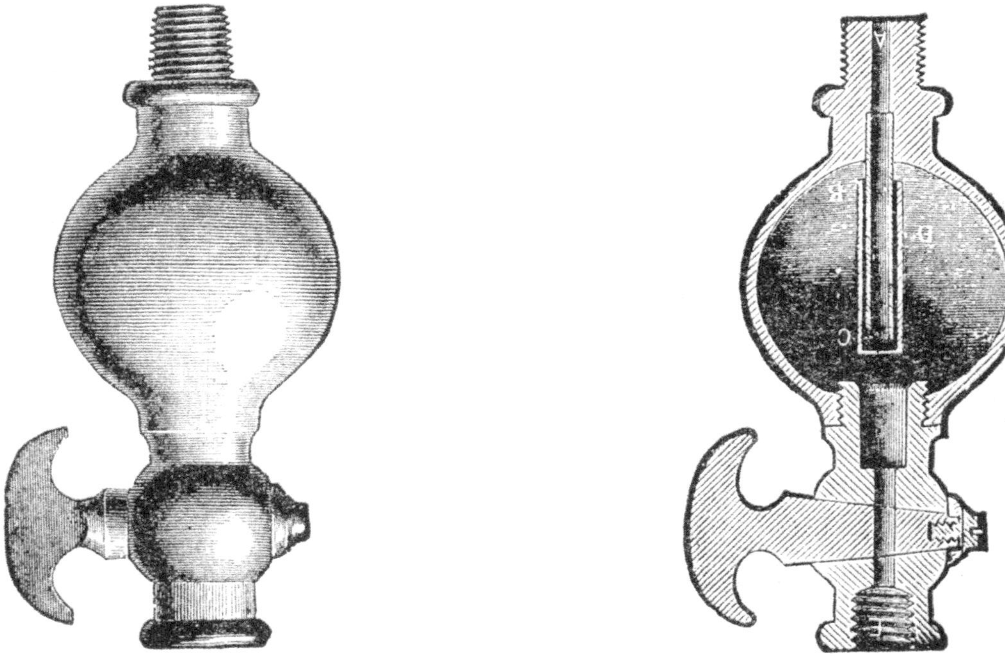

FIGURE 193

PRINDLE'S PATENT SYPHON COCKS

LUBRICATORS

In the "good old days," when tallow was the only known lubricant for the valves, it was one of the duties of the fireman on approaching a station or drifting down a hill, to seize the tallow pot and make a rush for the front end to oil the valves. This, with the weather at zero or below, was not a very desirable job, to say the least, and in course of time somebody thought out the plan of running pipes from the steam chests back to the cab, where they were fitted with a sort of funnel with a shut-off cock attached, and the melted tallow was poured in there, and the vacuum in the valve chests when the throttle was closed caused the tallow to find its way to the valve seats. But, thanks to progress and invention, tallow has been displaced by refined cylinder oil, and the tallow cup has given way to the modern sight-feed lubricator; consequently the lubrication of the valves has been much simplified, for the reason that the oil is constantly entering the valve chests drop by drop as it should. Of course the lubricator may get out of order at times, or the oil pipes become clogged, but these are troubles that are easily overcome as a rule. A few of the leading types of lubricators will be illustrated and described.

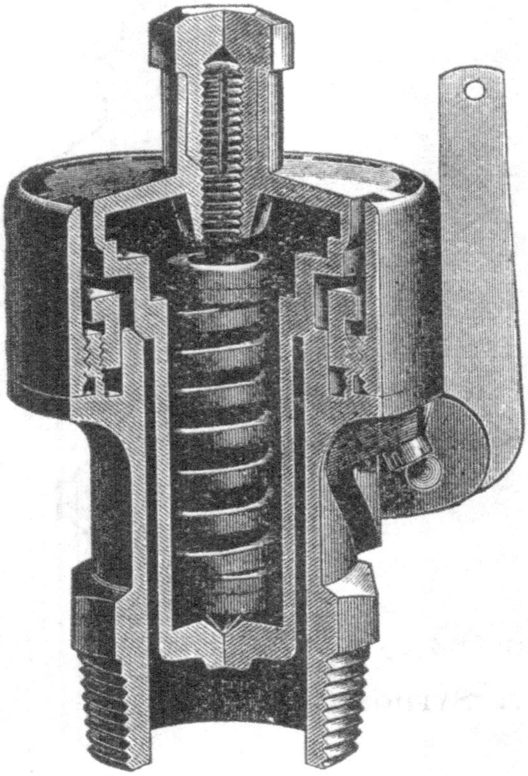

FIGURE 194
KUNKLE POP VALVE
FOR PORTABLE AND STATIONARY BOILERS

The Nathan New Bull's Eye Lubricator. Fig. 196.

General Features. The oil reservoir of the lubricator is of cylindrical form, which is generally acknowledged to be most suitable for high pressure.

The lubricator is provided with hand oilers for the cylinder feeds, and with gauge glasses which indicate when the reservoir is nearly empty.

The lubricator carries a reserve glass, packed in its casing, ready for use whenever occasion requires.

All glasses are packed in casings which screw into

the body, making their removal for inspection or repairs very convenient.

Directions for Application. 1. Secure the lubricator to boiler head or top of boiler, in the usual manner.

2. Connect for steam to fountain or turret, if large enough, otherwise direct to boiler. The steam pipe must not have less than ¾-in. I.D. when iron pipe is used, and not less than ⅝-in. I.D. when copper pipe is used. Steam valves and their shanks must have openings fully in accordance with these dimensions.

3. Oil pipes must have a continuous fall towards the steam-chest, without any "pockets" in them.

Directions for Operation. Fill the cup with clean, strained oil through filling plug A, and immediately after filling, open water valve D. Open steam valve (not shown), wait until sight-feed chambers are filled with water, then start and regulate the feed by opening regulating valves C, more or less, according to the feed desired.

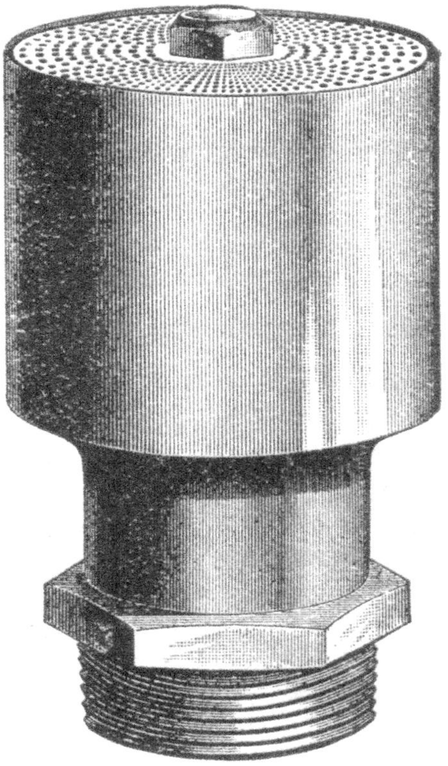

FIGURE 195
THE KUNKLE POP VALVE FOR LOCOMOTIVE BOILERS

To stop either of the feeds, close the respective regulating valve C.

To renew supply of oil, close all valves marked C and valve D, draw off water at waste cock W, then fill the cup as before, and open water valve D, immediately after filling, whether the feed is started again, or not.

FIGURE 196

FRONT VIEW

To oil by hand, close the steam valve. Fill the hand oilers O, open the hand oiler valves, and when all the oil has entered the tallow pipes, close hand oiler valves and open steam valve wide.

Notes. 1. Always open the steam valve before the engine begins to do any work whatever, whether the feed is started right away or not, and keep it open as long as the engine is doing service of any kind.

2. Keep the water valve D always open, except during the period of filling the cup, as per direction.

3. Once in two weeks, at least, blow out the cup with steam, opening all valves wide, with the exception of the filling plug, which should remain closed.

4. When putting on the lubricator for the first time,

or after it has been off for repairs, "follow up" the packing nuts of the glasses, when the lubricator gets hot, so as to take up any "slack" caused by expansion. This will tend to keep the joints tight.

The Detroit No. 21 Triple Feed Locomotive Lubricator, with auxiliary oilers and gauge glass. Fig. 200.

This device is simple in construction and simple of operation.

The oil is maintained at a uniform temperature and will not chill.

The feed is absolutely regular.

All feeds are visible from two sides.

An additional valve has been placed at the top of the lubricator to control the supply of steam from the boiler, making the device self-contained.

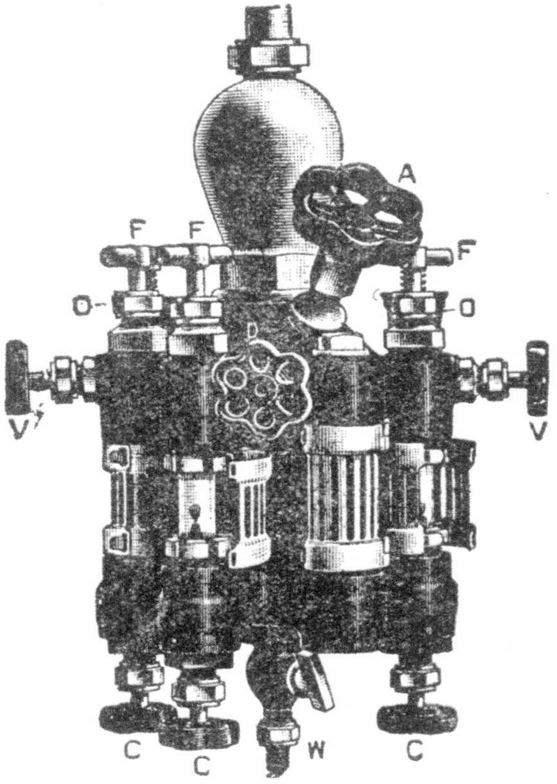

FIGURE 198
NEW NATHAN LUBRICATOR, FRONT VIEW

Directions for Operating. When the lubricator is first applied, blow out thoroughly, then close all the valves.

To Fill. Remove filler plug O and fill the reservoir with clean, strained oil.

Steam Valve. The regular boiler valve should be left wide open, and the steam valve B at top of condenser must also be kept wide open while the locomotive is in service.

To Start Lubricator. 1. Be sure that the regular boiler valve is open. Then open steam valve B at top of condenser gradually until wide open and keep wide open while lubricator is in operation. Allow sufficient time for condenser and sight-feed glasses to fill with water. 2. Open water valve D. 3. Regulate flow of oil to right and left cylinders by valves E E, and to air pump by valve L.

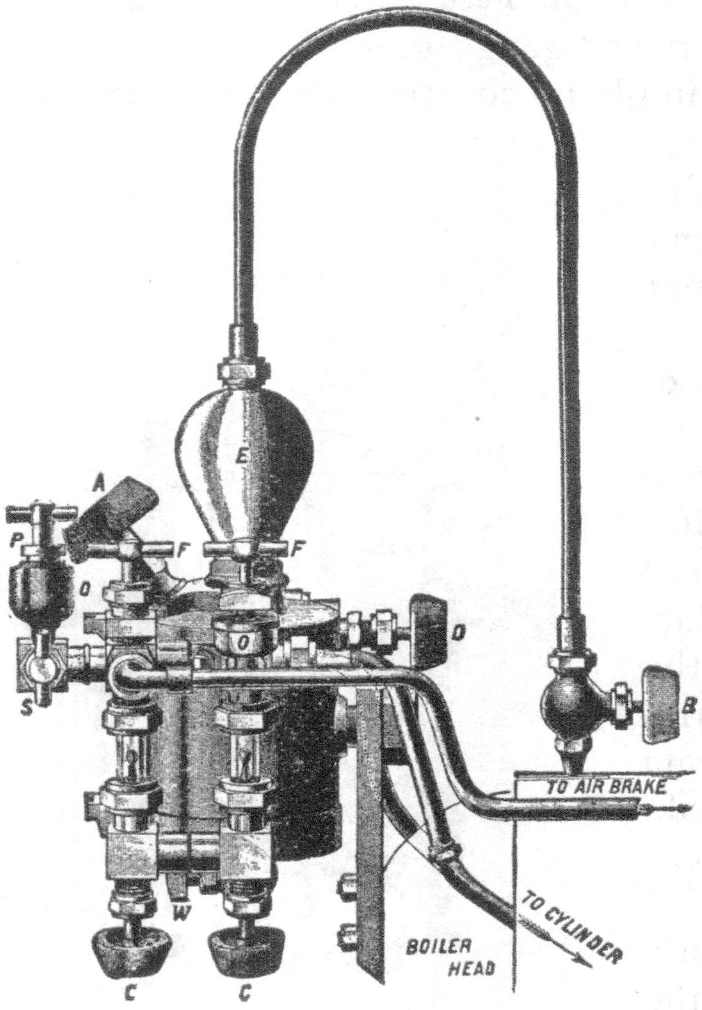

FIGURE 199

NEW NATHAN LUBRICATOR

SIDE VIEW WITH STEAM AND TALLOW PIPES

To Operate Auxiliary Oilers. See that valve H is closed. Then open valve X and fill body of oiler. Close X after filling and open valve H.

To Refill. Always close valves E E and L in advance of valve D. Open drain plug G, then filler plug O. Refill and proceed as before.

Immediately after filling the lubricator do not fail to open the water feed valve D, in order to prevent any excessive pressure due to the expansion of the heated oil.

Getting New or Rebuilt Locomotive Ready for Service. In getting a new or rebuilt locomotive ready for service, disconnect oil pipes at steam chest, and blow out thoroughly both oil pipes and automatic steam chest valves; also disconnect coupling to air pump and see that choke is free.

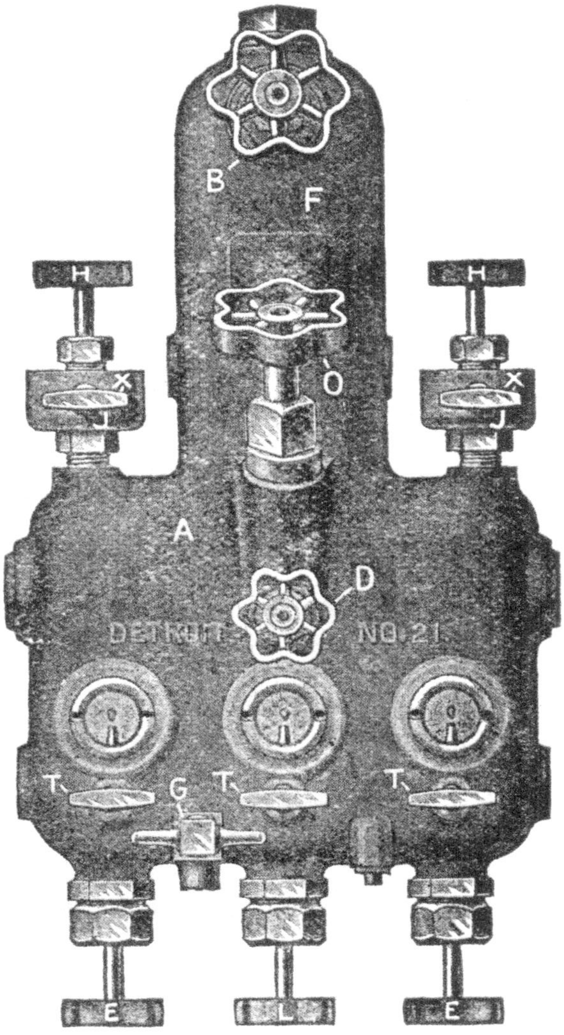

FIGURE 200
THE DETROIT NO. 21 LOCOMOTIVE LUBRICATOR

Steam for lubricator should be taken from turret if large enough, or from dome through an independent dry pipe of 1-in. iron pipe size or its equivalent.

When the No. 31, or four-feed lubricator is applied to the Vauclain type of compound, the two outer feeds are intended for the high-pressure cylinders. Two automatic steam chest plugs are furnished, stamped and tagged "H.P.," having $\frac{3}{32}$-in. chokes. The two remaining feeds lead to the low-pressure cylinders, and two other automatic steam chest plugs are furnished to be used with them, stamped and tagged "L.P." and having $\frac{1}{16}$-in. chokes.

Helpful Hints. *Blowing Out.* Blow out lubricator once a week.

Filling. If there is not sufficient oil to fill the lubri-

FIGURE 201

DETROIT 4-FEED LUBRICATOR

cator, always use water to make up the required quantity. This will enable the feeds to start promptly.

The steam valve B must be opened wide when the locomotive is in service to allow condensation to enter the condenser; otherwise condensation will be diverted to equalizing tubes. The feeds will gradually slow down as the water of condensation decreases.

When getting a new or rebuilt engine ready for service or when using soda ash boiler compounds, or when running in bad water districts, impurities will be carried over into the condenser and will gradually accumulate at base of water valve until the water is completely shut off. While this is taking place the feeds are affected the same as before described, and when this passage is finally closed by the sediment the feeds will cease altogether.

How to rectify while locomotive is in service: Close all feeds and water valve; open drain cock 2105 and allow about ½ pint of water to drain off; close drain cock and open water valve quickly. The condenser pressure will then force this sediment into bottom of lubricator, where it can be blown out in the usual manner, when the lubricator is empty.

Do not screw up too tightly the feed glass follower, as this will only serve to injure the packing. There is no danger of leakage at this point, as the glass and packing are so designed that the greater the pressure the better the joint.

Small Drop of Oil. The cause of a small drop of oil or the variation of size of drop during a trip. In alkali, salt water or oil well regions through which railroads pass, the water supply becomes impregnated with saline matter. This saline matter is carried over in the lubricator mechanically by the steam, so that the water in the sight feed glasses contains considerable of it and the amount increases as the locomotive proceeds on the trip until it crystallizes around the feed cones, thus gradually diminishing the size of the opening for the drop. Should the engineer undertake to force the feed it will result in the oil flowing in a very slender stream, scarcely perceptible. If this condition is not corrected the salt crystals will completely close the feed cone orifice.

How to Rectify. Close all feed stems; open all sight feed drain stems and blow out thoroughly. The action of the steam on feed cones will dissolve the salt crystals. Allow reasonable time for condensation; start the feeds, and the drop of oil will be normal.

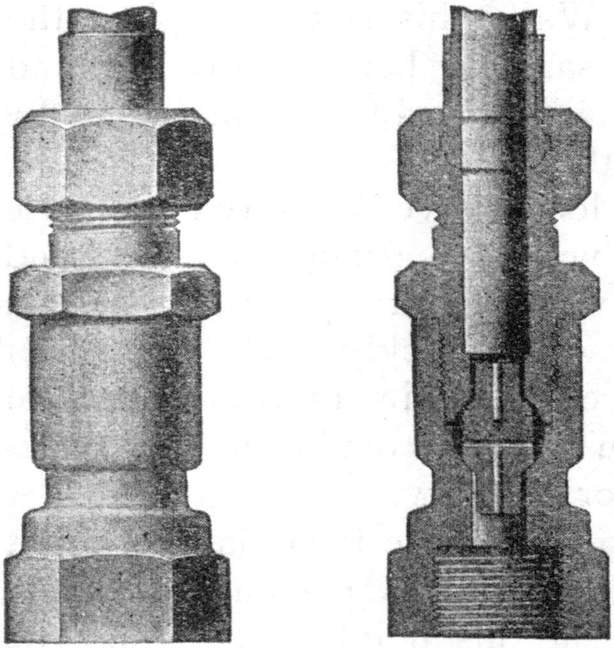

FIGURE 202
AUTOMATIC STEAM CHEST PLUGS AND VALVES

Air Bound. This condition is almost invariably brought about whenever it becomes necessary to fill a lubricator on the road. The temperature of a lubricator at such times is very nearly that of the steam pressure temperature. Sometimes the water feed valve seat may leak, and in order to fill the lubricator in this heated condition it is found necessary to shut off all steam pressure to the lubricator, including the air pump, and owing to the high temperature of the condenser the water flashes into steam, practically emptying the condenser and feed glasses of all water. The oil reservoir being very hot, the oil expands rapidly and the filler plug is usually put in before the reservoir is full. The steam and water pressures are hurriedly turned on and the feeds are opened before sufficient time has elapsed for sufficient condensation to accumulate. The feeds will not respond under such conditions, because the positive and negative pressures have equalized, and the lubricator is said to be air bound.

How to Overcome. Open all feeds and any one of the sight feed drain stems. This will allow the water in the oil tubes and the air occupying the highest space in the oil reservoir to escape to atmosphere.

The New Nathan Triple Sight-Feed Locomotive Lubricator. *Directions for Application.* 1. Secure the lubricator to boiler head or top of boiler by a strong brace, in some such form as illustrated. Fig. 197.

2. Connect top of lubricator to dome, top of boiler or bridge pipe by copper or brass tubing, which must not be less under any circumstances than $\frac{5}{8}$-in. inside measurement.

3. Connect the oil or tallow pipes to the union couplings on top brackets of the lubricator. The elbow on the top front bracket on right side is for the oil pipe to the air brake pump.

4 Remove valves from plugs over steam chests in order to maintain proper lubrication when steaming.

The sight-feed and gauge glasses of these lubricators are provided with proper shields and protectors, to prevent the flying of particles in case of a broken glass.

Directions for Use. Fill the cup with clean, strained oil through the filling plug A, and immediately after filling, open the water valve D. Open the steam valve B, and start to regulate the feed by opening the regulating valves C more or less, according to the quantity desired.

To stop either of the feeds, close the respective regulating valve C.

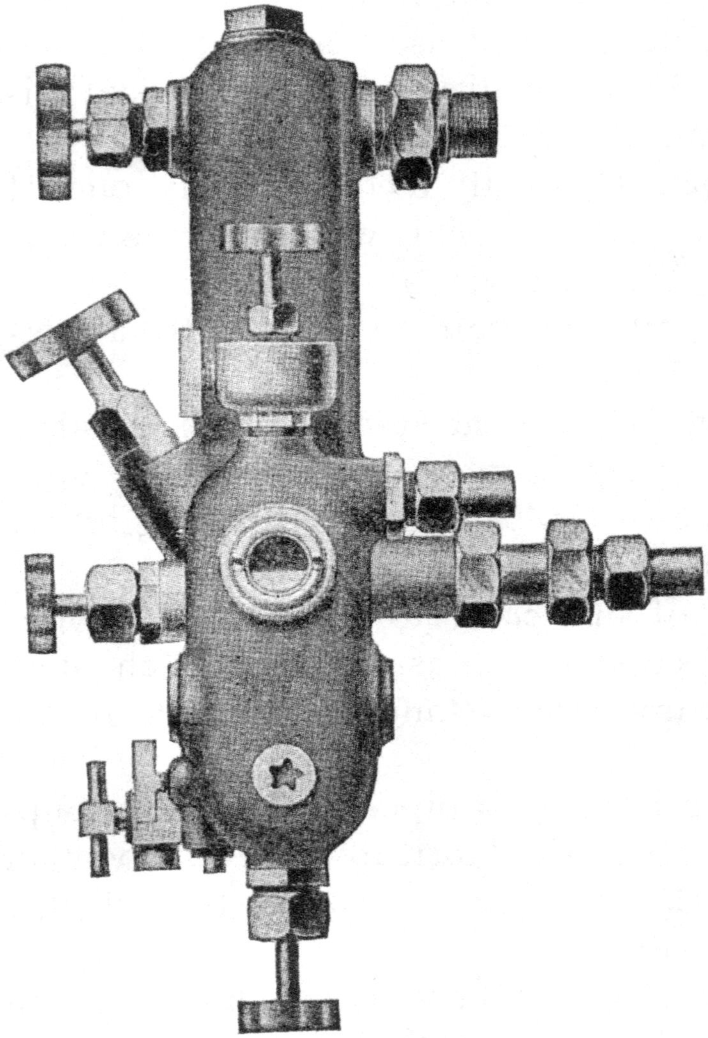

FIGURE 203
DETROIT NO. 21, SIDE VIEW

To renew the supply of oil, close all the C and D valves, draw off the water at the waste-cock W, then fill the cup as before and open the water valve D immediately after filling, whether the feed is started again or not.

Notes. 1. Valves F F F must be always kept open except when one of the glasses breaks. In such case close valves C and F belonging to the broken glass and use the auxiliary oiler O on that glass on down grades, as a common cab oiler.

The breaking of one glass does not interfere with the proper function of the others.

2. Always open the steam valve before the engine

begins to do any work whatever—whether the feed is started right away or not, and keep it open as long as the engine is doing service of any kind.

3. Keep the water valve D always open except during the period of filling the cup, as per directions.

4. Once in two weeks at least, blow out the cup with steam; open the valves wide, with the exception of the filling plug, which should remain closed.

MECHANICAL BELL RINGERS

Locomotive bell ringers are no longer a luxury—they are a necessity.

The duties of the fireman, who used to ring the bell, have increased with the increased size and speed of locomotive. A man furnishing coal to an up-to-date fire-box has little time to do much else when the machine is in motion.

The engineer's attention must not be taken from his work—cannot be with safety.

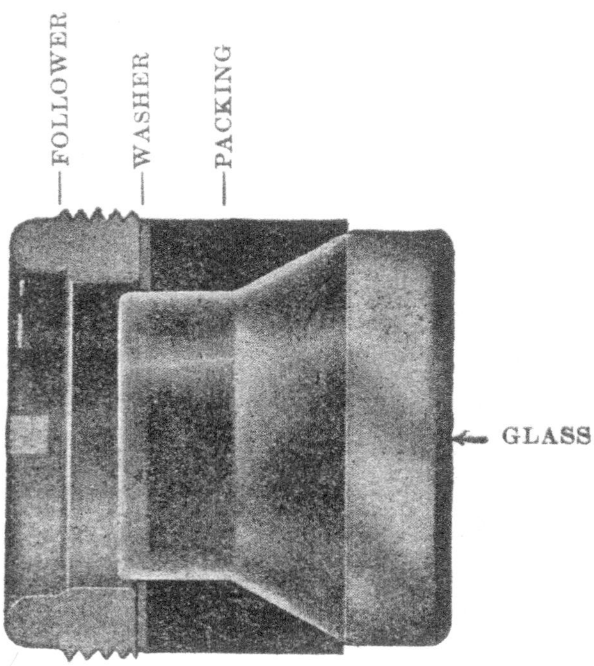

FIGURE 204.—DETROIT LUBRICATOR GLASS AND METHOD OF PACKING

DIRECTION OF PRESSURE INDICATED BY ARROW. THE HIGHER THE PRESSURE THE BETTER THE JOINT

The Sansom Bell Ringer, Fig. 205, lays claim to the following merits:

Has no packed joints except the piston, which has heavy, leather packing.

FIGURE 205
SANSOM BELL RINGER

Valve is a plug cock held to seat by a coiled spring.

Economical of air, because weight of bell compresses air in the cylinder almost to lifting point before admission.

Easily regulated to ring at any speed simply by control of air supplied.

FITTINGS

Patent Steam Sanding Apparatus for Locomotives. Fig. 206. (Nathan Mfg. Co., New York.)

Description. The ordinary methods for sanding the track in front of locomotive wheels being imperfect and unreliable, this apparatus has been designed with the view of obviating that difficulty and of greatly increasing the adhesive and tractive power of locomotive engines. This is effected by means of a combined jet of steam and sand projected to the point of contact between the driving wheels and the rails.

The advantages of the apparatus are:

1. The certainty of delivering sand at the proper point between the wheels and rails.
2. Saving in sand by being able to regulate the quantity delivered according to existing necessities.
3. Dispensing with rods and levers for working sand gear.
4. Capability of being applied to present sand boxes without any alteration.
5. Simultaneous delivery of sand at the point of contact for both wheels, whereby the liability to injure the crank, driving axles and coupling rods is greatly reduced.
6. Additional train resistance avoided, as no superfluous sand is left on the rails as by the old method.

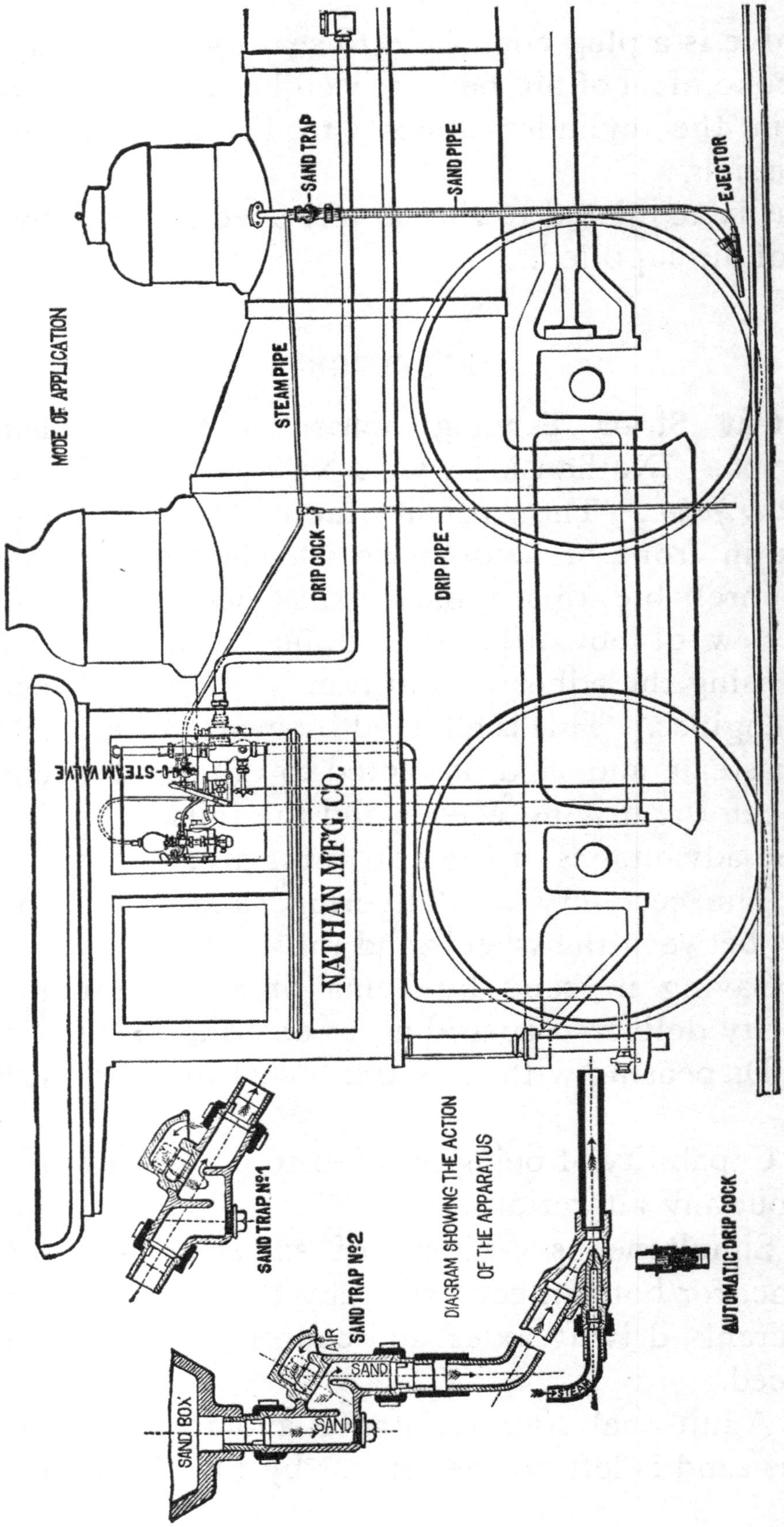

Figure 206
Steam Sanding Apparatus

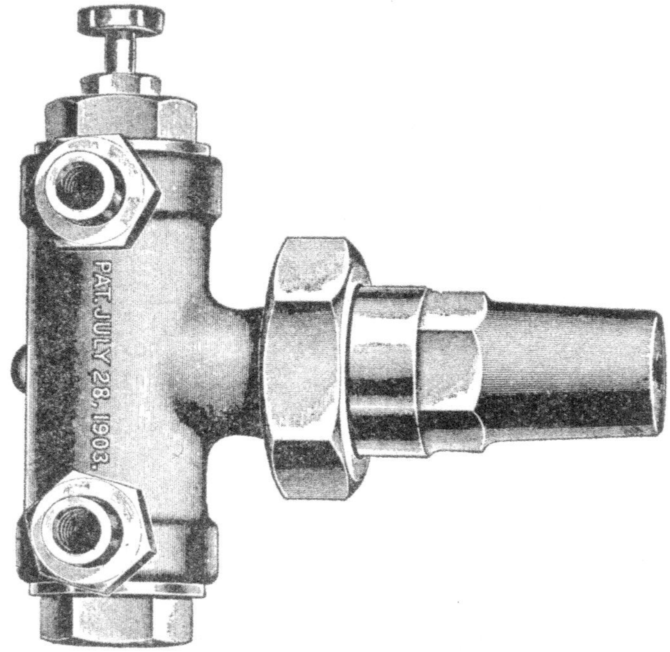

BACK VIEW

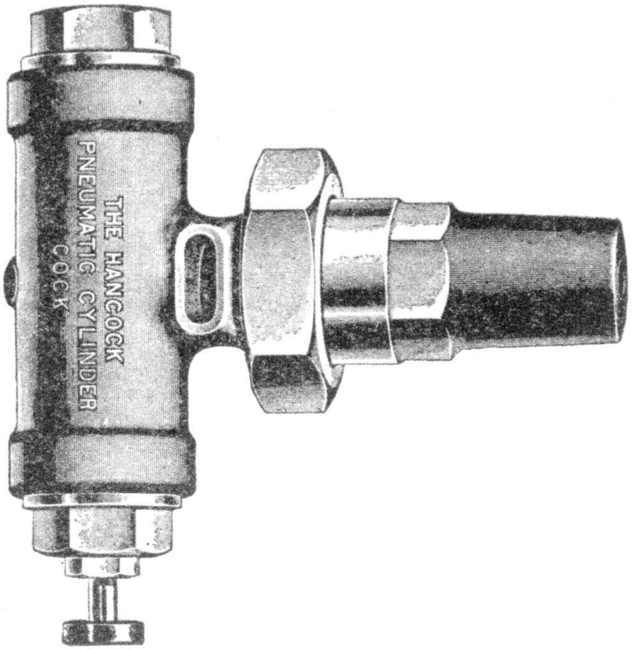

FRONT VIEW

FIGURE 207

HANCOCK PNEUMATIC CYLINDER COCK

7. Economy of steam; there being no slipping or unnecessary revolving of locomotive wheels in starting engine.

To obtain these advantages the following directions must be strictly adhered to:

1. The sand must be absolutely dry and finely sifted. Stones cannot pass through the apparatus, but will stop its proper function.

2. The cover of the sand box must be water-tight, to prevent the sand from getting wet by rain or snow beating in.

SECTIONAL VIEW

FIGURE 208

HANCOCK PNEUMATIC CYLINDER COCK

Directions for Application. Attach the sand trap to sand box or sand pipes by means of the union couplings in the most convenient position; the plug P must point straight downward and the air snout S must point in the direction of the delivery.

Connect the ejector to sand pipe in such a way that the blast will strike in about the center between tire and rail.

Attach the steam valve to dome, bridge pipe or other high place on boiler.

Connect the automatic drip cock to steam pipe and make the steam pipe to incline towards the drip cock in both directions, as illustrated. This will lead the condensed water, resulting from a leakage of the steam valve, down the drip pipe and thus prevent the freezing of the ejector pipe in cold weather.

Branch off the steam pipe to the ejector on each side of engine at a point between ejector and drip cock. The standard connections are made for 1-in. iron pipe for the sand traps. The steam pipe should be $\frac{3}{8}$-in., and connections on ejector and drip cock will be furnished blank or cut for $\frac{3}{8}$-in. iron, or drilled for $\frac{1}{2}$-in. copper, according to orders. The bottom of the drip cock is tapped for $\frac{1}{4}$-in. drip pipe.

The traps are made in two forms.

Trap No. 1, for oblique position, is most suitable for direct connection to sand box placed on top of boiler.

Trap No. 2, for vertical position is most suitable for connection to sand box placed on or below the running board.

Directions for Use. To start, open steam valve.

To stop, close steam valve.

One complete set of the single apparatus, for sanding in one direction, consists of one steam valve; one drip cock; two traps and two ejectors.

One complete set of the double apparatus, for sanding in both directions, consists of one double steam valve; two drip cocks; four traps and four ejectors.

The Hancock Pneumatic Cylinder Cock. Fig. 207. The Hancock pneumatic cylinder cocks are operated by air instead of the usual sliding bar with the connections to the cab. These cocks are composed of a small cylinder with a piston under each valve which is similar to the ordinary cylinder cock valve The piston is provided with an incline similar to the usual incline on the bar which lifts the valve, which is very similar to the ordinary cylinder cock valve. The movement of this piston back and forth opens and closes the valves. Air is admitted to each cylinder through a ¼-in. pipe, making four ¼-in. pipe connections to each cylinder of an engine. These are con-

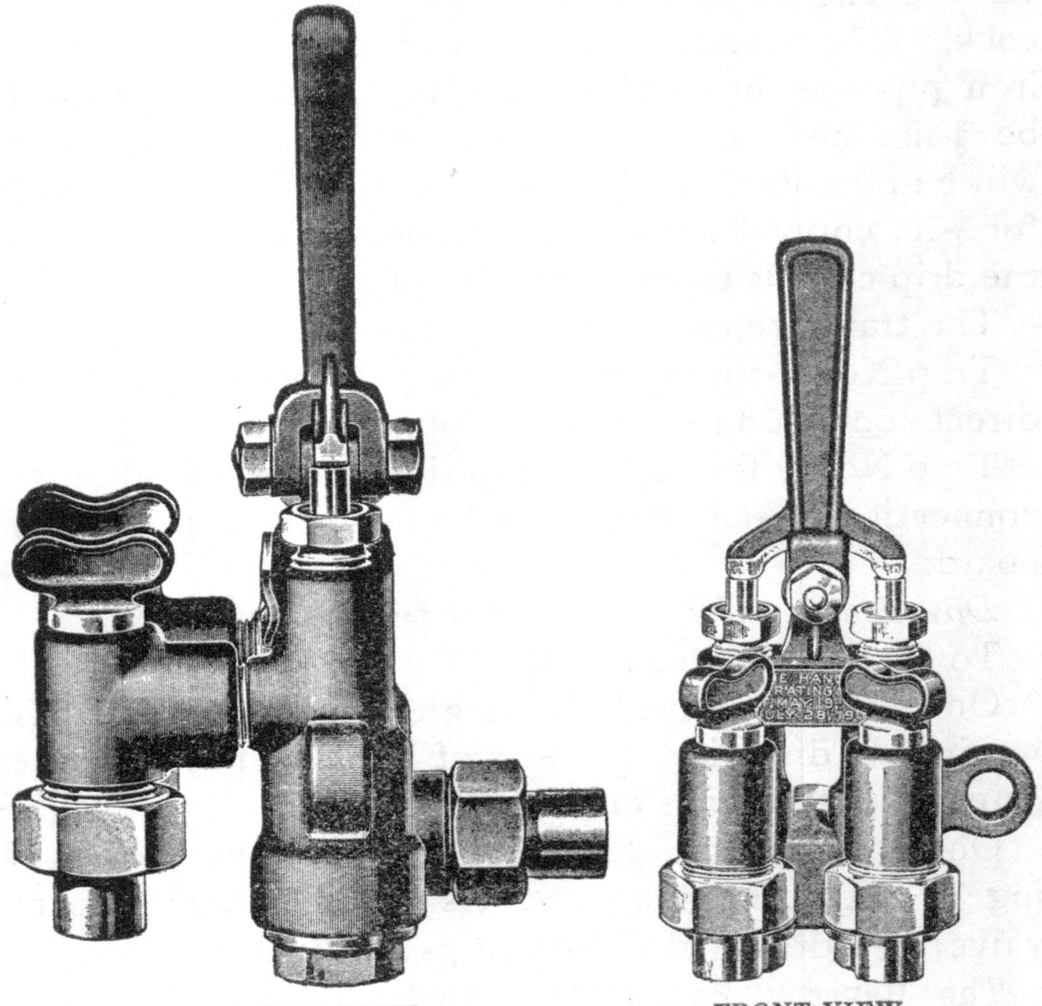

SIDE VIEW FRONT VIEW

FIGURE 209

OPERATING VALVE FOR CYLINDER COCK

nected to two ⅜-in. pipes run under lagging and connected to the operating valve in the cab. One of these pipes connects with the closing end of each cylinder cock. The other pipe connects with each opening end of each cylinder cock. The operating valve is also connected by one 3/8-in. pipe to the main reservoir.

To operate the cylinder cock, the operating handle or lever is pulled in one direction to open, the opposite direction for closing.

Each end of each cylinder cock is provided with a small hole which allows for the exhausting of all the air after it has thrown the piston in either direction. A hole also in center allows for any condensation or drip that may enter the cylinder when the cylinder cocks are open.

The operating valve is a very simple device, opening with a lever against a pressure of air and closed by pressure of air aided by a spring under each valve, and it has been found in practice that by opening and closing the operating valve, no matter how quickly, it will effect the purpose, and a very small quantity of air is consumed.

The two pipe connections on the operating valve leading to the cylinders are provided with a plug, which can be unscrewed by hand if occasion requires, and a small amount of kerosene or signal oil introduced, the plug screwed up and this oil blown through the pipe and valves for the purpose of cleaning, and will do it very effectually.

The illustration shows how the cylinder cocks and operating valve, Fig. 210, are located and piped.

Boiler Washing and Testing Apparatus. The new Rue washing and testing apparatus, will wash out, fill, and apply pressure to a boiler, with hot water. It has a capacity of 5000 gallons per hour.

When this apparatus is used, the boilers are washed much more effectually than can be done with cold

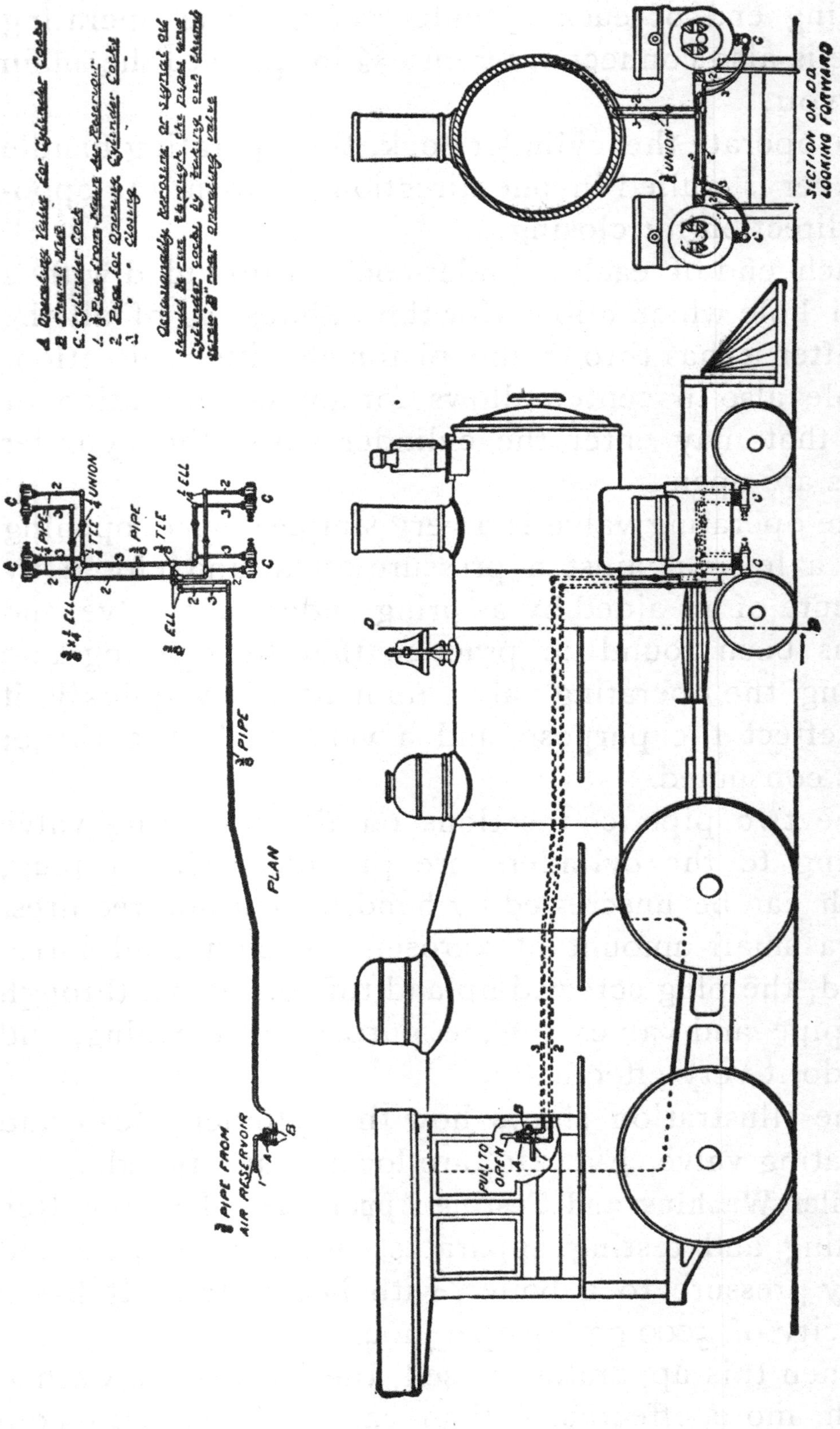

FIGURE 210
THE HANCOCK CYLINDER COCK AND OPERATING VALVE

water, and their temperature is not materially reduced.

It enables one to blow out, wash and fill with hot water and have engine ready for service within one hour, without injury to the boiler.

When applying pressure, this apparatus will produce and maintain from three to five times the amount of steam pressure used in operating it

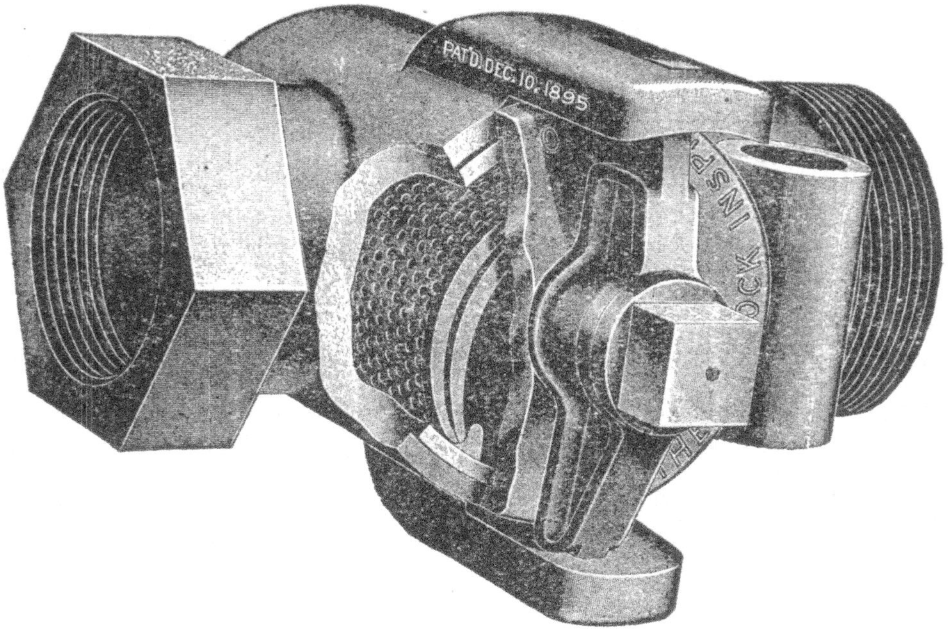

FIGURE 211
HANCOCK LOCOMOTIVE HOSE STRAINER

Keeping cock to the pressure gauge partly closed will prevent the hand from unduly vibrating.

One of the Many Ways It May Be Located. Take cold water to the apparatus out of, and put hot water back into, the pipe that supplies water for washing with cold water, always putting in a stop valve or cock between the connections. The hot water from the apparatus will pass with great force through the same pipe, hose, nozzles, etc., as are used with cold water.

This apparatus has connection for 2-in. pipes, and most be located where the water will flow to it.

THE ELECTRIC HEADLIGHT

A good headlight is a very necessary part of the equipment of a locomotive. It matters not how well the engine and train may be equipped with the air brake and other safety devices, if the man at the throttle does not or cannot see the danger ahead in time to apply the "air" and make the stop, the result is going to be a disaster of a more or less serious nature. Of course, by daylight there is practically no excuse for the engineer not seeing what is in front of him, but in hours of darkness the situation is reversed and unless the engine is equipped with a headlight that will project the light a far enough distance ahead to enable the engineer to see the impending danger in time to stop, or at least to considerably slacken the speed of the train, the responsibility for a smashup should not rest wholly upon the shoulders of the engineer. He may know every foot of track from one end of the division to the other end, but he cannot know intuitively whether or not the track ahead of him is clear and safe, except he can see it.

The ordinary oil head lamp falls far short of the requirements of a good headlight, and, until inventors shall succeed in perfecting a headlight that will illuminate the track ahead of the engine for a sufficient distance to cover the number of yards distance required in which to make a stop, the problem of the safe running of trains at night will remain in an unsolved state. There is but one device that appears to offer a solution of this most important question, and that is the electric headlight. Inventors have been striving for quite a numbers of years to bring out a reliable and efficient light of this kind, and although the process of development has been slow, still progress has been made, especially since the steam turbine has become available for operating the small dynamo or generator required to furnish the electric current needed. There is no doubt that the electric headlight will in time take its place as an element of safeto, second only to the air brake in the running of railway trains. In order that the reader may get at least an idea of what is being acccomplished along this line, the author presents the following information, descriptive and otherwise, together with illustrations.

The Pyle National Electric Head Light Company of Chicago are the makers of a light that appears to be deservedly working its way to the front, there being now about five thousand of their headlights in use, with a growing demand for more.

The device may be located either on the front end of the engine forward of the stack, cr the generator may be located on the left side just in front of the cab, while the lamp will occupy the usual position in front of the stack. Figure 212 shows the complete outfit on the front end ahead of the stack, and figure 213 shows the generator located on the boiler just in front of the cab.

Instructions for Applying the Equipment. If it is to be placed on the front end as shown in Fig. 212, measure the depth of the headlight case to be used and add 18 inches to this measurement. Then have a baseboard made, using these measurements for the length of board, and make the breadth the standard width for oil headlights. Bend a piece of $\frac{3}{4}$ by 3 in. iron flat ways "U" shaped, making center of iron come over the center of brackets on engine and make the length about two inches shorter than the base-

FIGURE 212

FIGURE 213

board. Place this iron "U" on the brackets within about two inches from the stack and mark for bolts through brackets.

Put baseboard on the iron "U" so the back edge will just clear the stack, and bolt to brace. Place the equipment on the board, with dynamo on the left side of locomotive and as near the back edge of board as possible, and bolt to board. Then bolt the headlight case to board in front of equipment. Put a $\frac{3}{4}$-in. angle valve in the highest part of the boiler in the cab with dry pipe, and run a $\frac{3}{4}$-in. pipe from this valve under the jacket on right side of the boiler to the equipment. Use $1\frac{1}{2}$-in. pipe for the exhaust, running it through the arch and have end of pipe about flush with the top of the nozzle tips of the locomotive. Use 2-in. pipe inside the arch, as it reduces noise and does not act on the fire.

If the equipment is to be located on the boiler just in front of the cab take up back sheet of jacket and fasten two brackets to boiler suitable for holding the equipment crosswise of boiler. Have a baseboard (iron or wood) made and bolt to braces, then place equipment with the dynamo on left side of locomotive and bolt to base.

If steel or iron baseboard is used, the equipment should be insulated from base by wood or asbestos sheet. Place a $\frac{3}{4}$-in. angle valve in the highest part of boiler in the cab (with dry pipe) and run a $\frac{3}{4}$-in. iron pipe to the equipment. Run the $1\frac{1}{2}$-in. exhaust pipe up above cab or down to running board and into arch. Run the lead wires through an independent pipe if possible or through the hand railing, being sure to insulate them carefully to prevent chafing, etc.

To apply the lamp to the reflector and case, remove the oil tank and all supports and guides from the reflector. Cover the board holding the reflector with tin, having the edges turned up about one-half inch.

This prevents sparks from the lamp from setting fire to the oil-saturated board. Secure the support for the back of the reflector to the lamp board on right-hand side, and adjust the screw until the reflector stands level. Remove both carbon holders from the lamp and set the lamp on the board, on the side nearest door of case, then put bottom holder in lamp and place the lamp on the board with copper electrode in center of hole in reflector, being sure to have base of lamp square with reflector slide. Mark for holes in center of large square holes in base of lamp and bore for $\frac{3}{8}$-in. bolts.

Secure the small wires for the incandescent lamps by the small screws at the right of brush holders. Run the incandescent wires through the hand railing to the cab, in wood strips. Run wires for cab in wood strips. Wind all the joints with tape.

The electric headlight equipment complete consists of engine, dynamo and lamp, and material for three incandescent lights in cab.

The Engine. The engine is known as the Pyle compound steam turbine. There are no wearing surfaces inside the engine requiring lubrication, hence they do not use any sight feed lubricator in the cab. Before starting the engine be sure the casing is thoroughly drained, and do not turn on steam too suddenly in starting the light, thus allowing time for the condensation to get out of the engine. It must have dry steam.

Remove plug in top of engine once each week and

pour in a little black oil. This will prevent corrosion of parts. The inside bearing only needs enough oil in the well for the loose ring to touch the oil and carry up on top of the shaft. If there is too much oil, it will be thrown out of the ends of the cellar by the motion of the locomotive, which may ruin the armature. The oil well for the outside bearing should be filled each trip. Use valve or cylinder oil in these bearings.

The Dynamo. The dynamo is constructed on the latest scientific principles, and the electrical balance is so perfect that no sparks should be seen at the brushes. The armature is held in place on the engine shaft by one screw which can be easily taken out if occasion demands. The brush holders are fixed, and the brushes can be taken out and replaced without changing the tension of the springs. A graphite brush is used for the top and a carbon brush for the bottom, and a few moments' care each trip is all that is required.

The mica between the copper strips of the commutator should always be a trifle below the surface. If it gets too high, file it down with a small file. Do not get it too low, as it will collect dirt, etc., and cause a short circuit. The commutator should be cleaned each trip with a damp piece of waste (not wet), rubbing endwise so as to keep the creases clean where mica is filed out.

Be sure and have the brushes fit perfectly on the commutator. If there is poor contact brushes will spark. If commutator is running out or has the appearance of getting rough, clean it up. To do this nicely, remove brushes and hold a strip of No. 0 sandpaper (not emery paper) on the commutator whiie running (see Fig. 214). Don't press the sandpaper on, for if there are any low spots they will increase in size.

If the brush tension spring is too tight, it creates friction, heat and unnecessary wear, both on the commutator and the brushes. If too loose it will spark and commutator will not run clean. Have it just tight enough to prevent sparking. In this case a little judgment must be used, for if the brushes are not in the proper condition, or commutator smooth and true, there will be sparking at the brushes, no matter how much pressure is used. Do not forget that the commutator is the vital part of all dynamos, and none will run successfully without regular care and attention. The voltage of the dynamo is entirely too low to force a current through any portion of the human body, so it may be handled freely and without any possible fear of being injured by it. It only requires a few moments' attention each day to keep the plant in perfect condition.

FIGURE 214

If the commutator becomes rough or out of round, it should be trued up in a lathe. The tool used must be very sharp, and light cuts must be taken, then polish it with fine sandpaper. It must be carefully examined to see that no two sections touch, as the copper is liable to lag or burr from one section to the other, and before putting it back, it would be better to cut or file the mica (between each section) a little below the surface, for it does not wear away as fast as the copper, and if the mica is not cut away it may lead to sparking. After doing this, be sure no ragged edges of copper stick up, for this will cut away the brushes rapidly. The speed of the armature should be as near 1,800 revolutions per minute as possible, unless the copper electrode burns off, when it should be reduced until the copper electrode does not burn.

The Lamp. (Fig. 215)—is simple, durable and reliable, and after a few trials it becomes an easy matter to trim the lamp in the dark, should occasion demand. In putting in the top carbon, it is much better to remove the carbon holder from the slide. After securing the carbon in the holder, take it between the thumb and forefinger and with the remaining fingers resting on the guide put it in place. If desired to clean the reflector remove only the top guide by loosening the thumb nut at the end of the upper arm, then remove the guide carbon and carbon holder.

FIGURE 215

The tension spring in the lamp is for two purposes. It brings together the points of the carbons, so as to establish the arc when the dynamo is set in motion, for there must be a complete circuit before there is any current. If the carbons are separated only a small fraction of an inch, the lamp will refuse to work, because the current will not jump across the separation. Sometimes there will be a deposit of scale on the point of the lower copper electrode which prevents the top carbon touching the copper, and as the current will not go through this scale there will be no light until this is removed. See that the point of copper is clean before each trip. Suppose all wires are connected and the lamp properly trimmed, turn on steam and set the armature in motion. The current enters the lamp and passing through or around the solenoid magnet draws down the iron armature. This in turn separates the carbons, thus forming the arc or light. It will be noticed that the spring is secured to the end of the lever toward the carbons, or on the opposite end from the magnet and pulls against it. This prevents the solenoid from pulling the carbon too far apart. The volume of light will depend largely on the way this tension spring is regulated. It may be so tight that the magnet will be unable to separate the carbons, consequently there will be no light. If the dynamo is run too long with the lamp in this condition it will burn out the armature or the fields, for the current becomes very heavy.

If the tension spring is very loose, the lamp will flash and go out, for the magnet will be drawn down too far. When the light goes out the current is broken, and there being no strength in the magnet, the spring will again bring the carbons together, then the current is instantly re-established. Adjust the spring so the lamp will flicker just a little, when the locomotive is at rest.

The wires leading back to the incandescent lamps may come together, causing a short circuit. This will put the light out. When this occurs the dynamo will be generating a heavy current, the speed will be quite low, and there will be a small light in the lamp. In this case just disconnect one of the small wires from connecting screw, then look for the cause of the trouble.

Most of the troubles are traceable to the adjustment of the lamp.

The magnet yoke may travel too far sometimes and strike the small lug on frame of lamp before carbons are separated sufficiently to make a proper arc. In this case the wire should be shortened so that the magnet yoke is about half way down before the clutch grips carbon.

If the wire is too short the lamp will jump or the carbon will stick in the clutch.

If the carbon feeds too fast the top clutch spring is too weak and should be given more tension. To do this, remove cotter pin and get top-clutch spring out of casing. Then pull it out a little, thereby giving it more "set."

If the light burns green the dynamo is running too fast and the speed should be reduced.

This can be stopped on the road at once by throttling the steam in the cab. There is another reason for light burning green. The main wires from the dynamo to the lamp may be connected wrong, therefore one wire should have a sleeve on each end large enough to prevent its going into the binding post with the small hole. The other wire should have plain ends.

The lamp can be moved in all directions for focusing. To get the proper vertical focus on the track, either to have the light close to you or to strike the track far ahead, loosen the set screw on the side, and by turning the adjusting screw the lamp can be raised or lowered as desired. To move it sideways, backward or forward, loosen the hand nuts and the lamp is free to move.

When once in focus there is no need of changing it again. Tighten all screws.

The back of the reflector is supported by an adjustable step, with a screw to raise or lower it, so the volume of the light will come out in parallel lines.

Focus Lamp. 1. Adjust back of reflector so front edge will be parallel with front edge of case.

2. Adjust lamp to have point of copper as near center of reflector as possible.

3. Have carbon as near center of chimney hole in reflector as possible.

4. Have locomotive on straight track and move lamp until best results are obtained on track. The light should be reflected in parallel rays and in as small a space as possible.

To lower light on track, raise lamp. To raise light on track, lower lamp.

If the light throws any shadows it is not focused properly.

If light is focused properly and does not then strike center of track do not change focus, but shift entire case on base board.

Point of copper should be about one inch above top of holder. If it is higher than this there will be too much heat on clutch.

To Roundhouse Men. A centrifugal brake is placed on back side of spoke of wheel and should be set so as to act at about 100 revolutions per minute more than where the governor acts, so that, if for any reason, the governor fails to act, this brake will check the speed and hold it at any speed at which the brake is set. The application of this brake commences with equipment No. 2,600, but cannot be applied to equipments with serial number lower than that

To adjust centrifugal brake, remove the armature and cap to the engine, pull out wheel and shaft when there will be free access to brake. If it is desired to adjust brake so that it will act at a higher speed, turn nuts to right, being sure to adjust both brakes the same, then tighten up jam nuts. One-half turn of the nut will change the speed at which the brake will act about 150 revolutions.

The governor should be examined once each month, and if the plungers are found cut they should be ground in or faced off as the case requires. If plungers are cut, the engine may run away and be broken by centrifugal force. If plungers are faced off, the ends of governor yoke should be bent further out from face of wheel, thereby allowing plungers to again seat firmly before governor weights are thrown out further than at right angles to face of wheel. If the speed is too high, adjusting screws should be turned back half a turn each, being careful to adjust all the screws the same. Half a turn of these screws should change speed about 100 revolutions per minute. If by turning back governor spring adjusting screw the speed is not reduced, the plungers do not seat, and should be faced off.

Suggestions. Have a few strips of No. 0 sandpaper about $1\frac{1}{2}$ in. wide to clean up the commutator.

If the light fails to burn when turning on steam, see that all screws are tight, and that point of copper electrode is clean. Push down on lever and see if carbon lifts up and falls freely. Put a carbon across both binding posts, and if there is a flash when it is removed, dynamo is all right and the trouble is in lamp. If there is no flash when carbon is removed take out brushes and clean commutator with sandpaper (not emery paper), put the brushes back and try the carbon again.

Keep all screws tight.

After putting in a new carbon, always push down on lever, and notice if carbon lifts and falls freely. If it does not lift, it is not in the clutch. If it does not fall down freely, turn it partly around and find the freest place.

The carbon should burn from eight to nine hours.

Engineers should be held responsible for the proper care of the equipment unless some one is appointed to examine and care for them at roundhouses.

Before leaving on a trip the equipment should be started and brushes examined, as to tension of springs, and adjusted if necessary before getting out on the road.

These equipments are not automatic, and as there are quite a number of enemies to electricity on the locomotive, such as grease, dirt, jar, heat, etc., it is necessary to give it a few minutes' attention each day.

Don't attempt to remove reflector from the case until the top carbon holder is removed by loosening thumb nut.

If the copper electrode burns off, equipment is running too fast, and the speed should be reduced by turning governor spring screws to the left until the trouble is stopped. Be careful and adjust all screws the same as nearly as possible. One-half turn of screws will change speed about 100 revolutions per minute.

Be sure and adjust tension spring as loose as

possible and not have the light go out while locomotive is standing still.

If light dies down when locomotive is running fast, the tension spring may be too tight, which prevents solenoid from separating carbons sufficiently to form proper arc, or top-clutch spring may be too loose, allowing back edge of clutch to be jarred up and release carbon.

If the light goes out when the locomtive is standing still, the tension spring may be too loose or carbon may not feed freely.

If light burns green on the road, throttle steam at once.

If electrode does not come in line with the carbon, the holder should be bent until electrode comes in line with top carbon.

Both ends of one lead wire should be doubled about one inch so it cannot go into binding post with the small hole, and thereby prevent crossing of wires.

Special motor brushes should be used with the improved brush holders, top and bottom brush being of the same quality. The graphite and carbon brushes used with the old style brush holders should not be used with the improved brush holders nor special motor brushes in the old style holder.

The Edwards Electric Headlight Co., of Chicago, whose factory is at La Porte, Ind., are manufacturing an electric headlight which, in addition to sending a powerful ray along the tracks in front of the engine, also projects a powerful vertical beam. The vertical beam makes a very decided illumination in the heavens, so much so that it is possible not only to detect the presence of an engine, but also in many cases to follow its path and determine in which way it is heading. An engineer is by this means placed in touch with the movements of other trains in his vicinity and is enabled to detect their presence where, if they carried ordinary horizontal beam headlights, he would

FIGURE 216

FRONT VIEW SHOWING SHADE DRAWN

be unaware of their location. The horizontal ray of light projected in front of the engine, highly illuminates the track and right of way for a distance of half a mile to a mile ahead of the train. Among the objections that have been raised to the use of electric headlights on locomotives is this one, that owing to the very brilliant light the engineer would not be able to distinguish the different colored signals, but practical experience seems to refute this idea, and the claim made for the Edwards light is that it does not in any wise decrease the efficiency of the signal lights, but on the contrary that they show up in their true colors.

Another valid objection is, that on a double track road there is danger of blinding an approaching en-

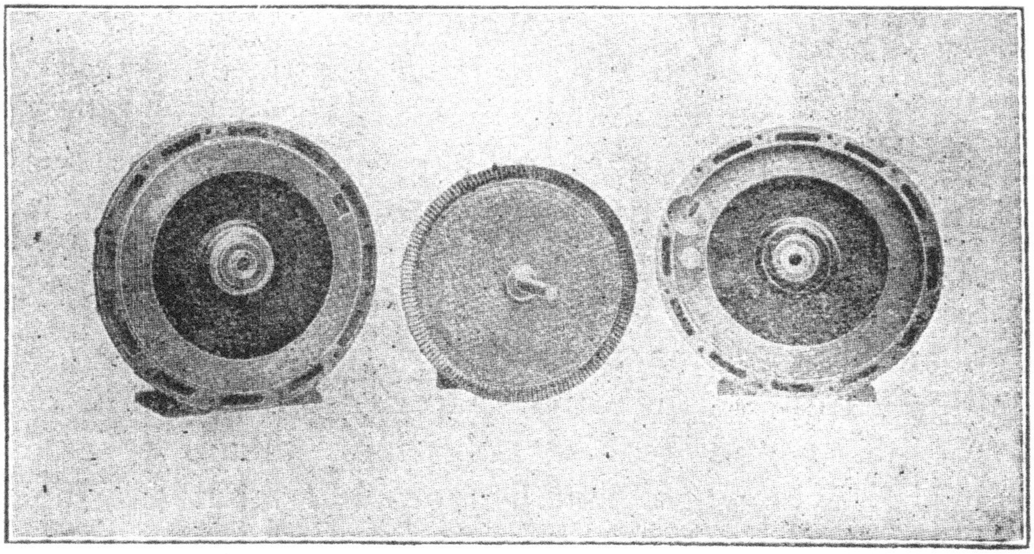

FIGURE 217
THE TURBINE ENGINE DISASSEMBLED

gineer. There is good ground for this latter objection and to guard against this contingency the apparatus is provided with a translucent shade, within the goggle, which may be drawn at will by the engineer when he is at the proper distance from an approaching engine. This shade destroys the strong glare of the light, giving the effect of frosted glass. As soon as the approaching train is passed the engineer releases the shade and again gets the full value of the light.

The Edwards equipment, a front view of which is

given in Fig. 216, consists of four parts: first, the motor, a simple-acting steam turbine; secondly, the dynamo, mounted on the same axle with the turbine and designed to yield to the arc light a current of from 30 to 33 amperes and from 30 to 33 volts; thirdly, the lamp, including the arc, the deflectors and the case; and fourthly, the bed-plate on which the whole apparatus is mounted.

The steam turbine, shown disassembled in Fig. 217, is provided with a propellor wheel, which is wholly constructed of rolled steel. It has a factor of safety of about 7, for, while the normal speed of the engine and dynamo is about 2,000 r. p. m., the wheel will withstand successfully a speed of about 14,000 r. p. m. The speed of the engine is held constant, or practically so, regardless of change of load or initial pressure, by a simple and efficient governor, which is so arranged with relation to the other parts of the engine as to be easily and readily accessible, should occasion demand. The wheel shaft is journaled in ball bearings, and the coefficient of friction is so low that the turbine will operate, running to its full speed, under a pressure so slight that a pointer upon a 180-pound steam gauge will not leave its stop, the gauge being connected between the governor valve and the nozzle. All the moving parts are encased in a cast-iron housing so designed as to thoroughly protect it from the elements dust, dirt, etc. The lubrication is automatic and is provided by loose rings feeding the oil to the ball bearings from the oil wells.

The dynamo is of peculiar construction, designed for the particular purpose for which it is used. The field is differentially wound, and the electric circuits so arranged that a burned-out armature is impossible.

Should a short circuit occur on any point of the circuit, the current is neutralized, and no matter how long the engine may run or the armature rotate, there will be no production of current whatever until the short circuit is removed. As soon as this is done the dynamo performs its proper functions and operates as usual. The current densities throughout the whole machine are very low, so that a minimum heat effect is produced, regardless of extremes of temperature or other conditions which might affect the resistance of the machine. Low-resistance carbon brushes are used, and many months of constant wear show very little deterioration of these brushes. Very large and long journal bearings are provided, and profuse lubrication is secured through the medium of loose rings dipping into the oil wells. An important feature of the equipment is the arc lamp with its parabolic reflector. It is strongly made, and care has been taken to insure a steady and constant light, free from flicker.

The vertical beam is caused to project upwards by an auxiliary plane deflector, placed outside the goggle at an angle of 45 deg. and in such a position as to intercept about 40 per cent of the whole volume of light issuing from the parabolic reflector and direct it vertically. This vertical beam forms a constant warning signal. Reaching to a great height, and on cloudy nights striking the clouds, it can be seen for many miles. In fact, upon the Big Four road it has been seen for a distance of 21 miles, and on the Chicago, Milwaukee and St. Paul road it has been seen for a distance exceeding 16 miles.

The whole apparatus is generally mounted upon one cast-iron bedplate, and it is the work of only six or ten hours to apply the equipment to the locomotive.

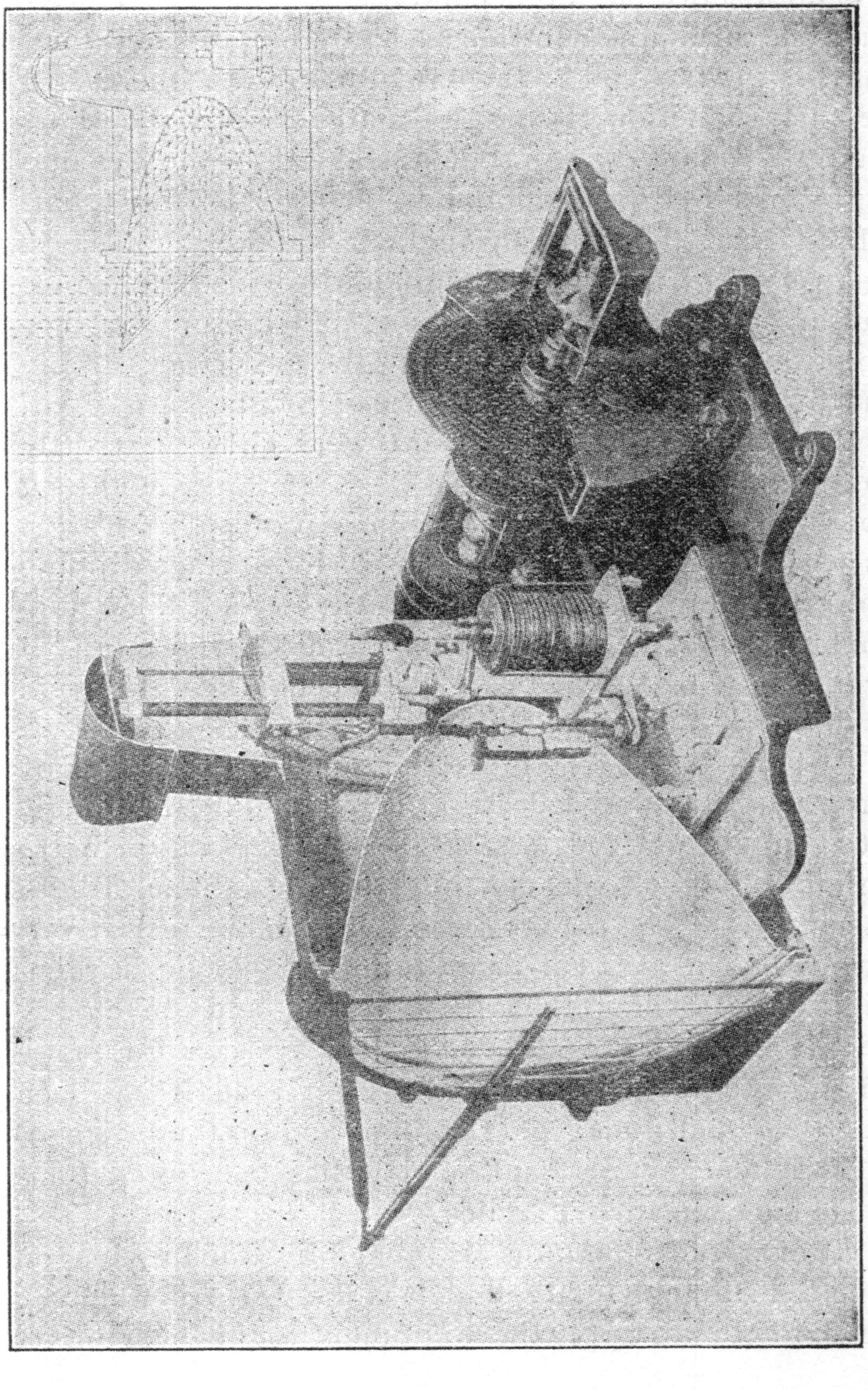

FIGURE 218

A SECTIONAL VIEW OF THE EDWARDS ELECTRIC HEADLIGHT

All that is necessary is to secure the bedplate at the proper place on the smoke arch by means of brackets bolted thereon, the running of a three-quarter-inch live steam pipe from the cab, and the passing of a one and one-quarter-inch exhaust pipe into the smoke arch.

Figure 218 gives a sectional view of the equipment. The light is fitted with a carbon positive pole and a copper negative. The focus is maintained by special mechanism, an adjustable guide being furnished for the carbon. This guide bears in three points on the carbon and permits of perfect adjustment regardless of small differences in diameter of the latter. Fig. 219 presents a view of the assembled turbine and generator direct connected. The steam is led to the engine through a $\frac{3}{4}$-in. copper pipe from the cab and the exhaust is led into the smoke arch by means of a $1\frac{1}{4}$-in. wrought iron pipe. The steam exhausts almost opposite the admission port of the engine, through the various exhaust ports seen in the cut, Fig. 217.

To Remove the Armature. To remove the armature, first take out the brushes from their holders, back of the set screw on the brush holder yoke; then remove the four nuts and two cap screws. The end plate may then be taken off, after which the armature may be withdrawn. The field coils are held in place by $\frac{1}{4}$-in. square keys. The coils may be removed by driving out the keys and removing the pole pieces. In reassembling the dynamo, be sure the oil rings are raised to permit the shaft to pass through the bearings, then replace the end plate. The field coils and armature are protected by a circular sheet steel casing fitting into grooves in the end plates. The armature is mounted in self-adjusting bronze bearings. These bearings are lubricated by means of rings, the shaft having spiral grooves to carry the oil throughout the bearings. Overflow oil holes are drilled in both oil wells to prevent too much oil being placed in the bearings. These overflow holes must be kept open, and the oil wells occasionally cleaned of sediment by removing the plugs, and cleaning the reservoirs

FIGURE 219

THE MINATURE TURBINE AND DYNAMO

with kerosene. Care must be taken to have the oil rings in the slots in the bronze bearings.

To Adjust the Lamp. To adjust the lamp, first have the correct speed on the dynamo, and the commutator and brushes working properly, then set the adjusting screw on the shunt side of the lamp, so that the pawl will clear the escapement wheel about $\frac{1}{100}$ of an inch; then raise the brass tube, or oil cylinder, which carries the carbon by means of the carbon holder support, and permit it to fall, then adjust the limit screw so that when the arc is established it will hold over without breaking the circuit. If the arc breaks when the light is started, or if, while the light is in operation, the equalizer sets up a pumping action, giving a vibrating or flickering light, the limit screw is set too high. On the other hand, if this limit screw is set too low, there will not be sufficient separation, when the light is started, and there will be only a small red light. This also may occur when the light is in operation and the locomotive in motion, which defect may be easily and quickly corrected by slightly raising this limit screw. When the arc is properly established the spring should be adjusted by means of the nuts so that the carbon will feed as it burns away without breaking the arc. If "tack head" deposits form on the top of the negative, and, breaking off, interrupt the light occasionally it is evidence that the spring is too weak, thus not allowing the arc to be drawn out long enough. A slight increase of the spring tension by adjusting the nuts will correct this fault.

It is necessary that the brass tube, or oil cylinder, guide rod, and valve rod should be kept perfectly clean. For this purpose a soft felt cloth should be used, never using sandpaper or emery paper or waste, as the sand and emery will roughen these surfaces, and the lint from the waste may clog the rack.

The Brushes and Brush Holders. The brushes and brush holders must be kept perfectly clean, and the brushes must always slide freely in the holders. The brushes must always occupy the proper position on the commutator, and the screw firmly set. Unless they are kept in this position there will be sparking between the brushes and the commutator, thus reducing the light and burning the commutator. The ends of the brushes which bear upon the commutator are slightly beveled, and in replacing them in the holders see that their full end surface has complete contact with the surface of the commutator. The brushes are self-lubricating and no oil or other lubricant must ever be put on the commutator.

Use only sufficient spring tension on the brushes to prevent sparking between the brushes and the commutator.

The Commutator. The copper bars of the commutator are separated by mica insulation.

The copper will wear more than the mica and if the mica is allowed to project above the surface of the copper, even slightly, it will prevent perfect contact between the copper bars and brushes, which must, at all times, be maintained. To prevent the mica from thus interfering with perfect contact between the brushes and the copper bars, the mica should always be a little below the surface of the copper bars. This is accomplished by filing out the mica to a depth of about $\frac{1}{64}$ of an inch by the use of a small file. This process will raise a slight burr on the edges of the copper bars which must be removed by using a strip of No 0 sandpaper (never use emery). In this operation run the machine slowly, meantime working the sandpaper back and forth lengthwise of the commutator so as to cover the whole surface, until it is perfectly smooth, then wipe the commutator clean, working a clean cloth or waste lengthwise to clean out the shallow grooves between the copper bars.

Use sandpaper as above described whenever the commutator becomes slightly rough. Should the commutator become too rough, or out of round, the armature should be removed, and the commutator dressed off in a lathe, using a diamond-point tool, and removing only enough metal to make a perfectly true and clean surface. After turning, polish with sandpaper (never use emery), then file out the mica as above directed.

THE VICTOR LOCOMOTIVE STOKER

The Victor Locomotive Stoker is the successor to the Kincaid Locomotive Stoker, which was invented by Mr. John Kincaid, who was for many years an engineer on the Chesapeake & Ohio R. R.

Necessity is, and always will be, the "mother of invention." Conditions arise which can only be met with approved ideas and approved methods, and the increased traffic, both in the freight and passenger service of all the prominent railways, has necessitated heavy power engines of approved types and enormous size, in order to accomplish desired results. In fact, the engines have grown beyond the capabilities of the firemen, and as the latter could not be reconstructed or endowed with greater strength and endurance, mechanical means became the only logical solution of the problem.

To meet these conditions, and to relieve the firemen of the severe and exhaustive drudgery imposed upon him, the inventor, himself a Brotherhood engineer, and a former wielder of the scoop, designed his now famous stoker. It was not designed to degrade the fireman's position, but to exalt it; not to make his duties within the province of "cheap men," but to put a premium upon his intelligence; to give him greater opportunities to study and thus hasten the time of his promotion to the position of engineer.

The merits of Mr. Kincaid's invention are manifold, but prominent among them are—

Its ability to fire an engine without opening the door, thus relieving the fireman from the extreme heat of the firebox; to scatter the coal in small quantities over the whole grate area, just as the needs of the engine require, thus obtaining almost perfect combustion, the reduction of the smoke and spark nuisance to a minimum and to insure the greatest possible saving in coal. A higher and more uniform steam pressure can thus be obtained; less clinkers are formed; longer runs can be made without cleaning the fires; as the stoker fires without opening the door, the in-rushes of cold air from frequent opening of the fire-door are eliminated, and the fire-box sheets and flues are protected from sudden contractions and expansions; thus the chief source of leaks in the furnace are practically overcome.

In firing with the stoker a much lower grade of coal can be used than is possible in hand firing and still accomplish as good or better results. Nut and slack coal is always preferable when using the stoker.

These stokers have been in operation since January 1st, 1905, attached to locomotives with wide fire-boxes, and to locomotives with long fire-boxes, used on all classes of passenger service. The result of this work has proven beyond all question that the stoker will do its work efficiently and economically and that the fireman, having once become acquainted with the stoker and recognizing its labor-saving features, becomes its enthusiastic supporter.

The continuous feeding of coal has a very marked effect upon the amount consumed. Run-of-mine coal is used, but it has been found that a good grade of slack will secure even better results owing to the principle of feeding coal in small quantities widely distributed. Absence of dense volumes of black smoke is also very noticeable.

Figure 220 shows the stoker as it appears in the cab, the small controlling engine required to operate it being located on the boiler head on the fireman's side.

The following are the dimensions of the apparatus: Length over all, 47 inches; as three inches of the trough enters the fire door, the stoker extends back on the deck 44 inches; width over all, 24 inches; height on short legs, 28 inches.

Figure 221 shows the stoker as attached to the furnace of a locomotive. A represents the hopper which receives the coal; B the plunger-trough through which coal is propelled into the furnace; C the stoking cylinder; D the rotary valve; E the furnace door; F the controlling engine, and at the left of the furnace door, the steam pipe extending to the locomotive boiler; also the valves and choke-plugs of the stoker, arranged for the convenience of the fireman in regulating the operation of the machine.

The illustration in Fig. 221 shows the controlling engine connected underneath the stoker. Fig. 222 is a transverse view from Fig. 221, and shows the head of the plunger and its position in the trough at the end of each stroke; the conical deflector, attached to the inside of the door, which spreads the coal in the fire-box, and when the stoker is removed, may be turned up to close the hole occupied by trough. The rocking-shaft connected with the controlling engine, which transmits power to the rotary valve, D, and to the conveyors in the hopper; also the end of the trough,

FIGURE 220

in which the plunger travels, and the exhaust steam port are likewise shown in this illustration.

Fig. 223 shows the hopper raised, and reveals one of the twin spiral conveyors which carry the coal forward; it also shows the opening through which the coal drops onto the apron after leaving the conveyors, and, when apron is retracted into the trough, to be thrown into the fire-box by the next forward stroke of the plunger.

FIGURE 221

Description. The stoker consists of the following essential parts, viz.: First, a main cylinder and a trough in which reciprocates a piston and plunger which with a variable stroke, throws the coal to the different portions of the firebox. This variable stroke is given to the plunger by means of a rotary valve, three separate steam ports leading from said valve to the rear end of the cylinder, and three choke plugs, one for each of the said steam ports.

Second, a small controlling engine. It has been found by experience that the most desirable location for this engine is on the boiler head on the fireman's side. This removes the liability of condensation and consequent dryness of engine parts when placed on and beneath the stoker itself. The steam for the operation of this engine is taken directly from the dome.

Third, a hopper with two spiral conveyors journaled in the bottom of the hopper pan. The conveyors carry the coal to the front of the hopper, onto the apron of the plunger, giving a regular and uniform feed. The speed of the conveyors may be increased or diminished by giving more or less steam to the controlling engine, as may be required. This also increases the number of strokes made by the plunger, but does not affect the plunger's velocity, or in any manner affect the distribution of the coal in the firebox, the latter being governed by the three choke plugs.

Fourth, a small steam chest containing a rotary valve which regulates the number of strokes made by the plunger.

That portion of the stoker forming this valve chest is cast in one piece with the main cylinder and has three separate steam ports leading to the rear end of the cylinder for the admission of steam behind the plunger or piston. These steam ports terminate in one common port before entering the rear end of the cylinder, first through a small preliminary port at the end of the cylinder (which also acts in the form of compression by retarding exhaust on the last portion of the return stroke), and after the piston has advanced a short distance it uncovers the main port, which also leads from the common port, giving free passage to the steam.

A choke plug is placed in each of the three steam ports between the valve-sleeve and the common port. The office of the three choke plugs is to vary the amount of steam reaching the rear end of the cylinder through the various ports and thereby giving a variable stroke to the plunger.

The valve operates in a rotary manner, each of the ports stopping fully open in front of its corresponding steam passage in regular succession: Beginning with No. 3, (the port nearest the rear of the stoker) the steam, after leaving this valve, passes through port No. 3 into the common port and the rear end of the cylinder. Choking down this steam port until it is almost closed causes a very light stroke of the plunger, distributing the coal over the grate near the fire-door. The other two valves operate in the same manner, each taking its respective turn.

FIGURE 222

They are adjusted so that more steam is admitted on the second stroke than on the third, thus distributing coal over the middle portion of the grate, and more on the first than on the second, thereby scattering coal over the front end of the grate. By this adjustment of the choke plugs any range of distribution can be obtained that may be desired.

The rotary valve and cylinder are provided with suitable live steam and exhaust ports for the return of the plunger and the exhaust steam from each end of the cylinder. In the front end of the main cylinder is a very small live steam port, connected directly with the live steam supply, and its office is to return the plunger after its forward stroke and also to add volume to the steam retained after the piston has passed over the forward exhaust port; thus giving the desired compression to prevent the piston ramming the front cylinder head. By means of a valve this port can be enlarged to give increased compression necessary when expelling water from condensed steam in starting the stoker when it is cold

Fifth, the furnace door.

Each machine is supplied with a furnace door made to fit the standard door-frame of the locomotive to which the stoker is to be attached. This door has an opening to receive the stoker trough and is provided with suitable brackets for holding the machine in position. Cast upon its inner side are curved lugs, which serve the purpose of hinges for a deflector for spreading each charge of coal over the width of the firebox. The end of this deflector can be raised, if necessary, to aid in the distribution of coal, by means of a set-screw directly under its center. It also has a small vertical sliding door for inspecting the fire, and the deflector can be turned up vertically and held in place by a latch to close the trough opening when the stoker is removed.

To Operate the Stoker. *Directions for Firemen.*—Don't fail to be on your engine *at least* 30 minutes before leaving time.

Know for yourself, before starting, that you have the necessary tools with which to do your work. If they are not on the engine, report the matter to the round house foreman, and don't go out without them.

A fireman's outfit should consist of, 2 scoops; 2 hooks

(one 12 feet and one 7 feet); 2 torches; one coal pick, and one ash-pan scraper.

Before leaving the round house, examine the shaker rigging and know before starting that it is O. K. Also see that the ash-pans are clean.

To build the fire break every particle of bank. Be careful not to get green coal on the grates. Spread the fire evenly over the entire grate area, then attach the stoker.

Assist the engineer all you can in getting the engine ready, but *never* neglect *your* work to do his. Always manage to keep the oil cans and torches filled and ready for immediate use.

Be sure the steam is turned on next to boiler. If you have a reducing valve set the gauge at about 60 or 80 pounds; if not, about one half turn on the globe valve next to the boiler is sufficient. Before admitting steam to stoker *always* open the *admission valve* to front end of cylinder to prevent the piston from ramming the front head. Leave this valve open, until the condensation is thoroughly blown out, and machine heated up to steam temperature. Then close and *leave it closed.*

Regulate the speed of the stoking piston by the throttle valve to the *controlling engine.* Run the machine slowly at first until enough steam is held in the cylinder on which to cushion the motion of the piston.

Starting the Stoker. *Turn on the steam gently, allowing sufficient time for condensation to blow out.* (Turning on a full head of steam before condensation is exhausted will burst this cylinder the same as it would on any other engine.) This may require a little time, as the steam ports are small. If there are any drain cocks on the stoker, see that they are open before turning on the steam. If there are none, considerable water will appear at the mouth of the stoker trough when starting the machine. This comes from the exhaust, and is caused by the steam coming in contact with the cold metal of the machine, but will disappear in a few minutes, or as soon as the circulation of steam has warmed up the cylinders. Should the piston head hit the back cylinder head on the return stroke open choke plug No. 3 to let in enough steam to cushion. If water appears at the exhaust after the machine has been in operation for some time, it indicates too much water in the boiler. As soon as steam has been turned on at the boiler, open the globe valve in the steam pipe leading to the small engine. This is the throttle valve for the small engine, and the amount of steam admitted by it regulates the speed of the conveyors and the number of strokes made by the plunger. To run the machine fast or slow increases or diminishes the amount of coal fed into the furnace, but does not affect the distribution of the coal in the firebox. Should controlling engine be lazy about starting touch one of the tappet rods at either end of the floating valve.

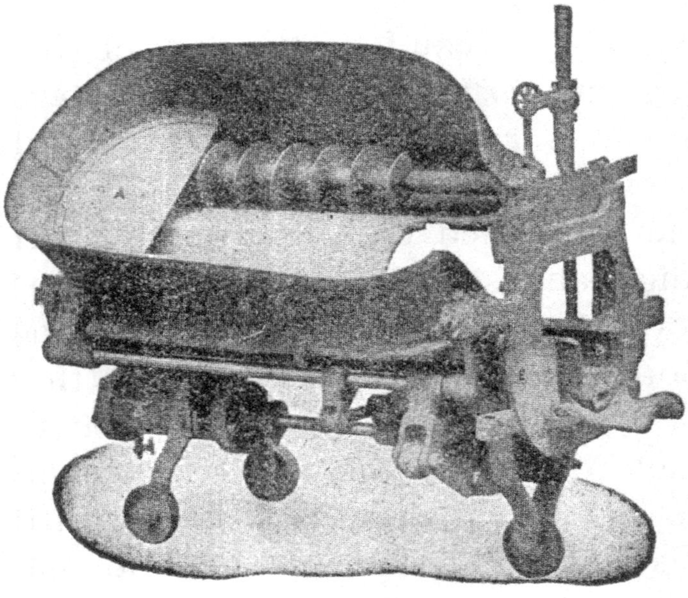

FIGURE 223

To Regulate the Choke Plugs. To regulate the choke plugs for the distribution of coal in the firebox, wait until the steam has reached the maximum pressure and machine is working; then with all the choke plugs wide open, turn on enough steam at the throttle valve to throw the coal near to, but not strike the flue sheet; then close down the middle choke plug (No. 2) until this charge falls near the center of the firebox; then close down the choke plug No. 3, next to the back end of the stoker, until the coal drops inside the furnace door. This gives a distribution over the entire length of the firebox. Don't fill the hopper with coal when regulating the choke plugs or you will get too much coal in the firebox, and waste fuel. Use only one or two scoopsful at a time and watch the distribution in the firebox.

In regulating the choke plugs when the engines are standing, be careful not to make the strokes too heavy, as the draft will assist in carrying the coal forward when the engine is working. Never try to regulate choke plugs with the front end admission valve open. If the steam drops 20 or 50 pounds while the stoker is in operation, open the throttle valve accordingly, unless some form of steam regulator is used. If, after running some time you find the stoker piston only making half strokes, examine front end admission valve and you will find it was not entirely closed or accidentally opened. Close this as far as you can and remedy the trouble.

Operating the Stoker. In operating the stoker on large engines with a heavy train, usually 20 to 30 strokes of the plunger per minute is sufficient, providing the hopper is kept full. The speed of the stoker can be varied from 12 to 40 or more strokes per minute. The speed of the stoker should be regulated to correspond with the work the engine is doing. Before starting out on a run, see that the entire surface of the grates is covered with fire. A heavy fire is not necessary, as better results can be obtained with a thin, bright fire. Avoid heavy firing. Remember the stoker is at work continually and there is little danger of the fire getting away, but don't neglect it when the engineer shuts off steam. Remember your fire is lighter than when firing by hand, and will die out quicker. *Notice the condition of the fire often.* This can be done by means of the small vertical sliding door above the stoker trough. If you can't see take your long hook and feel your fire occasionally. If the fire should become banked near the door, open up the choke plug No. 3 until this charge of coal goes beyond the bank; if the fire becomes banked near the flue sheet, close down the choke plug No. 1 slightly. If the fire is too light near the door or flue sheet, the reverse action should be taken; if the fire is too heavy in the center of the firebox, notice the choke plug No. 2, and if this is firing in the right place either open up the choke plug No. 1 or close down choke plug No. 3 slightly, or both, as this may happen by having the third stroke too heavy and the first too light. As a general rule, when the choke plugs have once been regulated, it is seldom necessary to change them. However, it is better to change the plugs than to hook the fire. If a bank is found anywhere in the fire, it can generally be burnt out without using the hook, by changing the choke plugs. With engines having a severe draft, considerable coal may be carried in by the draft before the plunger strikes it. This with some engines, is distributed over the back part of the grate, and with some engines it is distributed near the center. In either case the choke plugs should be regulated to suit these conditions. The coal should be broken as fine as possible, as much better results can be obtained. Railroads having the stoker in use should furnish nut and slack. It is much cheaper and the stoker will do better work with it.

In taking stoker apart be very careful not to lose nuts, bolts or other small parts, the loss of which might disable the machine.

Always keep the packing around the main piston rod tight. If this gets loose the steam used for cushioning the forward stroke escapes, causing the piston to strike front head.

Drifting Down Grades. While drifting down a long grade it is only necessary to feed enough coal into the firebox to keep up the heat and not let fire get too low. It has been suggested to build up a pretty good fire in the firebox before starting down grade and then to use the blower occasionally and feed in coal only just what is needed. Before reaching the bottom of the grade, if it is a long one, it would be well to start the stoker so as to have a good fire and a good supply of steam when the engine begins to use the steam again.

Care of Stoking Head. Always stop the stoker before hooking the fire as the hook might be caught by advance stroke of piston, and the stoking head or piston rod be broken. If at any time the stoking head should get a trifle loose, take off the nuts which hold it on and insert a sheet iron washer behind them to keep it tight. *Notice this every day.*

Shaking the Grates. The fireman should shake the

grates often but not too much at a time. Avoid getting green coal on the grates.

Dampers. Watch the dampers closely. On starting out it is occasionally best to keep the front one closed and the back one open, but over the last part of the run, or in any case where the fire is dirty and the draft obstructed, it may be better to open the front damper and close the back one. The regulation of dampers must be left almost entirely to the judgment of the fireman as the varying conditions require different adjustments of the dampers.

Black Smoke. The emission of black smoke indicates that the engine is being fired too heavily. It also indicates that coal is being wasted, and as the purpose of the stoker is to save, not to waste coal, these signs should always be recognized as an evidence of improper adjustment.

When the stoker is doing the duty it is capable of performing, the proper quantities of oxygen are admitted to the firebox to create nearly perfect combustion, which depends upon two parts of oxygen to one part of the combustible elements found in the coal and of course, a uniformly high temperature in the firebox Therefore, black smoke may be taken as a safe guide in determining a waste of coal, resulting from too heavy firing, and the stoker should be regulated accordingly.

Tools. With each stoker is furnished a 5/8 by 5/8 wrench, which will fit nearly all the small nuts and cap screws on the machine. The 9½-in. air pump spanner wrench, which is found on every locomotive, will unscrew the caps on the floating and reverse valve chests and will loosen the stuffing box on main cylinder, controlling engine and the rotary valve. The *thumbscrew* which holds the stuffing box on main cylinder from getting loose, is used to turn and draw out the reversing valve stem in the controlling engine. All the studs on stoker are $5/8$ x $2\frac{1}{2}$; all nuts are $5/8$ semi-finished; all cap screws are $5/8$ x $1\frac{3}{4}$.

All firemen should see that they are supplied with one 7 foot and one 12-foot hook, 2 scoops, 2 torches, 1 coal pick and 1 ash-pan scraper.

Coal. While the stoker will handle about as large lumps of coal as can be fired by the shovel, yet all railroads forbid the firing of coal in large lumps as it is not only to the advantage of the railroad but also of the fireman, for large lumps of coal start clinkers and in the end cause more trouble to the fireman than it would be to break up the lumps. With the stoker it is particularly desirable that all lumps be broken up to a size not larger than a man's fist.

DON'T FORGET TO FILL LUBRICATOR BEFORE STARTING

Oil. One of the most important duties of a fireman in operating a stoker is to see *that the lubricator is feeding oil, about three to five drops per minute all the time* stoker is in use, but shut off lubricator when standing on the side track. No steam engine will run long without oil, and the stoker is no exception to the rule. Therefore we desire to impress upon firemen the fact that a lack of oil in the stoker and on all its wearing parts will not only disable or damage the stoker, but will also brand the fireman in charge of the stoker as a careless man and not fit to handle a stoker or an engine.

Care of Stoker at Terminals. As different customs prevail on different railroads, it is probable that the same directions for handling the stoker at terminals cannot always be followed.

Our observation, however, has led us to believe that it is better for the fireman to detach the stoker at terminals and roll it back against the coal gate, leaving it in such a position that it cannot be damaged by dumping coal into the tender, or be in the way of the men cleaning or repairing the engine.

The fireman should instruct hostlers to have a good supply of steam by the time he has reached his engine, say 100 pounds pressure or more, and he should make it a point at all times to be at his engine in plenty of time to have everything ready before time for starting. When the fireman reaches his engine he should examine the stoker carefully to see that it is all right. Blow a little steam through the pipe before connecting the stoker, and see that there is no coal or cinders in the stoker end of the disconnected pipe. Some railroad men have suggested that the hostler should disconnect and connect the stoker at terminals. As hostlers, and particularly their helpers, would not be expected to be familiar with the stoker, we would discourage this idea as we believe it would be a source of trouble on account of misusage. Besides this, it does not take five minutes for the fireman to detach the stoker when coming into a terminal, and he can always have it unfastened and placed back out of the way before the engine reaches the stopping place, and but a few minutes to attach stoker when ready to start out; but the fireman should insist upon the hostler or his assistants having a good fire ready for him when he reaches his engine before starting out. Always try shaker bars to see if they are properly connected and in good working condition.

Accidents. Should an accident occur to the stoker while on the road, and you are sure that it cannot be readily repaired, cut off the steam from the stoker, loosen the connections, and if there is no room on the engine deck, run the stoker back out of the way; turn up the deflector to cover the hole in the door occupied by the trough and fasten it, firing the engine by hand until such a time as the permanent door can be put on. By removing journal caps on right-hand conveyor, the hopper can be removed and placed on back of tank. Should the engine be a small one, having no room to get the stoker out of the way, let it remain in position and fire through the small door until a stopping place is reached where it can be examined. Do not attempt to put on the regular door belonging to the engine until a station has been reached or until the conditions are such that it can be done without losing your fire. Then consider the matter very carefully as to what is best to be done.

If the stoker can be repaired in the shops of the road, have the repairs made as soon as the shops are reached, and if the repairs are of such a character that they must be made by the manufacturers of the stoker wire the shops at the first stopping place to have the necessary parts ordered. In dispatches or letters always give full information, leaving nothing to be guessed at. A few cents more expense by adding a few words more to a message should not be considered if there is the least possibility of a misunderstanding as to your meaning In a letter, if there is any doubt of misunderstanding or misinterpretation as to what is wanted, make a sketch of the part or parts required.

In repairing the stoker, should any nuts or bolts become rusted or unremovable, soak them thoroughly in kerosene, which will, in most cases, loosen them in a short time. Be very careful not to lose any bolts, nuts, keys or other small parts, the loss of which might disable the machine.

When out on the road, make note of any parts that may become broken or out of order as soon as discovered, and upon reaching the terminal look carefully over the stoker and see if any other repairs are needed. Make it a point to have the machine in perfect condition so that it will be ready to start out on the next trip. *This is imperative*, and must be attended to before leaving the roundhouse.

Do not lose tools and extra parts committed to your care, as such losses can only be attributed to carelessness.

BREWER PNEUMATIC FIRE DOOR OPENER

The occupation of locomotive firemen is one requiring a great deal of exercise, especially if the engine is on a long freight run, and consuming 15 to 20 tons of coal on a run. Some expert has calculated that for each ton of coal that is shoveled into the firebox with a No. 4 scoop (holding on an average 17 lbs. of coal), the fireman is required to make 585 distinct movements as for instance with each scoop full there are five movements divided up as follows:

1. Filling the shovel with coal.
2. Opening the door.
3. Picking up the shovel.
4. Throwing the coal into the firebox.
5. Closing the door.

Therefore the burning of ten tons of coal requires 5850 distinct movements on the part of the fireman,

Brewer Pneumatic Door Opener, Full View

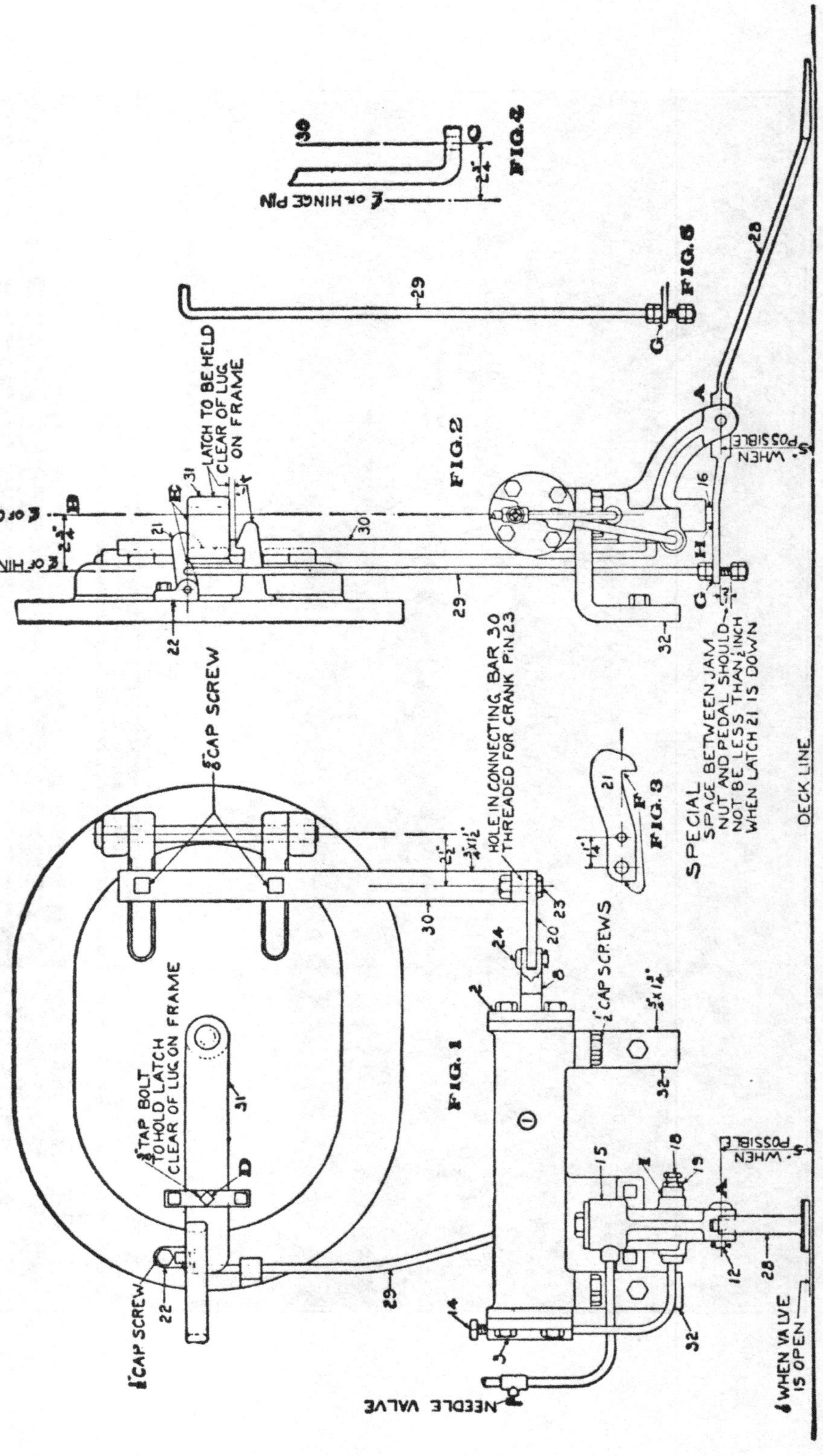

Brewer Pneumatic Door Opener, Sectional, Front and Side Views

and twenty tons would necessitate 11,700 movements in order to place it where it would do the "most good." Any device that will tend to save some of this hard labor, should certainly be warmly welcomed by the fireman. The Brewer Pneumatic Door Opener appears to be deservedly working its way to the front as not only a labor saver, but also a fuel saver, and by its action in opening and closing the firedoor almost instantaneously, it also protects the flues, as it prevents the large volume of cold air from entering the firebox with each shovelful of coal, which is unavoidable with the old style method of opening the door. The sudden cooling of the back ends of the flues is thus to a large extent prevented, while at the same time the single shovel system of firing is insured.

The apparatus is simple and durable, having very few parts, and therefore does not easily get out of order. It is clearly shown in the accompanying illustrations, which are self-explanatory. The quantity of air required to operate it is very small, almost imperceptible. It consists of a small horizontal air cylinder directly underneath the door. This cylinder is fitted with a piston, the rod of which is connected by means of a link and short arm or crank to the pivot upon which the door swings. The motion of the door in opening or closing is very rapid, but the door does not slam, as there is always a cushion of air to prevent this. The door is opened by simply placing the foot upon the treadle or trip shown on the deck. This action opens the air valve, admitting air to the cylinder, thus forcing the piston to move towards the right. To close the door, remove the foot from the trip, allowing the air behind the piston to escape. This permits the door to swing shut. This device is being used quite extensively on the Chicago, Rock-Island and Pacific, and other roads.

QUESTIONS

467. Upon what does the efficiency of a boiler depend in a large measure?

468. Theoretically what would be the correct way to supply a boiler with water?

469. Are these conditions practical?

470. What is the duty of engine men regarding the injector?

471. Is this an important subject?

472. When is a good time to use the injector?

473. How may the latent heat stored in the water be utilized?

474. Mention another "right" time and place to use the injector.

475. Should an injector have a wide range of capacities?

476. What can be said of the leadingty pes of modern injectors?

477. Who invented the injector?

478. How can an injector lift and force water into under pressure?

479. What two important qualities does steam possess that enables it to do this?

480. When steam comes in contact with a body in front of it what is the tendency?

481. What is momentum?

482. What is the velocity in feet per second of a jet of steam discharging at 180-lb. pressure?

483. What is the function of the combining tube of an injector?

484. What are some of the requirements of this tube as to construction?

485. Why does the combined jet enter the boiler?

486. What is the velocity of this jet in the delivery tube?

487. What velocity does the jet need to enter the boiler carrying 180 lbs. pressure?

488. To what then is the action of the injector due?

489. What can be said of the Sellers' Improved Self-acting Injector?

490. What is its range as to pressures?

491. How is the size of an injector determined?

492. What is the capacity per hour of a number $9\frac{1}{2}$ Sellers' at 200 lbs. pressure?

493. What are some of the things to be done in caring for an injector?

494. What about the Sellers' Self-acting Injector Class "P."

495. What should be the quality of the steam used in an injector?

496. Suppose that an injector suddenly stops working, what are some of the probable causes?

497. And what are the remedies?

498. What sometimes happens to the lifting tube?

499. Suppose the main check valve does not seat. What is to be done?

500. If there is an air leak in the suction pipe, what is the result?

501. What do lime and salts in the water do for an injector?

502. If the inlet valve does not feel cool what is the matter?

503. What type of injector is the Metropolitan?

504. What function does the lifting set of tubes perform?

505. How low a steam pressure will the Metropolitan start with?

506. How high a pressure will it work up to?

507. Is there any waste at the overflow?

508. Is it easily regulated?

509. What is the capacity per hour of a No. 9 Metropolitan?

510. How is the capacity of this injector regulated?

511. What are some of the causes of this injector not working properly?

512. What are line check valves?

513. What can be said of the "88" Monitor injector?

514. How should it be located with reference to the water level in the tender?

515. What kind of combining tube has Rue's Little Giant Injector?

516. How may it be used as a heater?

517. What type of injector is the "simplex"?

518. What is its throttling capacity?

519. How is it used as a heater?

520. What is the range in pressures for starting the Lunkenheimer Injector?

521. What is the capacity per hour of a No. 15 Lunkenheimer at 200 lbs. pressure?

522. Of what does the Hancock Inspirator consist?

523. How high will it lift water?

524. Of what does the lifting apparatus consist?

525. How is the combining tube of the inspirator made?

526. What is the range of working pressures of the Hancock Inspirator?

527. At what pressure is its maximum capacity?

528. At how high temperature will it take feed water?

529. How is it regulated?

530. What is the function of the intermediate overflow valve?

531. What is the capacity per hour of a No. 9 Hancock Inspirator?

532. How should the inspirator be located in order to obtain the best results?

533. What is a frequent source of annoyance in the use of the inspirator?

534. Mention other causes for the instrument not working.

535. What is a "composite" inspirator?

536. Mention some of the characteristics of the Hancock "Composite."

537. What qualities should a safety pop valve possess?

538. Describe the action of the Crosby Pop Valve.

539. How may the point of opening in the Crosby be changed?

540. What kind of a seat has this valve?

541. How is the American Pop Valve adjusted for blowing off pressure?

542. How is the American Pop Valve adjusted for blowdown?

543. What are some of the characteristics of the Crane Pop Valve?

544. What is a peculiar feature of the Kunkle Pop Valve Spring?

545. Mention another characteristic of this valve.

546. How often should a lubricator be blown out?

547. What is to be done with a lubricator when it becomes air bound?

548. Do the tubes and nozzles of a lubricator require frequent cleaning?

549. Does scale and sediment from the boiler settle in the lubricator?

550. If the oil pipes become clogged how may they be cleansed?

RED HAND-RESERVOIR
BLACK HAND-TRAIN LINE
CROSBY STEAM GAGE & VALVE CO BOSTON
0
10
20
30
40
50
60
70
80
90
100
110
120
130
140
150

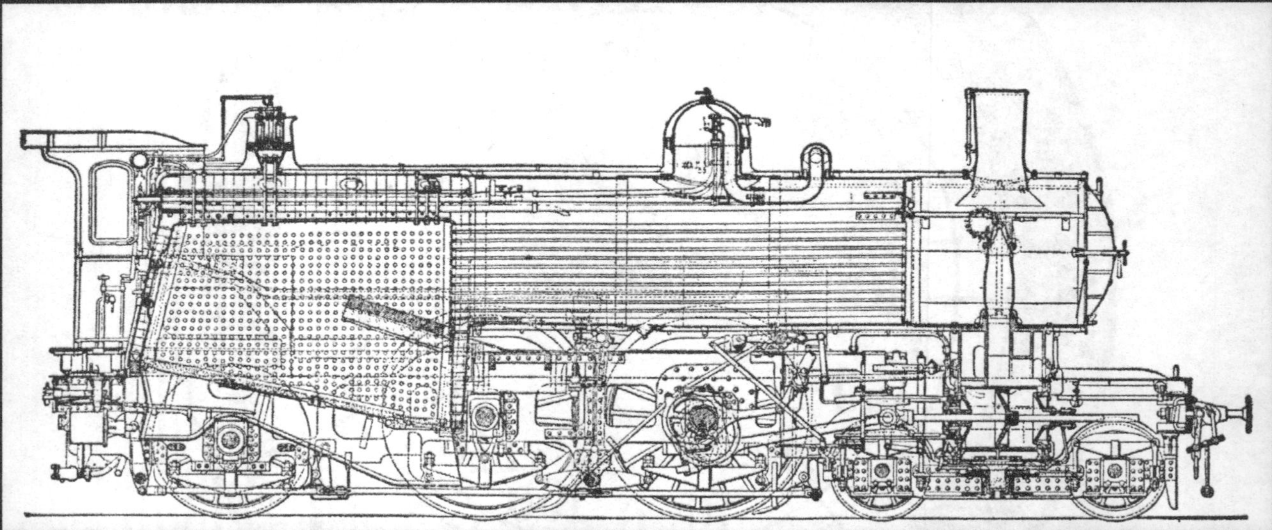

CHAPTER X

WHAT TO DO IN CASE OF BREAKDOWNS

The locomotive engineer, in addition to being competent to "run" his engine successfully over the route, according to the rules of the time card, keeping the water level in the boiler at the proper height, keeping the running gear of the engine properly lubricated, watching out for all signals, sources of danger, etc., should also be a man full of resources and determination; resourceful, in order that he may be able to deal successfully with any one of the many breakdowns that are liable to occur, and determined that when the breakdown does occur, he will if possible bring his engine in alive.

A careful engineer will always inspect his engine before starting out, and in this way be able very often to prevent a serious breakdown. By making regular inspections a man will become so familiar with his engine, and especially with those parts that are the more liable to become disabled, that he can tell at a glance if there is anything wrong. Before leaving the roundhouse he should test all the rod keys, by trying to drive each one back with a copper hammer. In this manner any loose set screws will be at once detected. Out on the road he will be able to detect any derangement of the valve gear, such as a slipped eccentric, loose strap bolt or blade bolts, broken valve yoke or loose valve stem key by the sound of the exhaust.

Although it is the fireman's duty to see that the engine is properly supplied with fuel, water, and the necessary fire tools, lamps, signal oil, and sand, yet it is well for the engineer to see that all of these have been provided. There should be a pinch bar and a pair of jacks, an axe, and a hand saw for use in blocking.

In fact, an engineer should know before starting on a trip that every thing about his engine is O.K. Then if a breakdown does occur out on the road he will have the satisfaction of knowing that he did his best to prevent it.

Derangement of Valve Gear. If an engine suddenly begins to go "lame," it indicates that one of four things has happened. Either a dry valve, a slipped eccentric, a loose strap bolt, or an eccentric blade is loose and has slipped. If, after stopping and looking her over, it is found that none of the three last mentioned things is the cause, and the engine is again started, and the exhaust sounds square, it shows that one of the valves was dry, and that when she was shut off, the oil remaining in the oil pipe was drawn into the valve chest by the vacuum. Sometimes the exhaust will in time wear a hole in the petticoat pipe, and this will cause the exhaust to sound "lame," or the tumbling shaft may be sprung, which will cause a longer cut-off on one side than on the other. In this case there will be two heavy blasts on one side, and two light ones on the other side.

Slipped Eccentric. Place one side of the engine on the center, as near as possible; either center will do. Put the reverse lever in extreme forward motion, and then, with a lead pencil or the point of a knife blade, make a mark on the valve stem at the gland. Now have the fireman put the lever in extreme backward motion, and if neither one of the eccentrics on that side has slipped, the mark will come back to very near the same position it was in when the lever was in forward motion. If it does not come within a quarter of an inch of its first position, the trouble is on that side, and if the original marks are on the eccentric and driver axle, the eccentric may be easily reset, but if there are no marks, the mark made on the valve stem at the gland will serve as a temporary guide in resetting the slipped eccentric.

There are several ways of getting very close to the center. Move the engine till the center of main axle, main crank pin and cross-head pin are on the same exact line on that side; or till the centers of axles and centers of crank pins are on the same horizontal line; or till a straight edge on top and bottom of the main rod strap comes the same distance each side of the center of the main axle. Or, go to the other side of engine, place her on the quarter, measure from the center of the main axle to the center of back crank pin, and from the center of back axle to center of main crank pin, move the engine if necessary till the distances are the same; the engine will be on the quarter on that side and the center on the other.

Broken Eccentric, Eccentric Strap or Blade, Broken Valve Rod, Broken Rocker Arm, Broken Link or Pin. Take off both eccentric straps and rods on that side, fasten the top end of the link by tying it to the link hanger and tumbling shaft arm so that it cannot tumble over and interfere with reversing the engine. Place valve to cover steam ports, clamp the valve stem so it cannot move, disconnect the main rod and block the cross-head. With a heavy engine, a better way is to take off the eccentric straps; tie the top of the link to the top end of link hanger; block the valve in such a position that it will admit a very little steam to the back steam port to lubricate the cylinder; have the lubricator feeding to that side. Take out the cylinder cocks or block them open on that side and any relief valves there may be in the forward cylinder head, leave the main rod up and proceed. If the engine gets caught on the center, close the cylinder cock opening in the back end of cylinder; steam leaking by the valve will soon move her off the center; then open this cylinder cock and go ahead.

Broken Reach Rod, or Arm of Tumbling Shaft. Put a very short block in the link on top of link block and a long one in the bottom end of link so that side will work full stroke. Do not block both links, only one. When the engine is moving, with both link hangers in position as they should be with a broken reach rod, at one point of the stroke one link tends to slip up on its block while the other link is slipping down. If both links are blocked solid top and bottom, the tumbling shaft must bend or spring. To reverse the engine, put the long block in top of link.

Broken Valve Seat. When a seat is broken the engine usually blows through on that side, how badly depends on what is broken and whether the valve is also broken. If the bridge or partition between the steam port to one end of the cylinder and the exhaust port is broken, when the valve uncovers that steam port live steam can get to the exhaust the full size of the broken place. If it is a false valve seat it may be broken so badly that steam will blow through in any position of the valve. When a valve seat breaks it usually catches the edge of the valve and springs the valve rod, the rocker arm or the eccentric blade, in which case an inspection of the engine should show the damaged parts that are outside the steam chest. If the valve catches so the engine cannot be reversed, it is an easy matter to locate the trouble by holding a hand on the valve rod while the lever is moved; if that side catches, it is soon felt.

After locating the trouble, take up steam chest cover and block over the openings to keep steam from passing through. A board covering both steam ports and the exhaust port will do this; in the case of a false seat taking out all the pieces, if they could not be fitted steam-tight. Usually the valve will have to be left out and a block fitted in between the board and steam chest cover to hold the board from rising up when engine is shut off and drifting. In the case of a balanced valve, the top of the valve comes so close to the pressure plate that the valve will not go in again with a board under it, nor can the broken false seat be taken out and the balanced slide valve be dropped on the cylinder casting, unless the top of valve is also blocked to keep steam out of the exhaust cavity of the valve. Some false seats are fastened to the cylinder casting by tap bolts going into the lands and bridges between the ports, in which case the broken seat cannot be taken out, but must be covered so that steam cannot get by it. After locating the trouble, disconnect the engine on that side, taking down the main rod and blocking the crosshead. It is usually necessary to take off both eccentric straps and rods, as the bottom rocker arm may be bent out so the link will be cramped on the block. If, after disconnecting, the reverse lever cannot move both links easily, uncouple the link hanger on the disabled side from the tumbling shaft arm.

Broken Valve Yoke. A valve yoke usually breaks off at the neck of the valve stem. It can be readily discovered in the exhaust by a tremendous blow. If the valve is pushed far enough ahead it will blow; if not, it is often mistaken for a slipped eccentric (examine the eccentrics first). It may be discovered in this way: Place the crank-pin on top or bottom quarter and reverse the engine; if the steam still continues to come out of the back cylinder cock it is usually the yoke. A great diversity of opinion exists regarding the best remedy for this kind of a break. The old and safest way is to raise the chest cover and block the valve central, replace the cover, remove the valve rod and main rod and block the cross-head at the back end. But this remedy requires much time and labor, and time is a very important consideration on the road, and there appear to be no mechanical objections to the other methods, providing the cross-head is securely fastened. Disconnect the valve rod and push the valve clear ahead, remove the stem if it would blow out, and use a gasket back of the gland, or hold the valve stem intact with valve stem clamp. Block the cross-head at the front end, and proceed; the pressure will hold the valve forward and if it should move it can do no harm, providing the cross-head is securely blocked. Another way is to remove the release valve, push the valve clear back, fit a block into the release valve long enough to hold the valve back, then block cross-head at back end. Still another way is to push the valve stem forward and clamp it by cocking the gland, then block cross-head at the front end. If the yoke is only broken at one side of the valve it will only affect one exhaust. When the yoke pushes the valve forward the exhaust will sound all right, but when it pulls the valve back the engine will be lame.

Broken Cross-Head. A slight break, such as a gib or plate, may sometimes be clamped, but be careful that the clamp does not strike the guide block at extreme travel of the cross-head. If it is a bad break disconnect the broken side. If the piston is not broken push it against the forward cylinder head and then block the cross-head in that position. If the cross-head is broken so that the cross-head cannot be blocked, the safest way is to remove the piston. If it cannot be taken out set the valve so as to admit steam to the back end of cylinder only, and clamp valve stem securely in this position.

Broken Main Rod or Strap. Disconnect on the broken side.

Broken Side Rod or Strap. Remove the broken rod and the parallel rod directly opposite to it. If it is a ten-wheeled engine and this cannot be done, remove all the side rods. If a front or back rod or strap on a twelve-wheeled engine, remove the broken rod and the one directly opposite to it, if this can be done, and leave the others up.

Broken Cylinder Heads. *Back Head.* Disconnect the engine on broken side. If it is necessary to remove the guides and broken head, then remove the piston also.

Forward Head. Disconnect that side of the engine. Another method advocated by many, but practiced by few, by which three-fourths of the power of the engine could be retained, is to remove the steam chest cover and plug up the forward steam port with wood and proceed working both sides. This method is impracticable, owing to the shape of the steam port cavity on most engines, and the time it would require, as time is usually the most important factor, besides the improbability of the block remaining intact.

Broken Guides, Blocks or Bolts. If any of the bolts break, try and replace them. See that all nuts are tight, or they may be the cause of springing the piston rod. If a guide bar is broken badly, disconnect one side.

Broken Guide Yoke. If a yoke is bent or broken and will not hold the guides secure, disconnect one side.

Disconnecting One Side. This necessarily implies that the engine is to continue its trip. Remove the main rod on one side and place the liners and brasses back in the straps. Secure the cross-head near the back end of the guides with a cross-head clamp, or with hard wooden blocks, securing the blocks with a rope so they cannot work out. Don't move the cross-head clear back to the striking point, as the cylinder packing rings may get down into the port or counterbore. Remove the valve rod and secure the valve stem with a valvestem clamp, set the valve central upon its seat and cramp the valve stem by tightening the gland on one side. Most engines that use metallic packing are supplied with a valve-stem clamp made to hold the valve central upon its seat; but the valve can easily be set to cover the ports by opening the cylinder cocks and giving the engine a little steam. Then adjust the valve stem until steam is entirely shut off from both cylinder cocks. Do not remove the eccentric straps or side rods unless it is necessary. Whenever the eccentric straps are removed on one side, the top of the link should be tied to the short arm of the tumbling shaft to keep it from tipping over, which would prevent reversing the engine. If it is necessary to take one side rod down, remove the one directly opposite to it; if this cannot be done, then remove all the side rods. Do not remove the eccentric blades, leaving the straps on the eccentrics, unless they will whirl and clear everything in all positions; otherwise they might punch holes in the firebox.

If the side rods have been removed from a ten-wheeled engine, or pony engine, see that the forward crank-pins will clear the cross-head in all positions; if not, take no chances, but disconnect both sides, blocking both cross-heads clear forward or wherever they will clear the crank-pins and have the engine towed in.

Broken Driving Spring or Hanger. If the engine is raised with jacks, block up the end of the equalizer that had been connected to broken part, so that it is a little higher than it was before, to allow for settling. It is customary also to block up between driving box and frame at the box where spring is broken. If it is the forward box, it puts the load on that box, which may be too much. It is better to block up over a back driving box, no matter which spring is broken, as the weight is carried there the best. If the engine is raised by running up on blocks or wedges, put block on top of the box that is under broken spring first, if possible, then run that wheel up on a wedge until engine is raised so that the equalizer can be blocked up level again; then put block over box, also, to carry what weight of engine the spring still at work on that side would not hold up; take out the broken spring or hanger if necessary. If equalizer is under frame and boxes, block under end that will hold it in proper place. If the reach rod is pinched so that the reverse lever cannot move the links, it may be necessary to take out the pin holding the reach rod to the tumbling shaft arm and handle the links otherwise.

Disconnecting Both Sides. This implies that the engine is dead and must be towed in. Remove both main rods and both valve rods, but it will not be necessary to block either, if the crank pins clear the cross-heads. Do not remove the side rods or eccentric straps unless it is necessary; and when it is considered necessary be sure to take the precautions previously explained.

In freezing weather, if the fire is down, all water should be drained out of the injectors, pumps, feed and branch pipes. If there are not frost plugs, slack the joints and let the water out. If there is danger of the water freezing in the boiler, run it out of both boiler and tank. See that all oil cups are well filled before starting. Almost all roads are very strict regarding the speed of dead or disconnected engines, as the engine is not then counterbalanced perfectly, and is therefore very injurious to the track. Some of the best roads limit the speed of all heavy engines which are disconnected on one or both sides, or which have the side rods removed, or dead engines hauled in a train, to twenty miles per hour.

Broken Equalizers. Raise the engine the same as for a broken spring or hanger when possible to do so. If an equalizer on a standard eight-wheeled engine, block on top of one box and block up the loose end of the equalizer, when possible, the same as for a broken spring or hanger; if it cannot be used, remove the equalizer and block on top of both boxes. If an equalizer is below the frame, do likewise, or chain it up. If forward equalizer on a ten-wheeled engine, block on top of the forward and main boxes, and block up forward end of back equalizer. If it is the cross equalizer on a mogul, block on top of both forward boxes and block on top of the back end of the long intermediate equalizer that goes to the truck. If the intermediate equalizer breaks, block between the boiler and the cross equalizer. If it is the cross equalizer on a four-wheeled pony, block on top of both forward boxes. When this equalizer is below or between the frames it is sometimes possible to block between the hangers and the frame. If a small equalizer that rides the back box, block on top of the back box and chain up the back end of the bottom equalizer. If it is a truck equalizer, block on top of truck boxes between the box and truck frame. Always remove or secure all loose parts.

Broken Equalizer Stands. If the stand breaks, then use the same remedy as for a broken equalizer, but if only the bolts break find some old bolts to replace them, or take bolts off some other part of the engine that will fit, and the loss of which will not impair the working of the other parts.

Broken Engine Truck Spring Hanger or Center Casting. If a four-wheel engine truck, block over the equalizers and under the top bar of engine truck frame close to band of spring, high enough so the engine will ride level with the other side. With a mogul, over the truck box. If the engine truck center casting breaks on a standard engine, block across under truck frame and center casting and over equalizers, from one side to the other; a couple of pieces of rail, 4½ or 5 ft. long, may come handy for this purpose. Or, put a solid block under the engine frame next to the saddle and on top of the truck frame on each side. This plan will give the use of the engine truck springs, although it does not always hold the center casting up against the male casting under the smoke arch, so the engine will track straight.

In case it becomes necessary to remove an engine truck entirely from a mogul or consolidation engine, proceed as follows: Block between the cross equalizer and bottom of the boiler; with the engine in this condition, she should be run carefully, as there is quite an additional load on the front driving boxes.

Broken Engine Truck Wheel, or Axle. If a piece is broken out of the wheel, it can be skidded to next side track by laying a tie in front of the pair of wheels. If an axle is broken or a wheel is broken off outside of the box, you can chain that corner of the engine truck up to the engine frame, being careful to chain so as to crowd the good wheel against the rail, and put a block between the top of the engine truck and bottom of the engine frame, on the other end of the same side of truck, in order to put the weight on that part of the truck.

Broken Back Spring on Consolidation Engine. Run the driver up on a wedge; pry up the back end of the equalizer and block between it and rail of frame; then run the back driver off the wedge and the next driver up on it, and block between the back driving box and the frame.

Broken Front or Back Section of a Side Rod on a Consolidation Engine. A consolidation engine has a knuckle joint between the first and second, and third and fourth pairs of drivers. In case of a section on either end breaking, remove the broken parts and the corresponding section on the other side. Be sure that the forward crank pin will clear the cross-head in all positions before moving the engine.

Broken or Loose Tire on Standard Eight-Wheel Engine. If a main tire breaks or becomes loose, raise the wheel center up off the rail a little higher than the thickness of the tire, to allow for the engine settling when blocked up; take out the oil cellar so the journal will not get cut on the edges of the cellar; put a solid block of wood between the pedestal brace and journal to hold the wheel center up clear of the rail; block up over the back driving box so the engine could not settle or get down so as to allow the wheel center to strike the rail. It will take a good deal of strain off the pedestal brace to put a block under the spring saddle on top of the frame. Taking out this driving spring makes a sure job; take off all other broken or disabled parts; if the rods are in good order, leave them up. If a back tire breaks, block up in the same manner as for a main tire, except that blocking comes next to the other journals and boxes.

Broken Main Rod. Remove the broken parts, block the cross-head back to within one-half inch of clearance to keep the cylinder packing out of the counter bore, disconnect cylinder-cock rod on disabled side and block the cocks open. Shift the valve in the same direction as piston if it is a slide valve or outside-admission piston valve, and in the opposite direction if it is an inside-admission piston valve. An easy way to remember and distinguish a direct from an indirect motion is in the position of the rocker arm. With the indirect, one arm is above and the other below the rocker box; with the direct, both arms are either above or below the rocker box. In moving the valve give it just enough opening to show steam at the cylinder cock, which will take the pressure off the blocking.

Broken Frame. For a broken frame ahead of a main

driver, disconnect the valve stem on disabled side, cover ports and leave up the main rod. Bring the engine in light with the good side. If the break is behind the main driver, take down the side rods on rear section only, if a consolidation engine. With a mogul type and the knuckle pin on forward section of side rod, take down all side rods.

Broken Draw Bar. If the engine has safety chains they will hold the tank, but not always a heavy train. If the engine is not equipped with safety chains, then secure a chain from the tank box or caboose and chain the tank to the deck. Safety chains should not have more than 4-in. of slack.

Broken Driving Brass. If a driving brass breaks and is cutting badly, run that wheel up on a thin wedge; then use an iron block between the top of frame and the spring saddle, which will take the weight off that box.

Broken Wedge Bolt. It is sometimes possible to screw the nut half-way onto each part of the broken bolt and thereby hold it up in place. If this cannot be done, then with a wire try to fasten a nut under the wedge to hold it up.

Broken Tender Wheel or Axle. Find a piece of a rail the proper length, or a cross tie will answer, place it across the top of the tank directly over the broken pair of wheels, block under the rail or tie to protect the flange on the top of the tender, jack up the broken pair of wheels to clear the rail and while in this position chain the truck to the rail above the tank on both sides.

Broken Crank-Pin. With a broken main crank-pin, on any class of engine, take down all side rods and be sure that the crank-pin on the forward wheel does not interfere with the cross-head in blocking the latter. With the back crank-pin on a consolidation or a ten-wheel engine, proceed as with a broken side rod, but if the crank-pin of an intermediate, otherwise known as driver No. 2, take down all side rods and run in light with the main rods up. Remember that taking down one section and not the other on the opposite side is dangerous; there is nothing to pull the wheel on the good side off the dead center. In only one case is this permissible—when the eccentrics are on the first or leading, and the main rod on the second or main drivers. In this instance if the forward section, with a solid end, breaks, the other side is to be left up so as to control the valve motion on the good side; but the valve gear on the crippled side must be disconnected.

Broken Cross-Head. If the break is with a four-bar guide or a Laird guide with yoke, block ahead and let the main rod rest in yoke; but the butt end brass and strap must come down, otherwise the rod would interfere with main pin. If the cross-head is of the alligator type and the yoke secured near the middle of the guide, block back and take down the main rod. It is always a good plan to allow enough port opening, in blocking valves central, to admit a little steam against the piston in the direction of the blocking. Remember also that an outside admission valve is pushed in the same direction as the piston, and an inside admission in the opposite direction.

Broken Eccentrics, Straps and Blades. With a broken go-ahead eccentric or blade, take down the back-up eccentric also. If the back-up eccentric is not disturbed the link lifter must be taken down. With a broken back-up eccentric, strap or blade, the go-ahead eccentric need not come down, but the engine must be run with a full cut-off and no attempt made to bring the lever back to the center of quadrant.

Disconnected or Broken Throttle Rod. This is generally regarded as a very serious mishap, but the seriousness depends entirely upon the nature of the break. If the throttle is open and cannot be closed, reduce the steam pressure to a point where the engine can be controlled with the reverse lever. It is a good plan to have some cars connected to the engine, in order to get the benefit of the brakes in case an attempt is made to run the engine in with a broken throttle.

Sometimes the valve becomes tilted or cocked and will not close. In such a case tapping the throttle rod with a hammer will sometimes bring it back to its seat.

If the throttle is closed and will not open it is very likely that the rod is disconnected inside the boiler, in which case the only remedy is to kill the fire, and prepare to be towed in, unless the company requires the engineers to make repairs.

Broken Whistle Stand. A broken whistle stand means a dead engine. Remove the broken part from the dome. A handy thing to have around an engine is a wash-out plug and several sizes of reducers. In the absence of a wash-out plug use the reducer in the dome cap, then take the nipple and angle cock off an air-braked car and insert into the reducer.

Broken Steam Chest or Cover. When the break is not a bad one, wedging between the chest and bolts is sometimes successful, but where the break is a bad one, remove the cover, block the supply ports, which on modern engines are at each end of the cylinder, with blocking of sufficient thickness to be held down by cover, disconnect the valve stem only, block the cylinder cocks open, and proceed on one side. The same method applies to a broken cover.

Metallic Packing Giving Out on the Road. Take off the stuffing box or packing case, or whatever it may be called, and if any of the old packing is left, leave a ring of it in the cone or cup; then make, out of wicking or old overalls, a ring of packing sufficiently large to fill the balance of the space in the cone, after which push the cone back against the "follower" on the end of the spring, put on the stuffing box and go along.

Broken Piston Gland Studs. On some of the old power in use on many railways, the fibrous packing is still used to quite an extent, and failure of the gland studs from one cause or another is not uncommon. When this occurs on engines having the four bar type of guides, the stuffing box should be partially filled, so that the gland would go well into it, then by driving a long taper wedge of iron, one on each side, between guides and gland, it (the gland) can be fastened securely and, if the wedges are driven carefully, set squarely. Wedges usually used to secure brake shoes to heads are very suitable for this purpose, but if they are not to be had, any piece of iron of the proper thickness and a little taper can be used. If only one stud is broken, fill the stuffing box as above, and if it is the top stud, use a piece of board to block from the oil cup on each guide to the top of the gland, driving a wooden wedge between back of the gland and cylinder head to prevent the gland wiggling and working blocks out. Never disconnect, or give up a train for this sort of failure.

Broken Piston Rod. Time can be saved when this occurs by doing just what is necessary and no more. If the rod breaks at the cross-head, as is usually the case, or near the piston, and the whole thing is blown out, cylinder head and all, just disconnect the valve rod and cover the ports, and go along.

Broken Driving Brass. Run the wheel upon a frog or wedge and block up between the frame and spring saddle, to take the weight as much as possible off the box.

With an engine having underhung springs there is no saddle to block under, and in a case of this kind place a jack under the equalizers nearest to the broken brass, then block the other end between the frame and the equalizer and remove the spring under the broken brass if possible.

Broken Driving Axle. This occurs usually close to the wheel and outside of the driving box. If it is a broken main driving axle, all rods on the disabled side and all side rods on the good side must come down. With any other driving axle, only such rods should come down as would give trouble to the rest of the rods.

To block up the axle on the broken side, remove the cellar and put a wooden block between the axle and the binder brace. If a hydraulic or screw jack is handy, raise the axle and driving box, if it has an overhung spring and block under the spring saddle above the frame to take the weight off the driving box. Use sponging on the sides of the blocking under axle or, better still, hot main-pin grease.

Water Glass Out of Order. If the water line in the glass is not in sight, and moving up and down when the engine is in motion, it indicates that the water glass valves are either stopped up or closed, and require immediate attention.

The blow-out cock at the bottom should be opened. If the water line now shows in the glass, and then suddenly rises out of sight when the blow-out cock is closed, it indicates that the water level in the boiler is higher than the top end of the glass. If only steam or a mixture of steam and water passes out through the blow-out cock it is evidence that the water in the boiler is too low, and if no water shows in the glass when blow-out cock is closed, the fire should be deadened at once. Every engine should be equipped with gauge cocks, and they should be tried every ten or fifteen miles passed over on the route.

If gauge cocks and lower water glass valve are stopped up, get engine and train off the main line onto a siding as soon as possible. Deaden or dump the fire, and report conditions. No engine should be worked in that condition.

If an Engine Works Water. Close the throttle a little at a time, until the water ceases to pass over into the cylinders. If it was foaming this would stop the trouble in the cylinders and the water level in the boiler would drop at once. If the boiler was pumped too full the water level would be above the gauge cocks.

When the Boiler Foams Badly. There is danger of knocking out cylinder heads, cutting the valves, stalling on some grade, or getting on some train's time because the engine cannot be worked to its proper power. There is also danger of burning the crown sheet when the water drops low enough to uncover it.

Broken Blow-off Cock, or Hole Broken in the Boiler. Either dump the fire or smother it with wet coal; get steam and water out of the boiler as quickly as possible. If the blow-off cock is broken off it may be plugged, but if it is blown out, it would be impracticable to plug the opening, and the only method is to treat such a case the same as with a hole knocked in the boiler, viz.: Disconnect the engine and be towed in.

If the water supply in the tender fails while out on the road away from a water tank, in warm weather, if possible, bail enough water into the tank to get to a water tank, and then fill up the tank. If in winter, and there is snow on the ground, shovel snow into the tank and melt it with steam from the boiler. If impossible to get water by the methods explained, draw the fire, disconnect the engine and be towed in.

If after getting a supply of water in the tank, the fire being dumped, and no steam in the boiler, it is found that the water level in boiler is below the crown sheet, the boiler may be refilled in the following manner by towing the dead engine with another engine. To do this, stop up all the openings where the outside air can get into the boiler or cylinders and steam chests (whistle, relief valves, etc.), open the throttle and steam and water connections to the injectors, put the reverse lever the way the engine is going and be sure to have her towed fast enough to create a vacuum in the boiler by the cylinders pumping the air out. The boiler can also be filled by connecting a hose to the overflow or delivery pipe of the injector on the live engine, and then to the suction pipe of the injector of the dead engine, or through the whistle or safety valve. A great many modern engines have wash-out plugs high enough to fill boiler to one gauge.

If an Injector Will Not Work. Be sure the injector gets a full supply of steam; the steam throttle may be only partially open. Some injector steam pipes are coupled to a turret; if the valve between the boiler and the turret is partly closed the injector may not get a full supply of steam and will not pick up the water. In this case a full supply of water is passing through the injector, but there is not sufficient steam to force the water into the boiler.

Examine the tank to see how much water there is in it; if plenty there, examine the hose, strainers and supply pipe to see if the injector could get the proper supply of water promptly; next, see if there were any leaks that would let air get into the supply pipe of a lifting injector; last, see if any foreign substance has got into the injector and choked any of the passages.

If an Injector Will Not Prime. The water may be all out of the tank, or the tank valve disconnected; air may be leaking into the supply pipe, the overflow valve stopped up or choked some; or the jet of steam may not pass exactly through the middle of the tube which exhausts air or starts the flow of the water.

If an Injector Primes Well, and Then Loses Its Priming When Steam Is Turned On Full. The boiler check valve may be stuck shut so the water cannot get away from the injector, the tubes may be coated with lime so they are too small or not of the proper shape, the tubes may be loose so they are not in line, or the supply of water may not be sufficient to condense all the steam; this may be on account of the feed water being hot.

When the boiler check valve sticks open or allows the boiler pressure to back up to the injector, jar the check case or delivery pipe a little so the check valve will settle into its seat. If it does not seat tight but leaks back, report its condition on arrival at terminal. Sometimes something will get into the delivery pipe and work under the check valve, holding it open; when the check is ground in, this foreign substance, which may be a part of the injector, will drop back into the delivery pipe and lie there till the injector is worked the next time, when it will get under the valve and hold it up again. Take off the delivery pipe and clean it out.

In Case of Bursted Flue. Dump the fire and lower the steam pressure as soon as possible in order to save the water in the boiler, then proceed to plug the flue with an iron plug if one is available. If no iron plug is at hand, use a wooden one, driving it into the flue for some distance. It will not burn, because no air can get at it. By putting on the blower slightly and putting a plank down on the grates, a man can often succeed in plugging a flue before the pressure is all out of the boiler.

Care should be exercised in driving an iron plug not to drive too hard, as there is danger of cracking the flue sheet.

In case of a boiler in which the flues are old and inclined to leak at the beads, pump the boiler as regularly as possible; have a bright, even fire; use great care that no cold air strikes the flues through the door or through holes in the fire near the flue sheet. Keep an even pressure of steam, as this means a steady temperature in the firebox. If possible, when going into the round house, leave two or three inches of fire on the grate after shaking out the old dead fire. This fire will die out slowly, so the fire box and flues will cool off slowly. The dampers should be closed after going into the house.

Be very careful in the use of the blower, that no cold air is drawn against the flues; this especially applies to the operation of cleaning the fire. Cold air contracts the flues, also the metal of the flue sheet, which causes them to begin leaking at once.

In Case the Lubricator Refuses to Work. First, see if

the steam valve to the boiler is open, then shut off the water valve from the condenser and open the drip valve in bottom of the oil tank; this would blow the water out of the glasses into the oil tank, with some makes of cups, and as soon as the glass filled up with water they might feed again. Or, shut off steam and water valve and open drip cock, then give engine steam and have steam from steam chest blow through the chokes, and clean them out. With the new style of cups having check valves, open the drip from the glass, blow it out clean and refill it with water. Lubricators usually stop feeding because some small openings are stopped up or something is wrong inside the cup.

The successful lubrication of valves and cylinders on long runs, with engines run at short points of cut-off, has been quite a difficult problem, and new ideas are being brought out all the time, and if the engineer is to be able to keep up with the rapid changes and developments that are coming along, it will be necessary for him to watch the lubricator question sharply.

Under certain conditions, the oil, as it is fed from the cup into sight feeders, does not pass to the steam chest direct, but this is no excuse whatever for running a dry valve, because a man who knows his business, in case of his valves getting a little dry, will ease off, almost shut off, for one second. This will drop the steam chest pressure, probably to about one-half of what it is ordinarily, at which time the strain at the lubricator will blow the oil and water from the pipe into the steam chest. The loss of time necessary to do this will not amount to as much in a trip as running a partially dry valve one mile would.

Engines that are being worked at a long cut-off do not, as a rule, give any trouble on account of valves running dry, because the pulsations of the steam chest pressure will render it impossible for water or oil to remain in one position in the pipe; it will work down, and in a few minutes after the lubricator is started the valves will receive the oil as regularly as it is fed from the cup.

The cases wherein the oil does not feed down regularly are when high pressure engines are run at high speed and short cut-off under full throttle. This can be explained in a few words, as follows: In the first place, when an engine is being run as above, the steam chests have almost boiler pressure and the steam that is admitted before the piston gets to the limit of its travel is, by the process of compression, to a certain extent, driven back into the steam chest, the effect of which is, owing to the necessary frequency of such action at high speed, that the steam chest pressure is stimulated to that extent that it is, perhaps, slightly higher, or at least equal to boiler pressure, in which case condensation will take place in the oil pipe near the steam chest and work upward; then the oil that leaves the lubricator will descend in the pipe, only until it comes in contact with the water in the same, and will remain there until the throttle is eased or shut off.

Broken Main Valve Rod. Put the valve on the center of the seat, so that it will cover all the ports on that side; disconnect the main-rod and block the cross-head the same as for simple engine.

BREAKDOWNS ON COMPOUND ENGINES, VAUCLAIN

Both High-Pressure Cylinder Heads Gone. If the stuffing-box could be made steam-tight, the steam valves would supply live steam direct to the low-pressure cylinders, but it would be doubtful if it could be done, owing to the lack of material for the same on an engine.

Broken High-Pressure Cylinder Head. Block the valve on the disabled side in its central position so as to cover the ports, disconnect the main rod and block the cross-head the same as with a simple engine. Run in with the other side, working compound if possible and simple if necessary, as it probably will be.

Broken High-Pressure Piston Rod or Piston Head. Remove the broken piston and rod and plug up the piston-rod hole in the cylinder-head from the inside with a wooden plug; then replace the head or heads and run in light.

Broken Low-Pressure Cylinder Head. Run with the starting-valve in compound. This would give the simple engine of the dimensions of the high-pressure cylinder. The exhaust from the high-pressure would necessarily go through the low-pressure cylinder and to the stack; but it would be very serious on the fire and would reduce the steam.

Broken Low-Pressure Piston Head or Piston Rod. Remove the piston and rod and plug the piston-rod hole and run light with the starting-valve in compound.

Baldwin Two-Cylinder Compounds. *Broken Main Rod, High-Pressure Side.* Remove the broken rod, block the cross-head at the back end of the guide, clamp the high-pressure valve in the center to cover both ports. Place the engineer's operating valve at point marked "simple." The steam will then pass through the reducing valve to the receiver and thence to the low-pressure cylinder.

Broken Main Rod, Low-Pressure Side. Remove the

broken rod, block the cross-head at the back end of the guide, disconnect the valve rod and close the plug cock in the pipe leading to the intercepting valve. Place the engineer's operating valve at the point marked "compound." The steam will then act in the high-pressure cylinder and pass to the exhaust without entering the low-pressure cylinder.

For Valve Stem on Either Side. The same remedy as for a broken rod.

For the Intercepting and the Reducing Valve. The intercepting and reducing valves are automatically operated by steam pressure. Access can be easily had to either of these valves by removing the head of their respective cylinders, and any failure of the valve to act can be readily ascertained and remedied.

Schenectady Cross Compound. *Broken Main Rod, High-Pressure Side.* Remove the broken rod, blocking the cross-head at the front end of the guides. Clamp the high-pressure valve forward to clear the exhaust port; the steam will then pass through the exhaust port on the high-pressure side into the receiver, thence to the low-pressure steam chest. The low-pressure cylinder then acts as a high-pressure engine. Open the throttle valve moderately on account of the large area of low-pressure cylinder.

Broken Main Rod, Low-Pressure Side. Remove the broken rod, secure the cross-head at the back end of the guides; clamp the low-pressure valve back far enough to clear the exhaust port. Exhaust steam can then pass through the low-pressure cylinder and out through the stack.

Broken Valve Stem on Either Side. The same remedy as for a main rod on that side; also remove the main rod and secure it as previously instructed.

Intercepting Valve Disabled. Should the intercepting valve become disabled, clamp the poppet valve open, if possible; if not, then remove the back head from the intercepting valve steam cylinder, and push the piston forward, putting in a block to hold it in that position; then put on the head, which will prevent the steam in the receiver closing the poppet valve. The same may be done with the small piston which moves the valve, admitting steam to the intercepting valve steam cylinder; this will also prevent closing the poppet. Live steam would then be admitted to both cylinders for starting.

Baldwin Tandem Compound. *Broken Cylinder Head.* This would be a very serious case, and it would be necessary to clamp the valves centrally over the ports on the damaged side, and go in on one side.

To Test for Leaks and Blows. The Baldwin tandem compound is tested in the same manner as other tandems.

Questions

551. What are the qualifications in general that a locomotive engineer should possess?

552. What will a careful engineer always do before starting out with his engine?

553. What should he do with the rod keys before leaving the roundhouse?

554. What special tools should be carried on the tender, to be used in case of breakdowns?

555. When an engine suddenly begins to go lame in her exhaust, what does it indicate?

556. How is a dry valve easily detected?

557. Mention some other causes besides those

previously enumerated that may be responsible for a lame exhaust.

558. If an eccentric has slipped, how may it be located and reset?

559. If it is necessary while out on the road to place the engine on a center, how may it be done in a hurry?

560. What is to be done in case of a broken eccentric, or a broken valve rod?

561. If the reach rod, or arm of tumbling shaft, should get broken, what should be done?

562. When a valve seat becomes broken, what should be done?

563. At what point does a valve yoke usually break?

564. How may a broken valve yoke be discovered?

565. Mention several remedies for this kind of a breakdown.

566. How may a broken gib or plate on the cross-head be taken care of?

567. If the cross-head is badly broken, what should be done?

568. What is the only remedy for a broken main rod, or strap?

569. In case of a broken side rod or strap, what should be done?

570. What is necessary in case of a broken back cylinder head?

571. In case a forward cylinder head breaks, what may be done?

572. What is to be done with broken guides, blocks, or bolts?

573. If a guide yoke is bent or broken, what should be done?

574. What is implied when but one side is disconnected?

575. How should the cross-head be secured when one side is disconnected?

576. What is to be done with the valve in case one side is disconnected?

577. If it is necessary to remove the eccentric straps, what must be done with the link?

578. If it becomes necessary to take down one side rod, what must be done with the one opposite to it?

579. If the eccentric blades are removed and the straps left on the eccentrics, what precautions should be taken?

580. If the side rods on a ten wheeler have been removed, what precaution must be taken?

581. In case of a broken driving spring or hanger, what method should be pursued, if the engine is jacked up?

582. If the engine has been raised by running her up on blocks, what is the best method of procedure?

583. When both sides are disconnected, what is implied?

584. In case the engine must be towed in, and the weather is freezing, what precautions should be taken?

585. How is a standard eight wheeler to be blocked up in case of a broken equalizer?

586. If the engine is a ten wheeler, how is a broken forward equalizer to be blocked?

587. If an intermediate equalizer on a mogul breaks, what is to be done?

588. If a cross equalizer on a four-wheeled pony engine breaks, how is it to be blocked?

589. If a truck equalizer breaks, how is it to be taken care of?

590. If an equalizer stand breaks, what is the remedy?

591. In case of a broken engine truck spring hanger

or center casting on a four-wheel truck, how is it to be blocked?

592. How should a mogul engine truck be blocked, if spring hanger is broken?

593. If engine truck casting breaks, what should be done?

594. How may a mogul or consolidation engine be blocked so as to run, in case it is necessary to remove the engine truck entirely?

595. What may be done with a broken engine truck wheel or axle?

596. What is to be done in case of a broken back spring on a consolidation engine?

597. How may a consolidation engine be fixed up so as to run with a broken front or back section of a side rod?

598. What is to be done with a broken or loose tire on an eight wheeler?

599. What is the best method to pursue in case of a broken main rod?

600. Mention an easy way to remember and distinguish a direct from an indirect valve motion.

601. In case of a broken frame, what should be done?

602. What may be used to replace a broken draw bar?

603. What should be done when a driving brass breaks?

604. What may be done in case of a broken wedge bolt?

605. When a tender wheel or axle breaks, what may be done to enable the engine to proceed?

606. What is to be done with any class of engine in case of a broken main crank-pin?

607. When the back crank-pin of a consolidation or ten-wheeled engine is broken, what is to be done?

608. If the broken pin is on an intermediate driver, what must be done?

609. What should always be remembered when disconnecting an engine?

610. Under what conditions is it permissible to disconnect a section of one side, and not the same section on the opposite side?

611. What is to be done with a broken cross-head if the guide is a four bar, or Laird guide?

612. If the cross-head is of the alligator type, what should be done?

613. What is a good thing to remember when blocking valves central?

614. What must be done when a go-ahead eccentric, strap, or blade breaks?

615. In case of a broken back-up eccentric, strap, or blade, what is necessary?

616. What may be said of a broken or disconnected throttle rod?

617. If the throttle is open and cannot be closed, what should be done?

618. What is a good plan to pursue in case an attempt is made to run in with a broken throttle?

619. If the throttle becomes tilted or cocked, how may it sometimes be seated?

620. If the throttle is closed and will not open, what is the only remedy?

621. What does a broken whistle stand mean?

622. What is a handy thing to have around an engine?

623. With a broken steam chest or cover, what methods may be pursued?

624. In case of metallic packing giving out on the road, what is to be done?

625. When a piston gland is broken, how may the break be remedied?

626. What should never be done in case of this sort of a breakdown?

627. What should be done when a piston breaks?

628. What is to be done in case of a broken driving brass?

629. If the engine has underhung springs, and a driving brass breaks, what should be done?

630. At what point does a driving axle usually break?

631. If a main driving axle is broken, what must be done?

632. If the broken axle is not a main driving axle, what should be done?

633. How is an engine with a broken driving axle to be blocked up?

634. If the water line in the glass is not in sight and moving, what does it indicate?

635. What should be done immediately in this case?

636. How is too high a level of water in the boiler indicated?

637. What indicates too low a level of water in the boiler?

638. If no water shows in the glass when the blow-out cock is closed, what should be done at once?

639. What should always be done with the gauge cocks?

640. If both the water glass and gauge cocks are stopped up, what should be done as soon as possible?

641. What should be done in case an engine works water?

642. What are some of the dangers incurred when a boiler foams badly?

643. In case of a broken blow-off cock, or a hole broken in the boiler, what should be done?

644. What may sometimes be done with a broken blow-off cock?

645. If the water supply in the tender fails while away from a water tank, what should be done?

646. If this should occur in the winter with snow on the ground, what may be done to obtain a supply of water?

647. How may a dead engine be pumped up, and the boiler refilled by another engine?

648. By what other method may a boiler that is dead, and partly empty, be refilled out on the road?

649. What is to be done in case an injector refuses to work?

650. What are some of the things to be looked after, when an injector will not work?

651. If an injector will not prime, what are some of the probable causes of it?

652. If an injector primes readily, and then loses its priming, what are some of the various causes?

653. If the boiler check valve sticks open, how may it sometimes be righted?

654. What other causes may be responsible for the check valve not working properly?

655. What is to be done in case of a bursted flue?

656. If no iron plug is at hand, what may be used to plug a bursted flue?

657. What precaution should be observed in the driving of an iron plug into a bursted flue?

658. What should be done with a boiler having old and leaky flues?

659. How should the blower be used on such a boiler?

660. What effect does cold air have upon the flues, and flue sheet?

661. What is one of the first things to do in case the lubricator refuses to work?

662. How may it be cleaned out?

663. What is usually the cause of a lubricator not feeding the oil properly?

664. Is there any excuse for running a dry valve?

665. What may be done towards forcing the oil to pass into the steam chest?

666. Under what conditions is it difficult to get the oil to feed regularly?

667. What causes tend to bring about such conditions?

668. What is to be done with a Vauclain compound, in case of a broken main valve rod?

669. What may be done with a Vauclain compound when both high-pressure cylinder heads are gone?

670. In case one high-pressure cylinder head on a Vauclain breaks, what should be done?

671. If a high-pressure piston rod or piston head should break, what is to be done?

672. What may be done with a Vauclain compound if one of the low-pressure cylinder heads is gone?

673. If a low-pressure piston head or piston rod breaks, what is the remedy?

674. If the main rod on a Baldwin two-cylinder compound breaks, how may the engine be run?

675. If the same break should occur on the low-pressure side, what should be done?

676. If a valve stem on either side of a Baldwin cross compound should break, what is the remedy?

677. If either the intercepting or reducing valve on a Baldwin two-cylinder compound fails to work, what may be done?

678. What is to be done with a Schenectady two cylinder compound in case of a broken main rod on the high-pressure side?

679. In case a similar break occurs on the low-pressure side, what should be done?

680. What is the remedy for a broken valve stem on either side of a Schenectady cross compound?

681. In case the intercepting valve becomes disabled, how may a Schenectady two cylinder compound be operated?

682. In case of a broken cylinder head on a Baldwin tandem compound, what is to be done?

683. How are Baldwin tandems tested?

684. Under what conditions is it allowable to operate a compound locomotive as a single expansion engine?

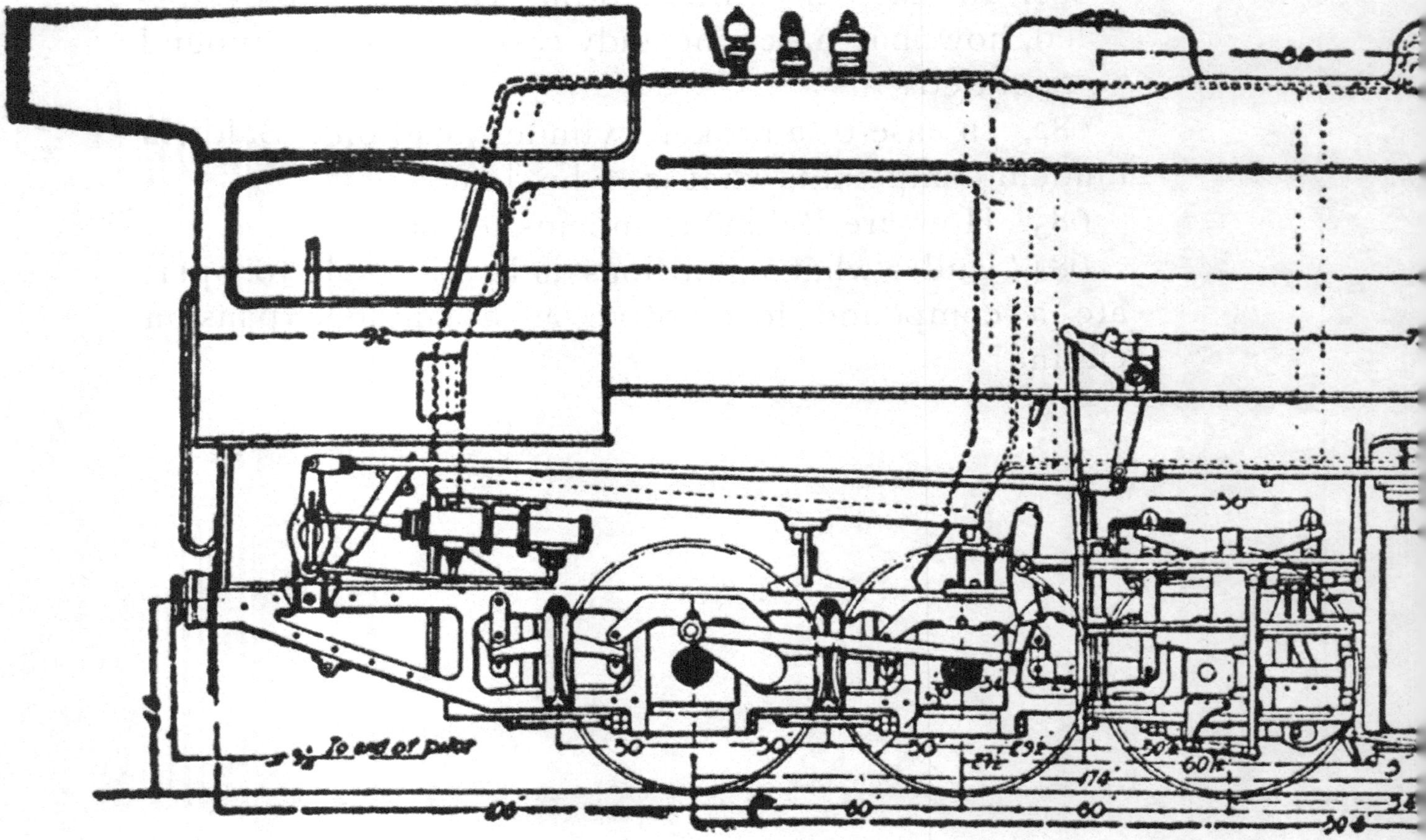
To end of pilot

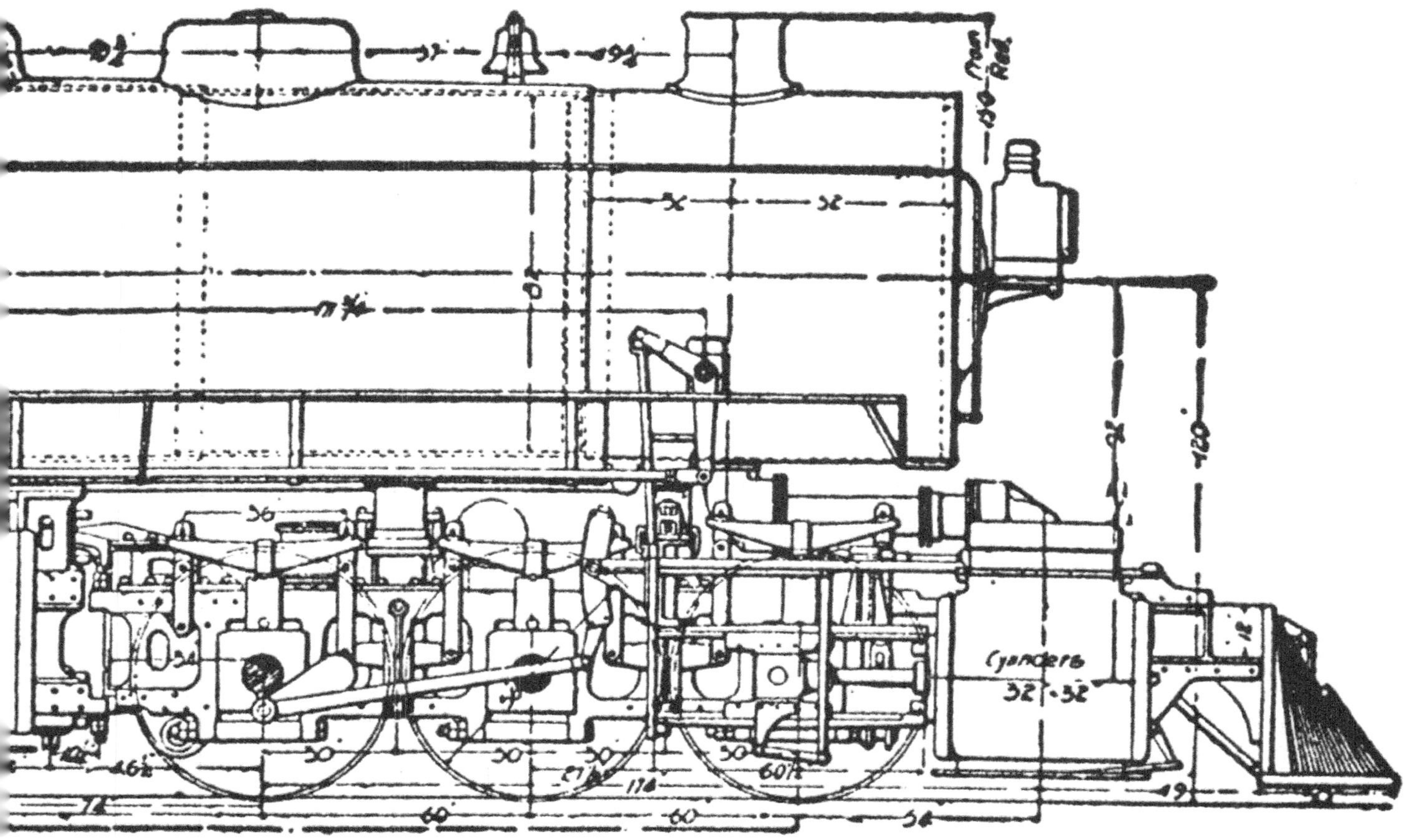

Cylinders
32 - 32
120
60
60
54

CHAPTER XI

WALSCHAERT VALVE GEAR—GENERAL DESCRIPTION.

The motion of the valve, as shown in Fig. 293, is derived both from the crosshead and the eccentric crank from a driving axle. The crosshead connection imparts the lap and lead at the extremities of the stroke, when the eccentric crank is in its middle position. The eccentric crank in this position imparts its fastest movement to the valve to give a very quick opening. The crosshead motion in advancing from the dead point effects an approximate uniformity in the combined motion given to the valve, as if it was derived from a single crank or eccentric set with an angle of advance corresponding to the lap and lead. The valve motion may therefore be graphically illustrated in the same manner as that of the Stephenson motion, with a circle representing the path of the eccentric, the diameter of which is equal to the travel of the valve, and the valve movements may be determined in the same way by any of the well-known methods of Professor Zeuner and others, as illustrated on another page. It will be observed that the only apparent variation due to this gear is that brought about by the invariable lead.

The Walschaert motion, as usually constructed, does

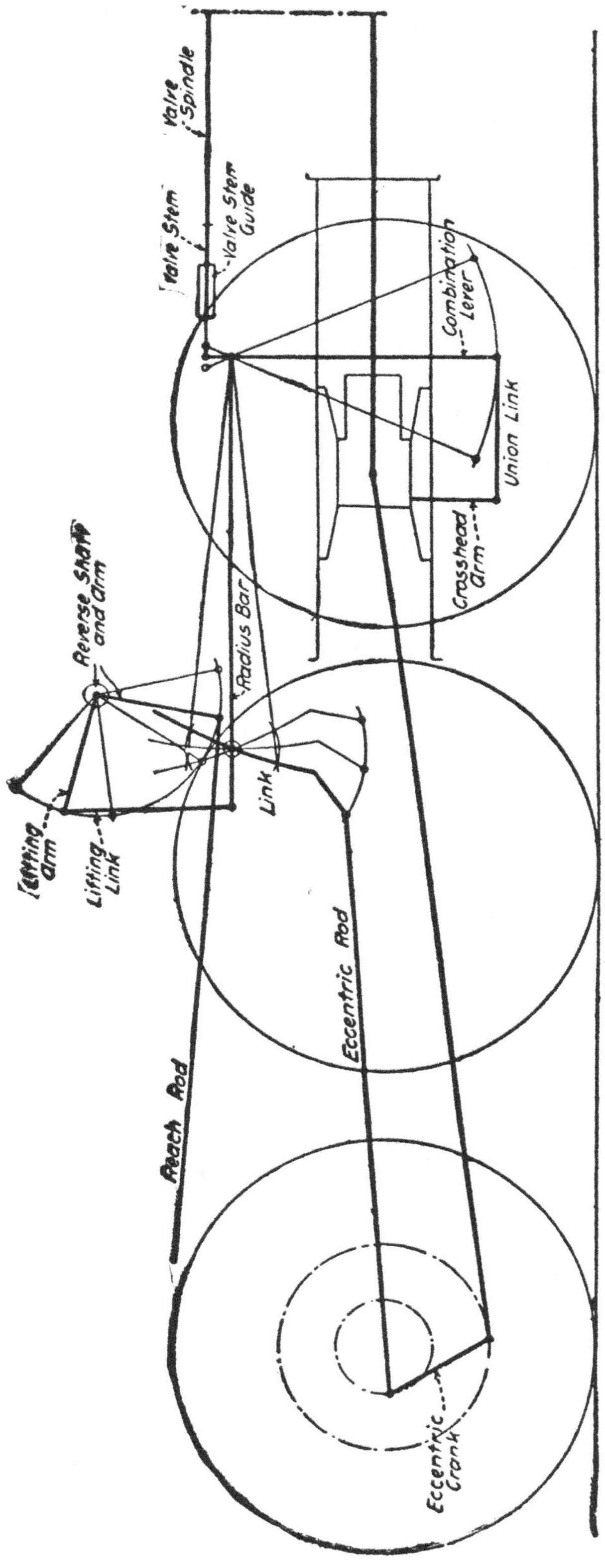

FIG. 293. OUTLINE OF WALSCHAERT VALVE GEAR AS APPLIED TO COMPOUND LOCOMOTIVE—BALTIMORE & OHIO RAILROAD

not lend itself as freely to adjustment as does the Stephenson motion with independent eccentrics, and for this reason it is not as liable to get out of adjustment. It must be correctly laid out in the design and correctly fitted up. The importance of this cannot be overestimated. The various points must be carefully plotted in order to give the best results in the combination movement of the parts of the motion. The movements of the motion involve such complications in plotting as to render the complete plotting of all too laborious for every new design, and for this reason the use of an adjustable model is very desirable in designing this gear. However, with complete knowledge of the nature of the gear, simple methods and formulas may be used to determine the locations of the various points covering the motion. One object of this description is to avoid the necessity of a model except to verify the results.

To entirely overcome the irregularities inherent in all motions transformed from circular into lineal, cannot for practical reasons be expected, but the errors may be reduced to such an extent that they do not affect either the power or economy of the locomotive. This remark is made to forestall the inference that the accuracy of the Walschaert motion as to the cut-off points is not superior to the Stephenson motion when the latter is turned out of the shop.

In the construction of the Walschaert gear the desired travel of the valve, the lead and the maximum cut-off which determines the lap of the valve, are selected. The stroke of the piston being given, the combination lever is proportioned so that a motion equal to the lap and lead is given to the valve when the crosshead is moved from one end of the stroke to the other. The link may be made of any approved design, and is so located that the radius bar will have a length of at least eight times (ten or twelve times is better) the travel of the link block, and the radius of the link should be equal to the length of the radius bar.

For outside admission valves the radius bar is attached to the combination lever between the valve stem and the crosshead connections, and for inside admission (piston valves) it is attached above the valve stem. The fulcrum of the link should lie as nearly as practicable upon a line drawn through the union of the radius-bar and combination lever, parallel with the center line of the valve stem. The suspension point of the lifter should have a locus which causes the link block to travel as nearly as practicable on a chord of the arc described by any point of the link wherever the block happens to be when the link is swung into one of its extreme positions.

This is most closely approached by a lifter through which the radius bar slides, and does not swing with the link. A properly suspended hanger will accomplish practically the same result, though the slip of the link bar will be somewhat more in the back than in the forward motion, but as the suspension point cannot be made to follow the theoretical locus, it should be made to do so as nearly as possible by favoring the position of the most commonly used cut-offs.

In locating the longitudinal position of the link fulcrum, consideration should also be given to the length of the eccentric rod, which should have a minimum length of 3½ times the eccentric throw, and should be made as long as circumstances will permit, with an approximate equal length of the radius and eccentric rods. The point of connection between the eccentric rod and the link should be as near the center line of motion of the main rod as this correction for rod angularities will permit, but this is often accompanied with the requirement of excessive eccentric throw. In such cases a compromise must be made to raise this point. The fore and aft position of this point relative to the tangent of the link arc must also be determined with reference to the angularity of the eccentric and main-rods, so that the link is exactly in its central position when the piston is at either end of the stroke. The angles through which the link swings on both sides of its central position should be as nearly as practicable equal, but this is subordinate to other conditions. Attention should be paid to the effect of the angularity of the main connecting rod upon the cut-off, to reduce this to a minimum, this having an effect upon determining the locus of the suspension point of the lifting link as well as that of the eccentric rod connection to the link.

It is evident that a proper design of Walschaert gear can only be laid out by a skilled draughtsman. In maintenance, care is required that all parts should preserve their original forms and position, and this should be checked by verifying the valve movements through turning the main driving wheels before the locomotive goes into service.

The chief point of difference between the Walschaert and Stephenson motions is that the former gives to the valve a constant lead at all cut-offs, whereas the latter produces an increase of lead which becomes excessive at short cut-offs.

METHOD OF ADJUSTING VALVES WITH WALSCHAERT GEAR.

The lap and lead are determined by the proportion of the arms of the combination lever and the stroke of the piston. The amount is found by turning the engine from one dead center to the other center in any cut-off position.

1. The motion must be adjusted with the cranks on the dead centers by lengthening or shortening the eccentric rods until the link takes such a position as to impart no motion to the valve when the link block is moved from its extreme forward to its extreme backward position. Before this change in the eccentric rod is resorted to, the length of the valve stem should be examined, as it may be of advantage to plane off, or line under, the foot of the link support which might correct the lengths of both rods, or at least only one of these should need to be changed.

2. The difference between the two positions of the valve on the forward and back centers is the lap and lead doubled, and cannot be changed except by changing the leverage relations of the combination lever.

3. A given lead determines the lap, or a given lap determines the lead, and it must be divided for both ends as desired by lengthening or shortening the valve spindle.

4. Within certain limits this adjustment may be made by shortening or lengthening the radius bar, but it it desirable to keep the length of this bar equal to the radius of the links, in order to meet the requirements of the first condition.

5. The lead may be increased by reducing the lap, and the cut-off point will then be slightly advanced. Increasing the lap introduces the opposite effect on the cut-off. With good judgment these quantities may be varied to offset the irregularities inherent in transforming rotary into lineal motions.

6. Slight variations may be made in the cut-off points as covered by the previous paragraph, but an independent adjustment cannot be made except by shifting the location of the suspension point, which is preferably determined by a model.

METHOD OF LAYING OUT WALSCHAERT GEAR.

Having presented a general outline of the gear, we may proceed in determining the more important points necessary to obtain a successful motion of the valve, and, as previously stated, the stroke of the engine is given, and the travel and lap and lead of the valve are selected to suit a desired cut-off. We have first to find the proportions of the combination lever. By designating the lap and lead with a letter C, the crank Radius with R, the cross-head end of the combination lever with *L,* and the valve end of the same with V, we have $R : C = L : V$, or $V = \frac{CL}{R}$, with the connection F of the radius bar as a fulcrum. The length of the combination lever must be determined from the height of the valve stem over the piston rod and a convenient angle of oscillation of 45° to 50°, which should not exceed 60°.

We have next to find the required travel of point F,

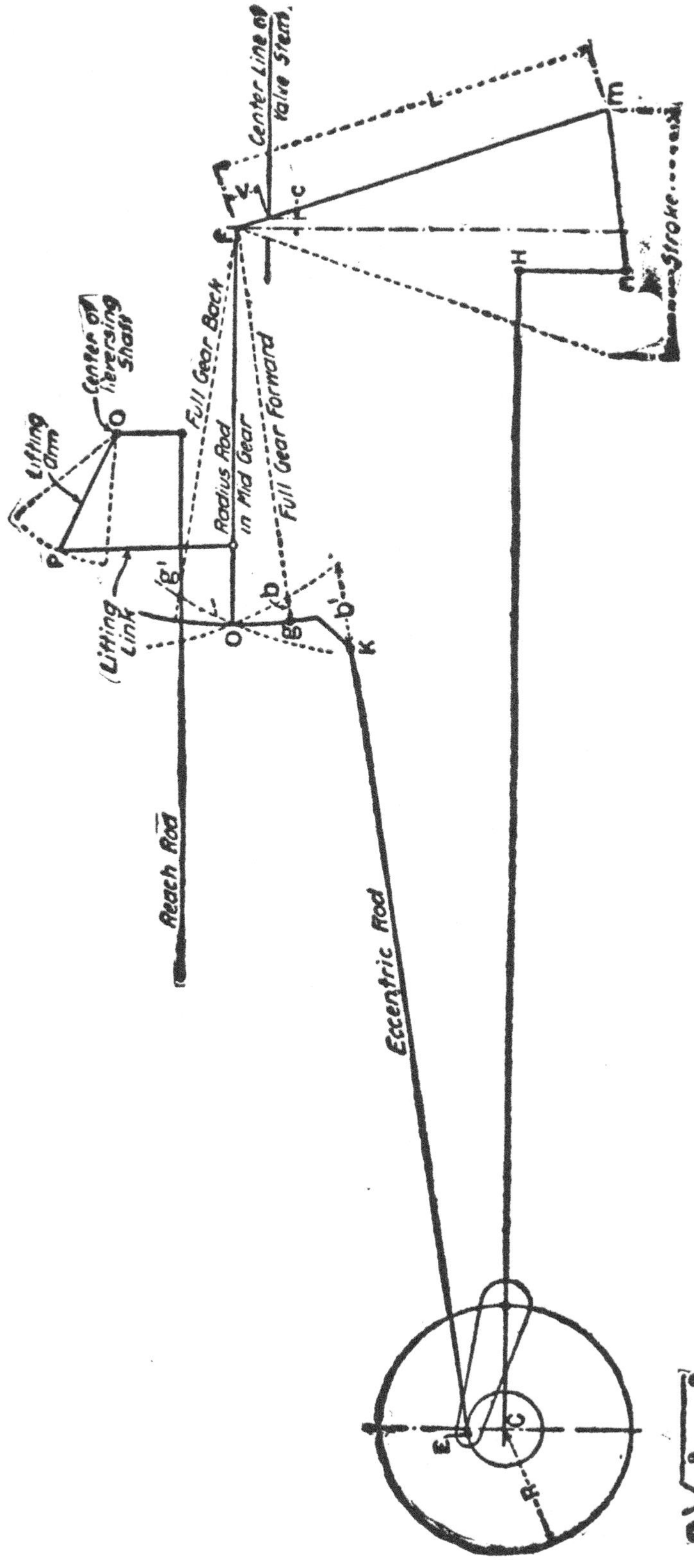
Center Line of Valve Stem
L
E
V
c
Stroke
H
Center of Reversing Shaft
Lifting Arm
O
P
Full Gear Back
Radius Rod in Mid Gear
Full Gear Forward
Lifting Link
g'
b
b'
O
B
K
Reach Rod
Eccentric Rod
R
E
C

NO. [illegible]

Fig. 294, to obtain the desired valve travel, which for convenience sake is taken on one side of the center position, or half its total travel in full gear, and which we will designate b, when we have:

$b=\frac{R\sqrt{a^2-c^2}}{R+c}$ for outside admission, and $b=\frac{R\sqrt{a^2-c^2}}{R-c}$ for inside admission valves, where a is the radius of an eccentric that would give the required travel of the valve, and c is as given above.

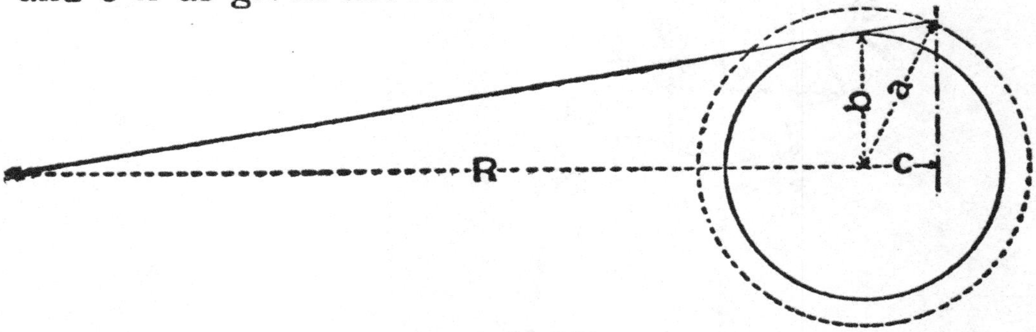

FIG. 295

This may be laid out graphically as in Figs. 295 and 296, when *a* is equal to one-half the travel of the valve and R and *c* the same as in the above formulæ.

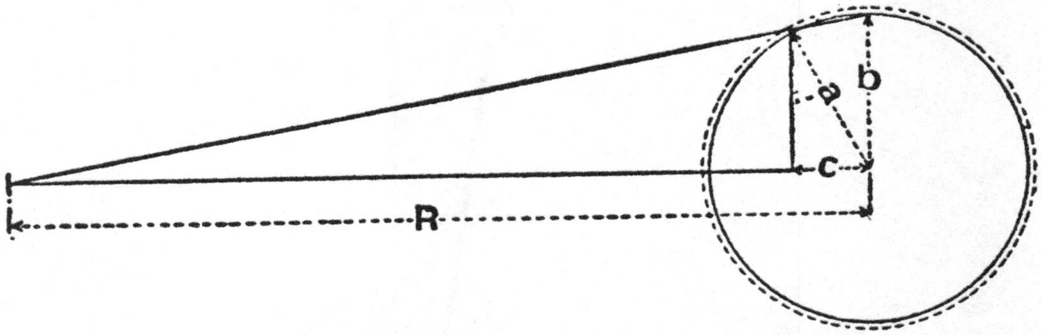

FIG. 296

With the limited amount it is advisable to allow in raising or lowering the link block in reversing the motion, we can without practical error consider the half movement of the link block g to be the same as that of point F, and by limiting the angle of the swing of the link to a maximum of 45°, we get the raise or depression of the radius bar and link block $Og = \frac{b}{\tan. d}$, where O is the link fulcrum and $d =$ half the angle of the swing of the link.

The location of the link and eccentric rod connecting point K, Fig. 294, cannot be determined with any practicable formula, but must, as already stated, be found by plotting to meet the requirements of the different cut-offs and corresponding crank positions. The same is also the case with determining the locus for the suspension point P of the lifting link, and in these two locations lies the principal success of the gear.

THE REAULEAUX AND ZEUNER DIAGRAMS.

Figure 297 shows a combination of two diagrams, namely, those of Reauleaux and Zeuner, which coincide exactly as to the different valve movements, which may be found as follows:

The distance AB represents the travel of the valve as well as the stroke of the engine, though in different scales, which makes no difference when the cut-off is always expressed in fractions or per cent of AB. The maximum cut-off is determined upon to be AR. Draw a perpendicular line RC from AB until it cuts the arc ACB. Next decide on a desired lead and, with that as a radius, draw an arc with A as a center. Draw a line from C tangent to the lead circle around A, when the lap of the valve is found to be equal to the perpendicular distance from the line CS to the center O of the diagram. The crank will then be in the position OS when the valve commences to open, or the angle AOS in advance of the

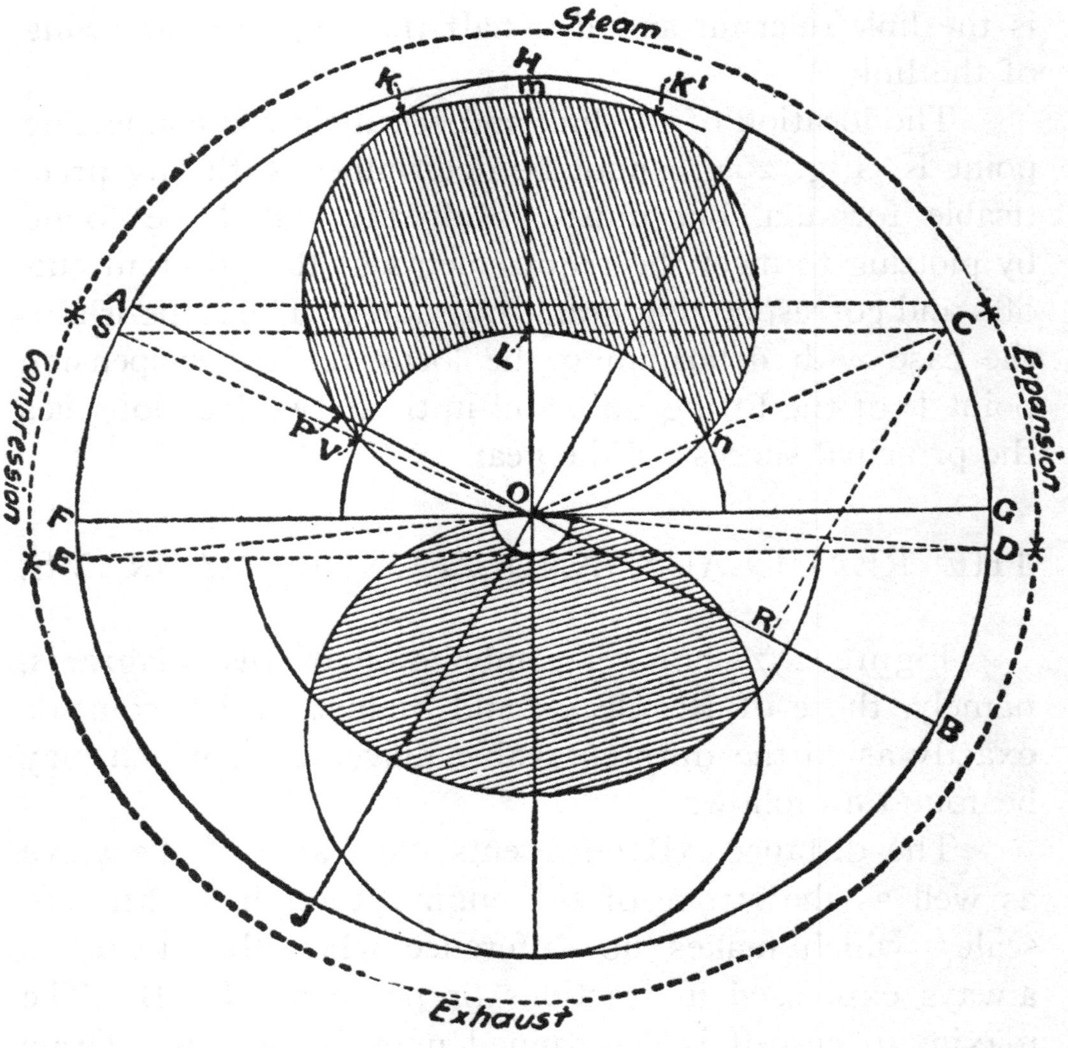

FIG. 297

dead center, and on OC at cut-off. Continuing, we find the valve in its middle position when the crank is on OG, which is drawn parallel to SC through the center O.

Extend this line to F, and with the exhaust lap as a radius, draw the exhaust lap circle on the opposite side of the line GF and draw DE tangent to this circle, when OD is the position of the crank at the release point. From this point the exhaust remains open until the crank reaches the position OE, when it closes and compression takes place until it again reaches OS for admission and one revolution is completed.

By placing the Zeuner diagram upon this, draw HJ perpendicular to FG, and with the radius OH of the eccentric circle as a diameter, draw the admission valve circle OVH*n*O and the lap circle with the steam lap as a radius and find the intersection occurs at V, both with the circles and the previously laid down admission line OS and the cut-off point at the intersections at *n*. On the line OH set off the width of the steam port from *L* toward H equal to L*m* and with O*m* as radius draw the arc K*m*K. The shaded figure enclosed by the letters VKK′*n*L represents the steam port opening during the admission period and the width of the port opening at any desired position of the crank is found by measuring the distance radially from O between the lap and valve circles on the port line, as the case may be, on the desired crank position.

The exhaust openings are determined in the same manner and are shown on opposite side of FG, where the crank passes through the arc DJE during the exhaust period with a positive exhaust lap of the size EF. When the exhaust edge of the valve is line and line, this arc becomes GJF, or 180°, and when a negative lap (clearance) occurs, the duration of the exhaust period exceeds the half revolution of the crank. The various events are indicated around the eccentric circle on the figure as they take place during a complete turn of the crank.

In Fig. 298 the eccentric and admission valve circles

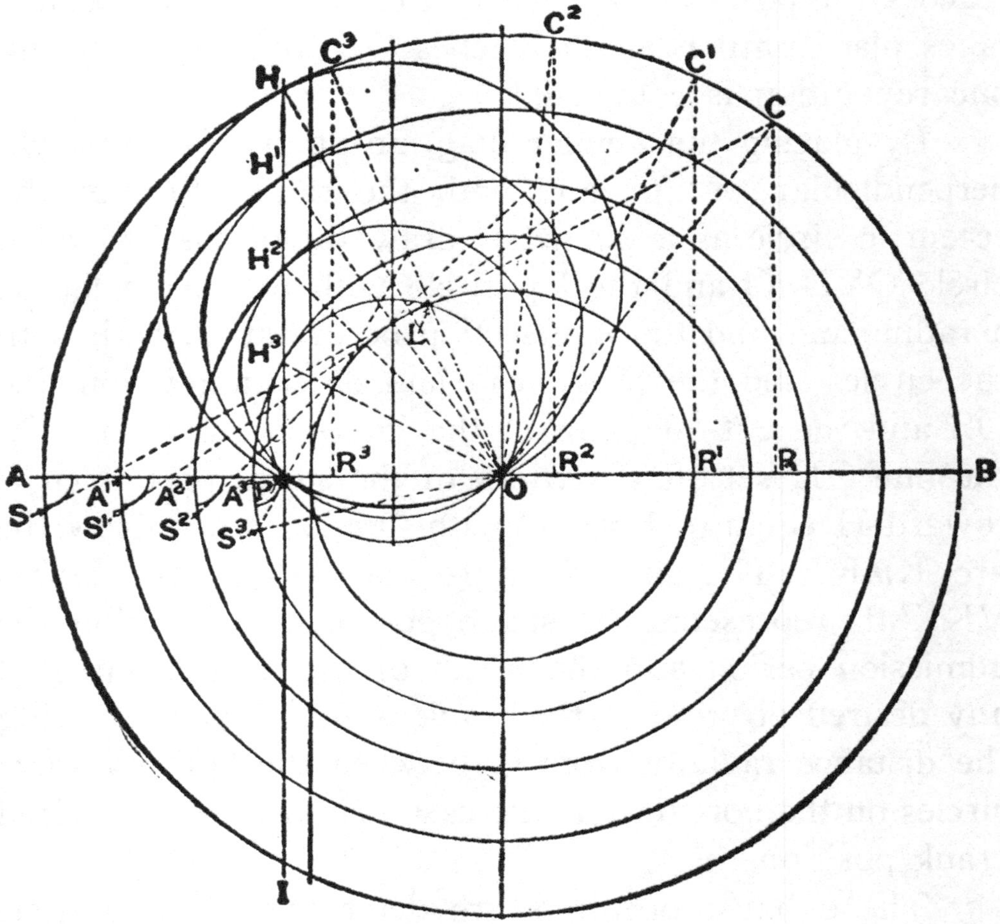

FIG. 298

are shown at different cut-offs where each set of lines and circles is governed by the same explanation as those of Fig. 4, where the admission points S, S^1, S^2, S^3, correspond to the closing positions C, C^1, C^2, C^3 cut-off points R, R[1], R[2], R[3], etc. On OH we have the full travel valve circle and OL the lap or radius of the lap circle,

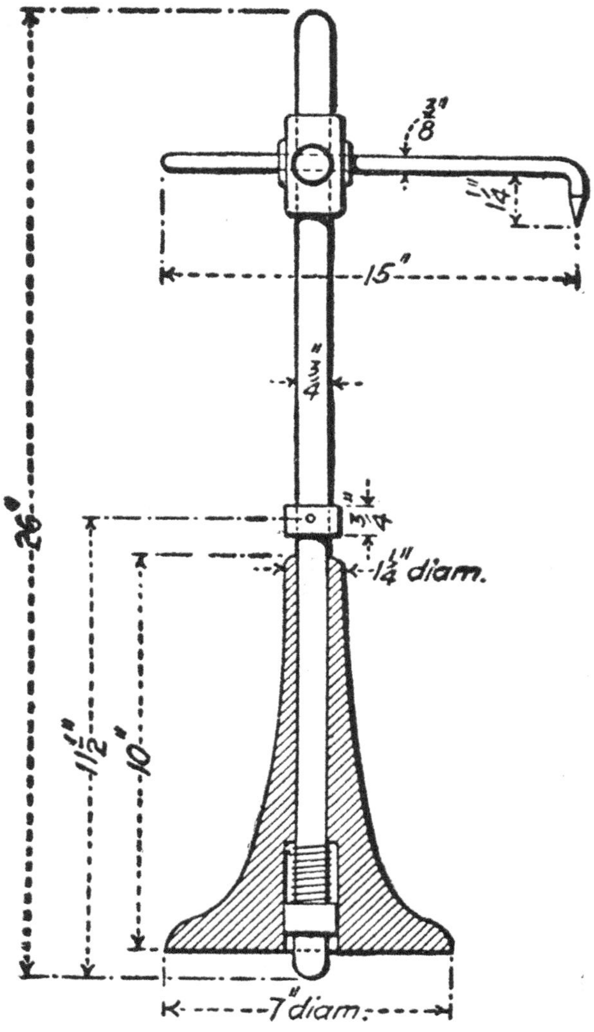

FIG. 299. A CONVENIENT GAUGE FOR SETTING THE ECCENTRIC CRANK

the latter being the same for all cut-offs as well as the lead, the radii H[1], H[2], H[3], etc., of the eccentric circles or diameters of the corresponding valve circles terminate

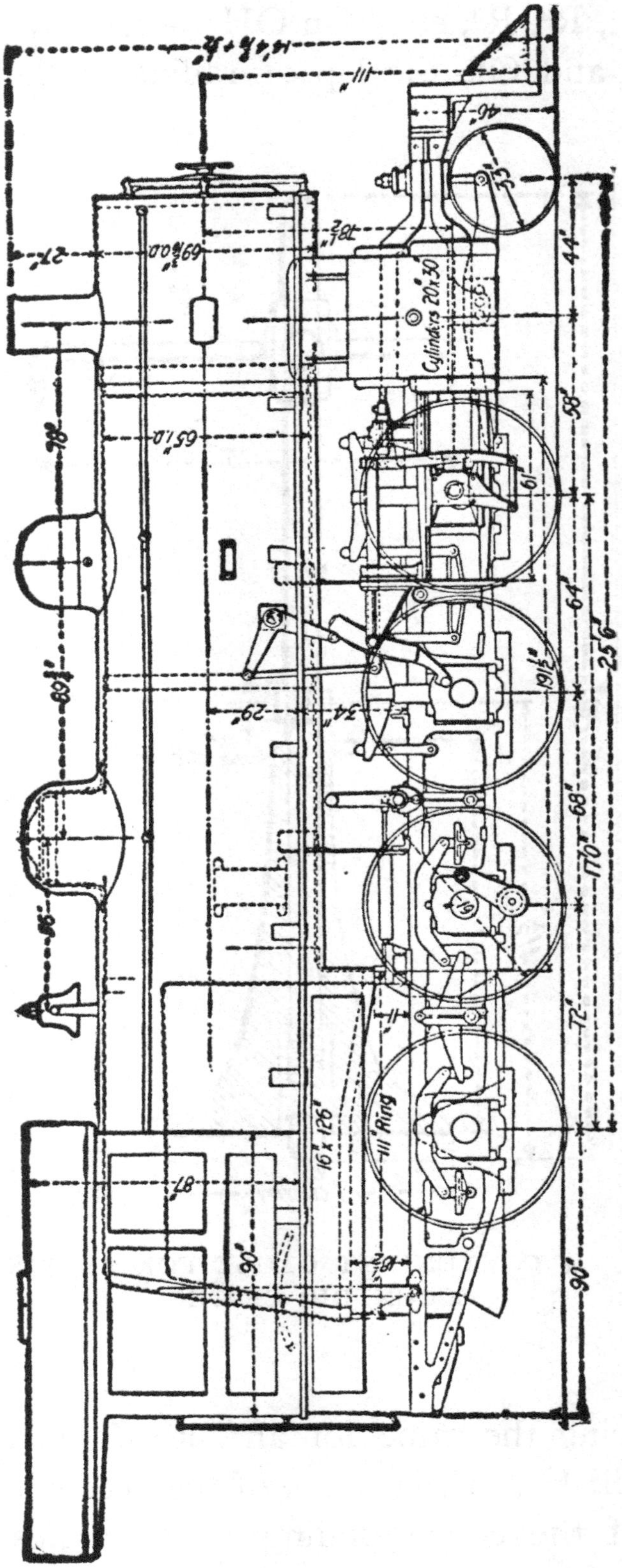

FIG. 300. WALSCHAERT VALVE GEAR—CONSOLIDATION FREIGHT LOCOMOTIVE—BOSTON & MAINE RAILROAD

on a line HI drawn perpendicular to AB and at a distance from O equal to that of lap and lead.

When the reverse lever is in its center position the diameter of the valve circle falls on the line AB and is equal to lap and lead. Continuing in back position we have the same method repeated and OI would be the full travel valve circle diameter, or the same as the eccentric radius for the valve travel. Any desired cut-off position may be laid out in the same manner as that in Fig. 297, which shows all the valve movements for a complete revolution of the axle.

The movements are in actual practice not so regular as the circles indicate, as it is impracticable to obtain the various loci in their theoretical positions; besides, we have the angularities both of the main rods and the eccentric rods to contend with and whereby irregularities are entering in the problem that must be compensated for, as referred to in the general description. It is not to be considered that a uniform circular motion is the best, but an approximation to it works with fewer shocks or jerks, and is therefore more desirable for so high-speed an engine as a locomotive. A few advantages can be taken, however, in selecting the suspensions and various connections, so that better results can be obtained than from a true circular motion, which are principally affected by three union points and are, first, the connecting point of eccentric rod and link; second, the locus of the lifting link suspension point; and third, the relative height of the crosshead connection point of the union bar to the corresponding point of the combination lever.

It is not necessary to lay out the valve diagrams except where a given cut-off per cent is wanted. This is the most convenient way to find the required lap.

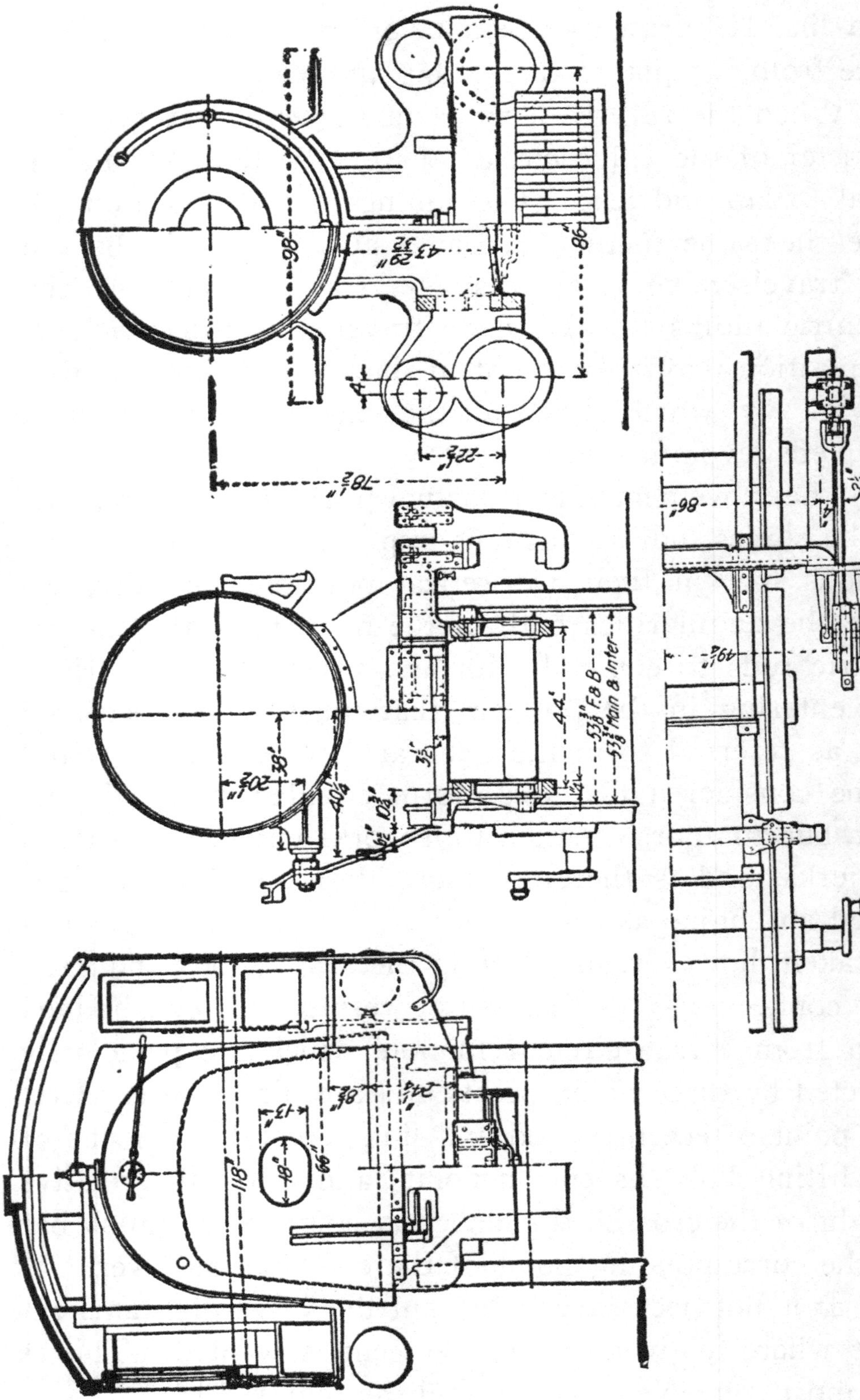

FIG. 301. WALSCHAERT VALVE GEAR—CONSOLIDATION FREIGHT LOCOMOTIVE—BOSTON & MAINE RAILROAD

WEIGHT OF WALSCHAERT GEAR.

In the matter of weight of parts, the details, in pairs, of the Walschaert gear of this engine (Lake Shore & Michigan Southern 2-8-0 type No. 912) are as follows: Crosshead arms, 60 pounds; vibrating rods, 220 pounds; eccentric rods, 220 pounds; links, 260 pounds; transmission bars, 140 pounds; valve rods, 70 pounds; eccentric cranks, 100 pounds; vibrating links, 70 pounds; valve stems, 72 pounds; and transmission bar hangers, 72 pounds. This means a total weight of 1,252 pounds for the entire valve gear of the Lake Shore engine, not including the valves. The weight of the corresponding valve gear parts of a recently constructed 20 x 28-inch, 4-6-0 engine, with Stephenson link motion, is 2,734 pounds. Such a weight, which must be moved and reversed for every revolution, imposes severe duty upon the eccentrics, and it is not surprising that they heat.

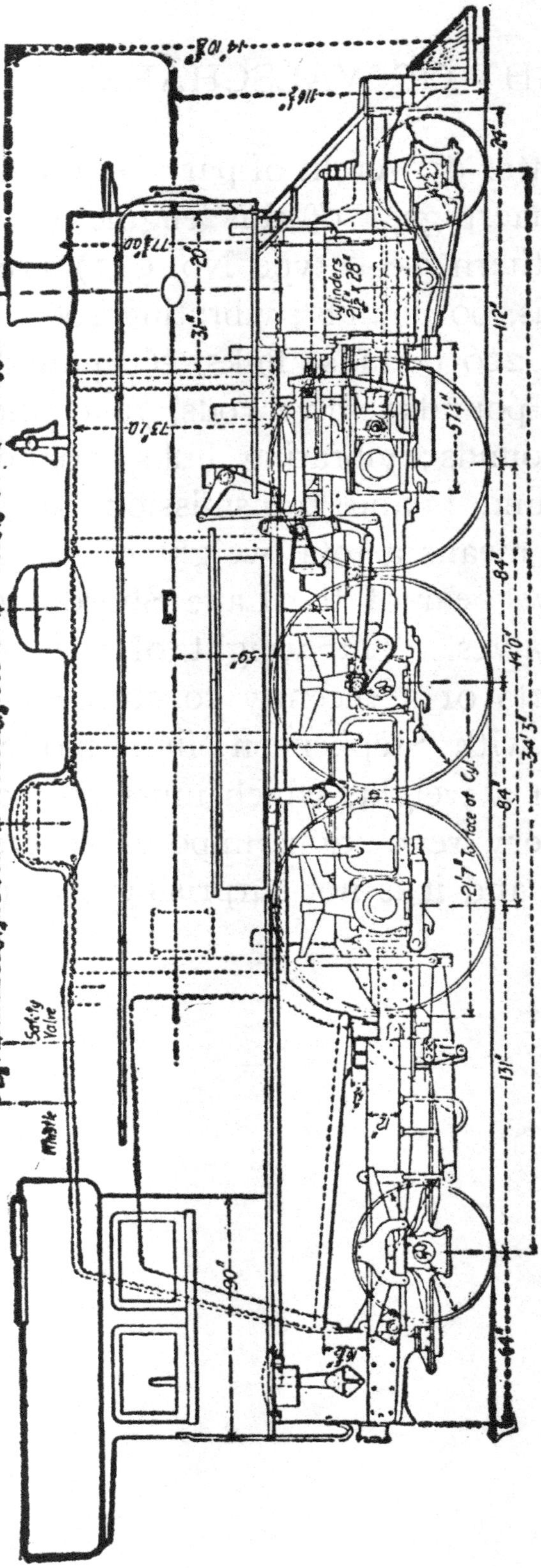

FIG. 302. WALSCHAERT VALVE GEAR—PRAIRIE TYPE PASSENGER LOCOMOTIVE—PENNSYLVANIA RAILROAD

SAVING IN WEIGHT WITH WALSCHAERT VALVE GEAR.

Comparative Weights, Stephenson and Walschaert Valve Gear.

Lake Shore & Michigan Southern Locomotives.

	Class and Type					
	Class D, 2-8-0		Class J, 2-6-2		Class J, 2-6-2	
	Stephenson	Walschaert	Stephenson	Walschaert	Stephenson	Walschaert
Crank pins, main	520	490	390	365	440	415
Crank pin arms	......	100	......	80	......	90
Crosshead arms	......	60	......	50	......	50
Eccentric	600	......	740	......	740	......
Eccentric strap	800	......	880	......	1,120	......
Eccentric rods	200	220	200	175	280	200
Link	280	260	260	240	300	275
Link support	......	280	......	250	......	260
Link lifter	45	......	120	......	120	......
Reverse shaft and arms	260	400	350	375	385	390
Rockers	260	......	280	325	280	350
Rocker boxes	240	......	300	300	300	325
Transmission bar	300	140	270	160	300	170
Transmission bar hanger	80	72	120	65	120	75
Transmission bar bracket	......	......	......	......	200	......
Valve rod	80	70	130	100	100	100
Vibrating rod	......	220	......	180	......	180
Vibrating link	......	70	......	60	......	60
Total, lbs.	3,665	2,382	3,940	2,725	4,685	2,940
Saving in weight, lbs.		1,283		1,215		1,745

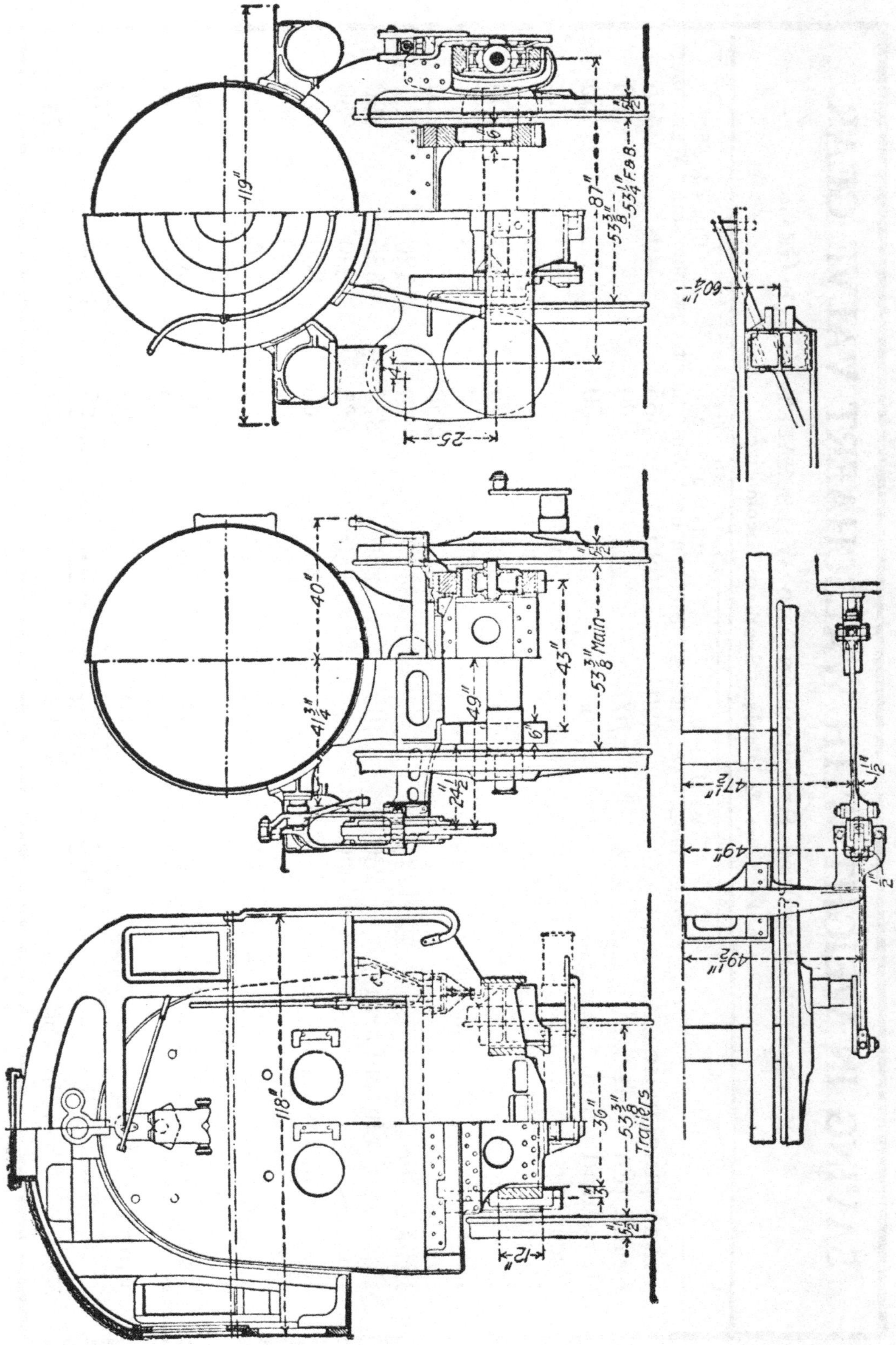

FIG. 303. PRAIRIE TYPE PASSENGER LOCOMOTIVE—PENNSYLVANIA RAILROAD—WALSCHAERT VALVE GEAR

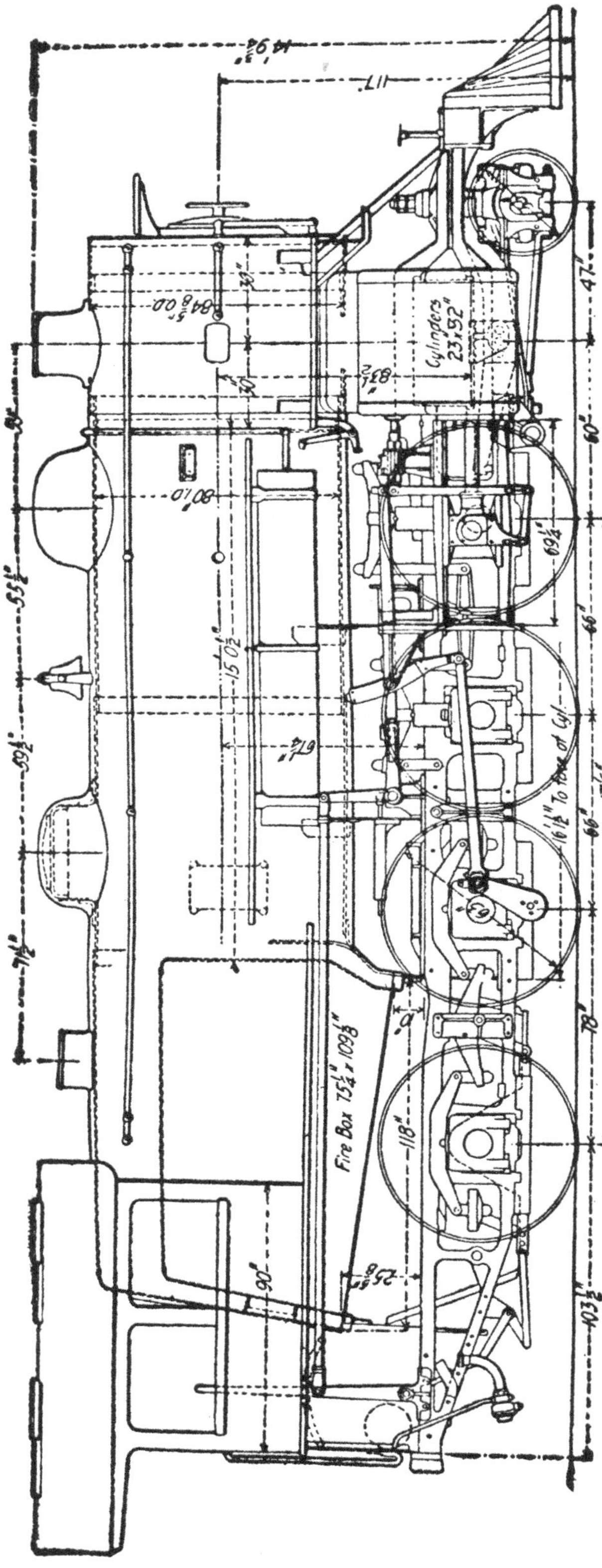

FIG. 304. CONSOLIDATION FREIGHT LOCOMOTIVE—NEW YORK CENTRAL & HUDSON RIVER RAILROAD—WALSCHAERT VALVE GEAR

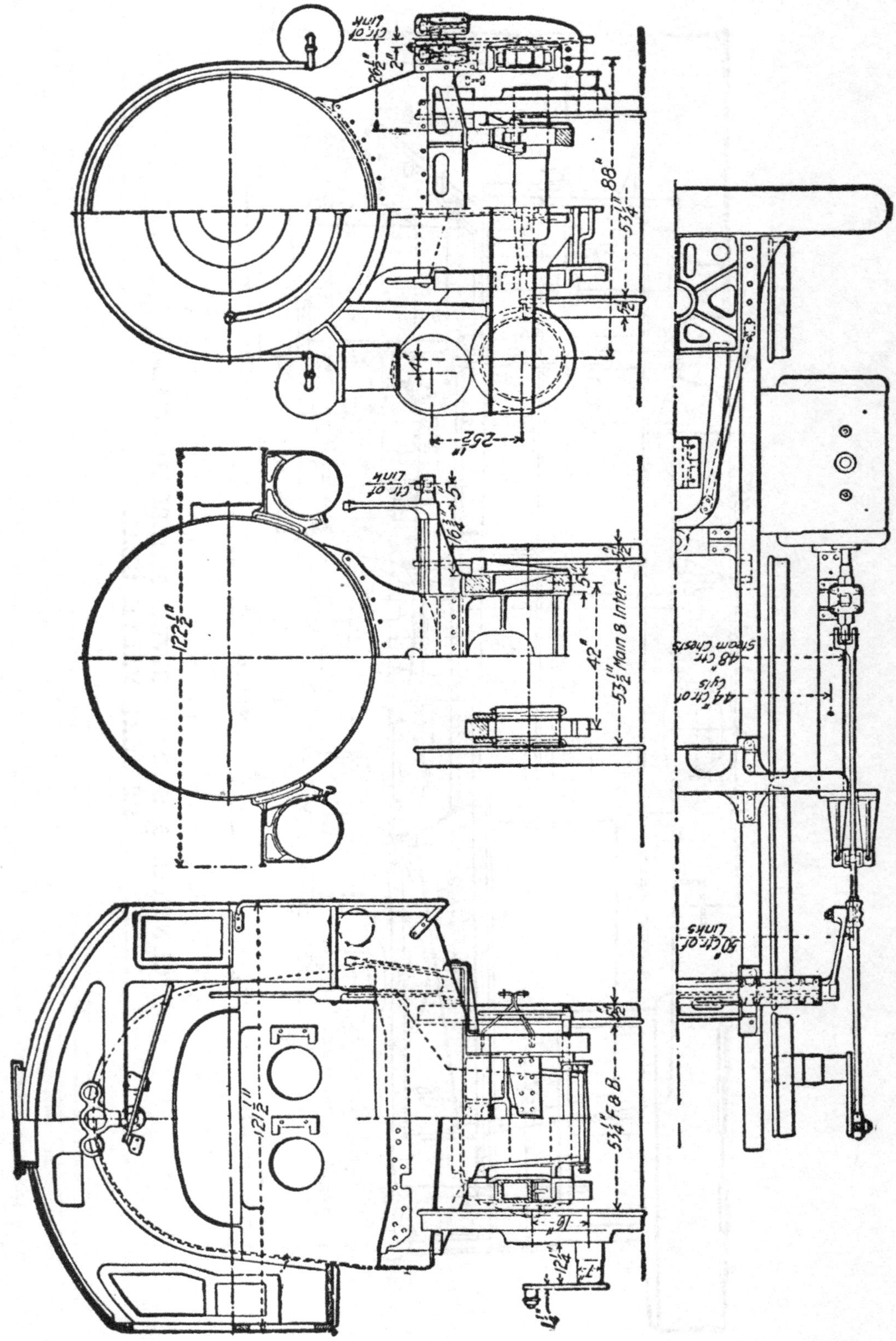

FIG. 305. WALSCHAERT VALVE GEAR—CONSOLIDATION FREIGHT LOCOMOTIVE—NEW YORK CENTRAL & HUDSON RIVER RAILROAD

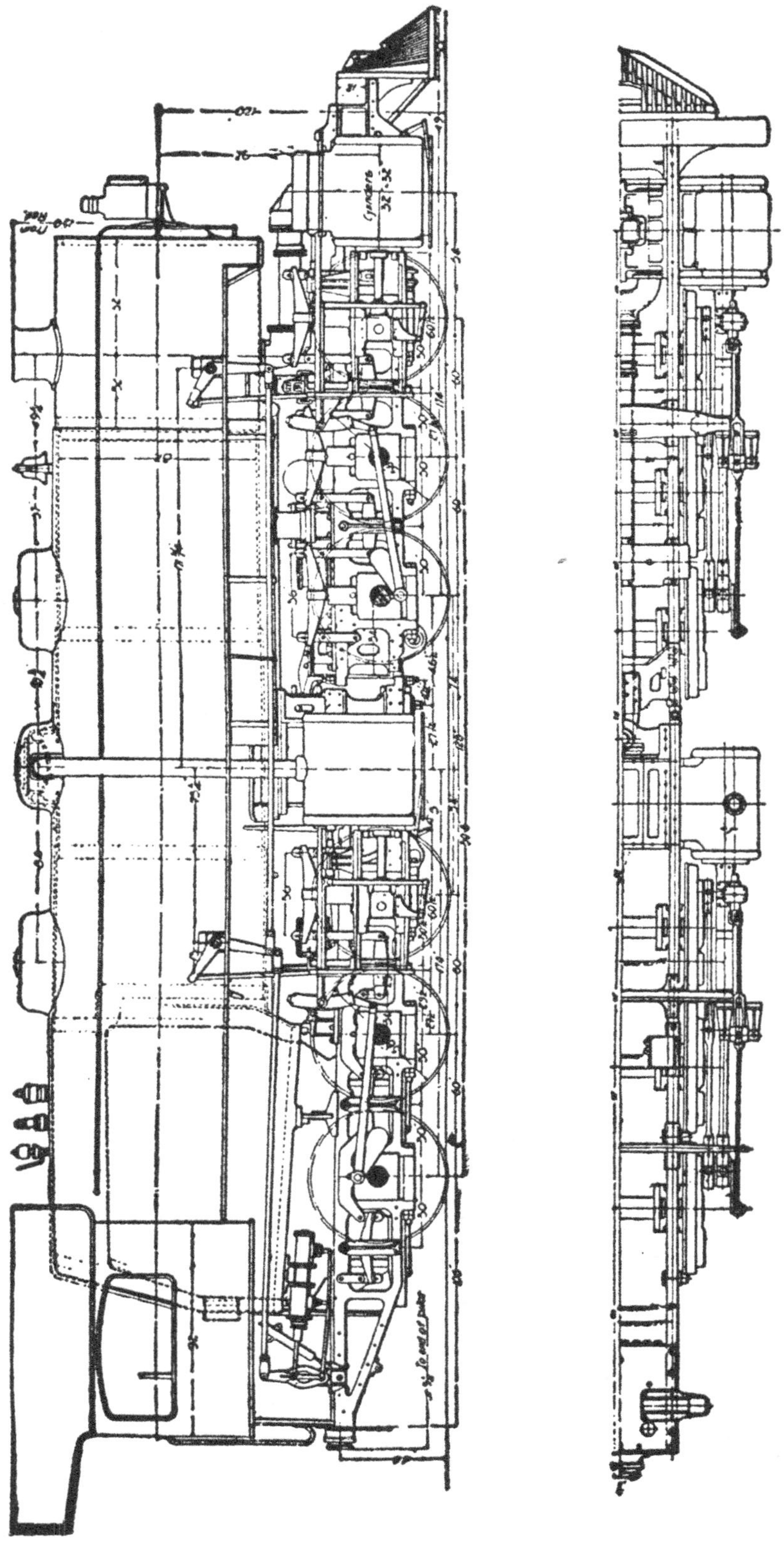

FIG. 306. WALSCHAERT VALVE GEAR—MALLET ARTICULATED COMPOUND FREIGHT LOCOMOTIVE—BALTIMORE & OHIO RAILROAD

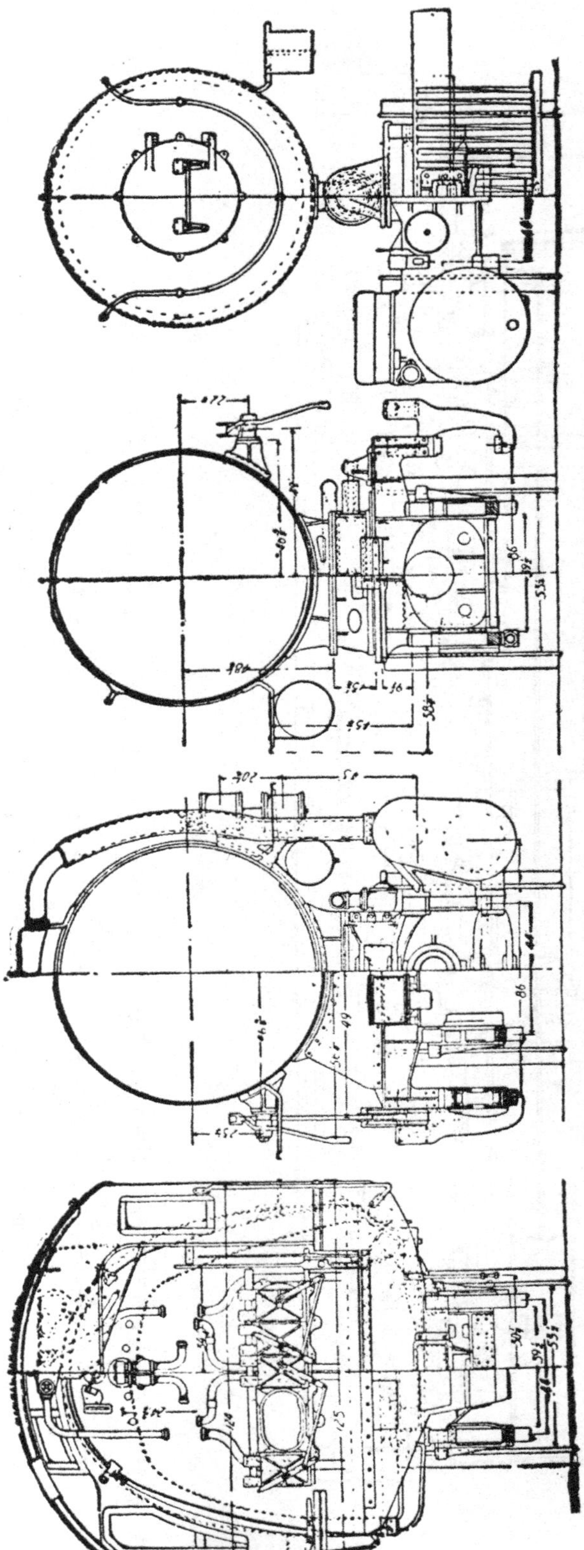

FIG. 307. MALLET ARTICULATED COMPOUND LOCOMOTIVE—BALTIMORE & OHIO RAILROAD—WALSCHAERT VALVE GEAR

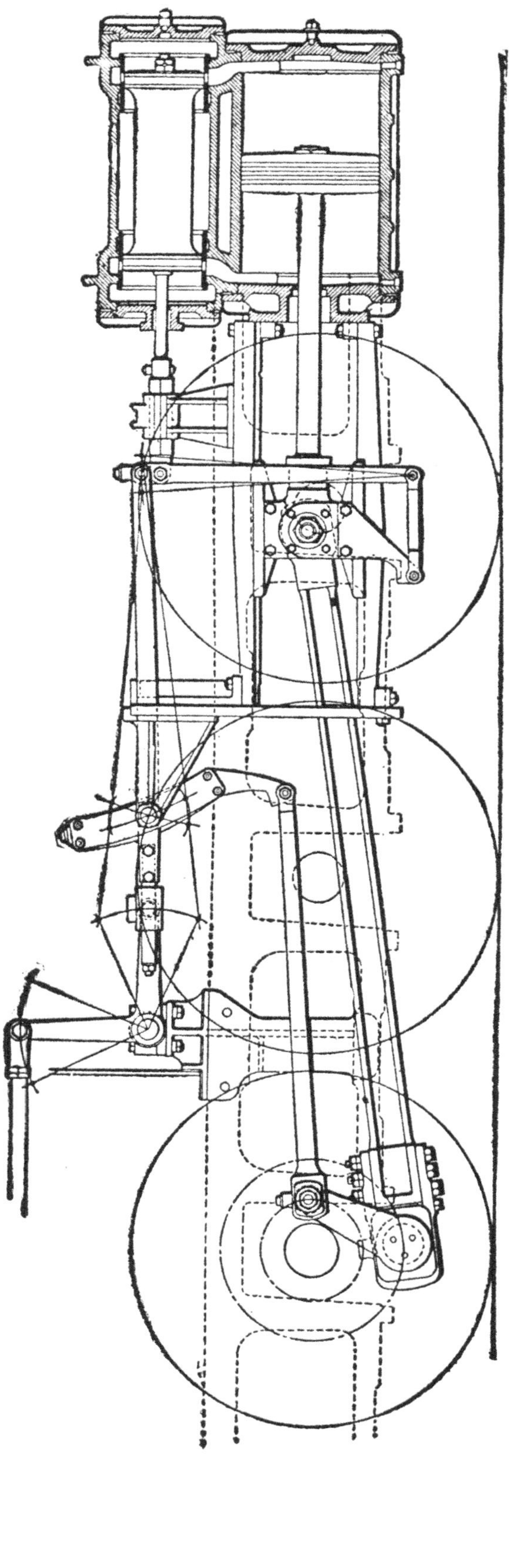

FIG. 308. WALSCHAERT VALVE GEAR AS APPLIED TO CONSOLIDATION FREIGHT LOCOMOTIVES—LAKE SHORE & MICHIGAN SOUTHERN RAILWAY

QUESTIONS.

1. From what two points is motion transmitted to the valve by the Walschaert valve gear?

2. From what point are the two important factors of lap and lead imparted?

3. What advantage is derived from the cross head connection as regards lead?

4. How may the movements of the valve as operated by this gear be diagrammatically illustrated?

5. Mention an important feature of the Walschaert valve gear as regards adjustment.

6. How is the combination lever of the Walschaert gear proportioned relative to the piston stroke?

7. What kind of link is used with this valve gear, and how should it be located relative to the length of the radius bar?

8. At what point is the radius bar attached to the combination lever for outside admission valves?

9. At what point is this connection made for inside admission valves?

10. Where should the fulcrum of the link be located?

11. Where should the suspension point of the link be located?

12. What should be the minimum length of the eccentric rod, relative to the throw of the eccentric?

13. What rule should be observed regarding the point of connection of the eccentric rod to the link?

14. In what position should the link be when the piston is at either end of the stroke?

15. What is the chief point of difference between the Walschaert and the Stephenson valve motions?

16. How are the factors of lap and lead determined?

17. In the adjustment of the Walschaert valve gear, what position should the crank be first placed in?

18. Before changing the length of the eccentric rod, what precaution should be observed regarding the valve rod?

19. What is the second rule to be observed in the adjustment of this motion?

20. Mention the third maxim for guidance in this adjustment.

21. Give the fourth rule relative to the adjustment of this gear.

22. According to the fifth rule, how may the lead be increased?

23. What effect does increasing the lead have upon the cut-off?

24. What is the sixth proposition relative to the adjustment of the Walschaert valve gear?

25. Can the location of the link and eccentric rod be determined by formula?

26. How then can these important points be found?

27. How is the location for the suspension point of the lifter determined?

28. Upon what two factors does the success of the Walschaert valve gear depend?

29. Are the valve movements in actual practice as

regular as the circles in the Reauleaux and Zeuner diagrams indicate?

30. Give some reasons for this discrepancy.

31. How may many of the irregularities due to the angularities of the main rod and eccentric rod be largely modified?

32. How does the Walschaert valve gear compare with the Stephenson link motion as regards weight?

SKETCH OF EGIDE WALSCHAERT,

THE INVENTOR.

Egide Walschaert was born in 1820 at Mechlin, then a little, retired village in the vicinity of Brussels. The railway line from Brussels to Malines was opened in 1835, and that decided the career of young Walschaert who entered the railroad shops of the state railways at Malines in 1842. He became chief superintendent of the shops of the Brussels Southern road, and at the early age of twenty-four had already acquired to an eminent degree all the qualities which go to make the successful engineer and which ought to have secured to him in a few years the position of technical director of the locomotive service of the system. It is humiliating to state that he remained chief shop superintendent all the remaining active years of his life.

On October 5, 1844, M. Fischer, engineer of the State Railways at Brussels, made an application in the name of Egide Walschaert for a patent of an invention relating to a new valve gear for locomotives. This Belgian patent was accorded by royal decree on November 30, 1844, for a term of fifteen years. The rules of the railway company did not allow the foreman of shops to advertise a patent in Belgium to his profit, which perhaps explains the mediation of M. Fischer, who never claimed to have any part in the invention. The mechanism described in the patent of 1844 presents a strong semblance to that which we are now familiar with, and

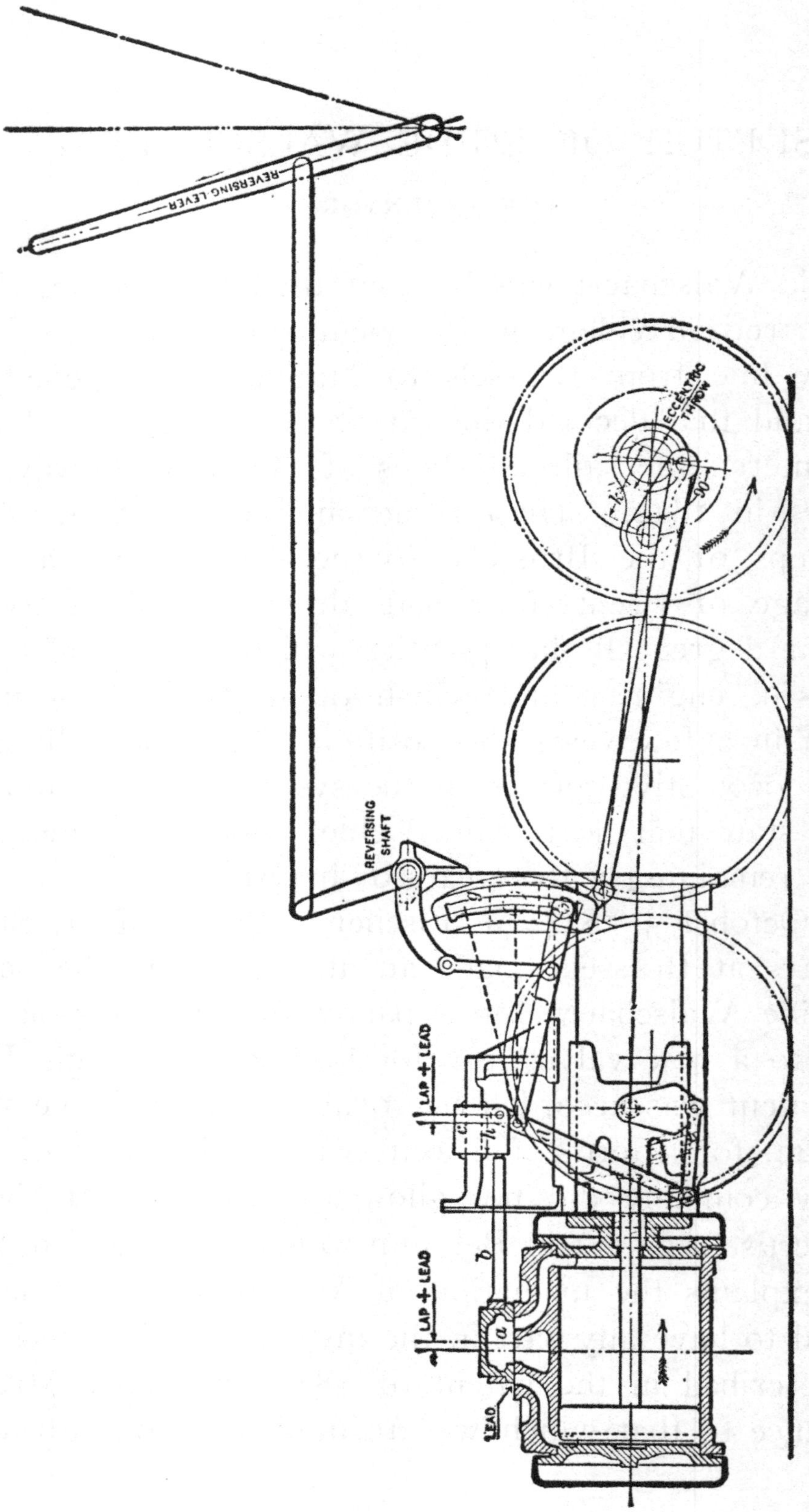

FIG. 309. WALSHAERT'S VALVE GEAR

the inventor constructed in 1848 a similar valve motion for application to locomotive No. 98. At this time the valve gear in use was that of Sharp, with two eccentrics and the usual forked rods. The shifting link attributed to Stephenson had been invented by Howe in 1843, and it is doubtful if Walschaert had ever seen it. The problem which today seems very easy was at that time an intricate one and to Walschaert, who then gave the correct and most elegant solution, is certainly due unreserved admiration. He also invented a valve gear for stationary engines, somewhat on the Corliss or Sulzer principle, and he built at Brussels a shop for the manufacture of such engines, and this was managed by his son.

At the Paris Exposition of 1878 a gold medal was awarded him for his engine, and in 1883 the exposition at Antwerp awarded him a diploma of honor, in which his locomotive valve gear was given merited praise.

Walschaert died Feb. 18, 1901, at Saint Lilles, near Brussels, at the age of 81 years. His reputation, however great, was accepted with singular modesty and his business relations were met with absolute disinterestedness. He gave his remarkable inventions to the world at a time when the study of steam distribution and valve gears was in its infancy, and he was deprived of the resources of a science which was not yet developed. On account of his great merits, which are here imperfectly recited, it is unfortunate that proper justice has not always been accorded Walschaert, for the ingenious mechanism which was original with him has been purloined through long years in the greater part of Europe.

ORIGIN AND ADVANTAGES OF THE WALSCHAERT VALVE GEAR.

Egide Walschaert patented the valve gear which bears his name Nov. 30, 1844, producing it because he saw the need of an improved valve gear. He wrought better than he knew, for this gear now seems indispensable to meet conditions recently arisen because of the development of the locomotive in this country, of which the inventor could not possibly have dreamed. His work now meets a need which did not exist during the lifetime of the inventor, and for this reason Egide Walschaert is entitled to credit, for solving a problem which was to confront a future generation.

The Walschaert valve gear has been in use since 1844. About 90 per cent of the main line locomotives in Continental Europe are equipped with it, and out of thirty-one locomotives exhibited at the Liege Exposition twenty-five had this valve gear. It has also been used in this country in a number of special cases. Valve gear has not been improved proportionately to the improvement in other factors in American locomotive development during the past twenty-five years. American locomotive designers are now turning to the Walschaert valve gear, because to meet recently arisen conditions it offers mechanical and structural advantages over the Stephenson motion. The reasons for applying this valve motion to large American locomotives are:

Accessibility. There is not room enough for the Stephenson gear under a very large passenger or freight locomotive. The eccentrics are crowded, and proper

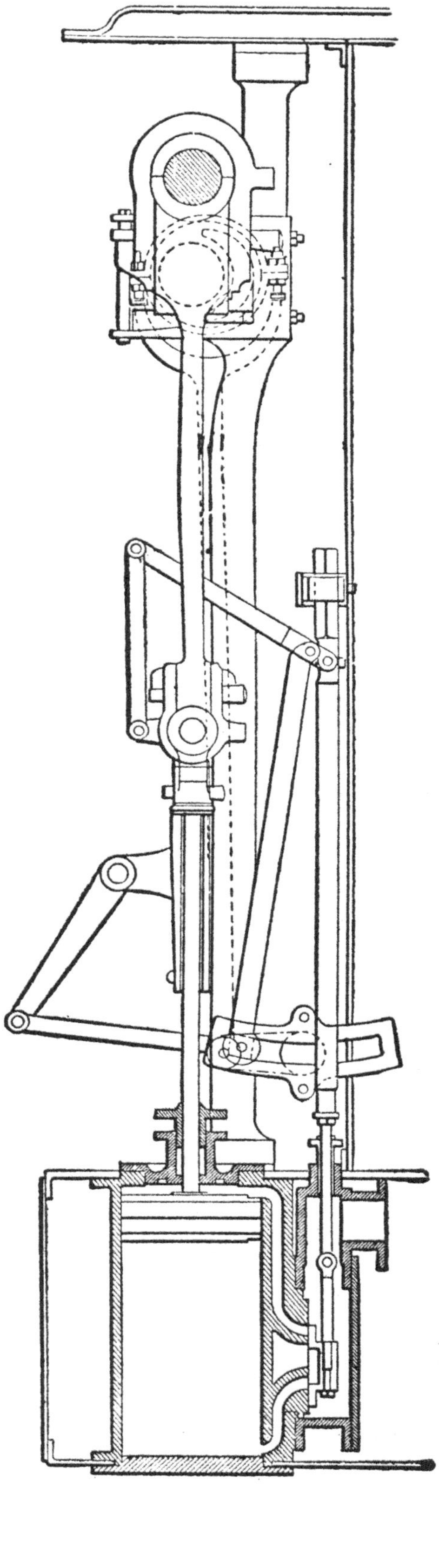

FIG. 310. THE VON WALDEGG GEAR OF 1849

inspection, not to speak of proper care, is difficult, except over a pit. Valve gear to be properly maintained must be accessible for inspection and lubrication. The accessibility of Walschaert gear should reduce engine failures.

Weight. A saving of 1,745 pounds is possible by using the Walschaert gear, in the case of very heavy passenger locomotive. Stephenson gear, weighing as much as two tons, is far too heavy to be satisfactorily reversed twice in every revolution on fast running locomotives.

Directness. The Walschaert gear transmits the moving force to the valve in very nearly straight lines, avoiding the springing and yielding of the rocker arms, rocker shafts and transmission bars, which cannot be avoided in these parts of the Stephenson motion, even if they are made very heavy.

Permanence of Adjustment. The advantage of permanence of adjustment lies with the valve gear which has no large eccentrics. All connections in the Walschaert gear are made with pins and bushings, which are designed specially to resist wear.

Wear. Large eccentrics, besides occupying too much space, wear unevenly, and lubrication is difficult with the high surface velocities of the largest sizes. With hardened pins and hardened bushings the Walschaert gear has an important advantage in maintenance.

Smooth Operation. The Stephenson link, under the influence of two eccentrics, moves through wide angles, resulting in a wedging action of the link block, which strains the gear when working hard, and produces lost motion. The Walschaert link oscillates through smaller angles, producing less lost motion. The effect of this angularity of the link is plainly discernible on the testing plant.

Frame Bracing. The removal of the valve gear between the driving wheels facilitates bracing the frames of the locomotive laterally.

OPINION OF MR. WILLIAM FORSYTH.

DELIVERED BEFORE THE INTERNATIONAL RAILWAY CONGRESS.

The use of the Walschaert valve motion on the Baltimore & Ohio Mallet articulated locomotive, and on the Pennsylvania Railroad's de Glehn four-cylinder compound, which are now on exhibition at Saint Louis, has again brought the merits of this gear to the attention of American designers. These two locomotives represent extreme conditions as to the speeds for which they are intended. The one is for slow speed on heavy grades, the other for high-speed passenger work. This would indicate that the Walschaert gear is well adapted to any kind of service, freight or passenger. It is fortunate that a well-designed gear of this type will soon be seen in operation in this country, and its performance on these locomotives will be watched with interest.

A valve gear outside of the frames is conveniently inspected and repaired, while one inside of the frames is certainly in an awkward position for either operation. With inside cylinders and crank axles there is little room for eccentrics and links, and if all this be removed it allows ample length for main pin bearings,

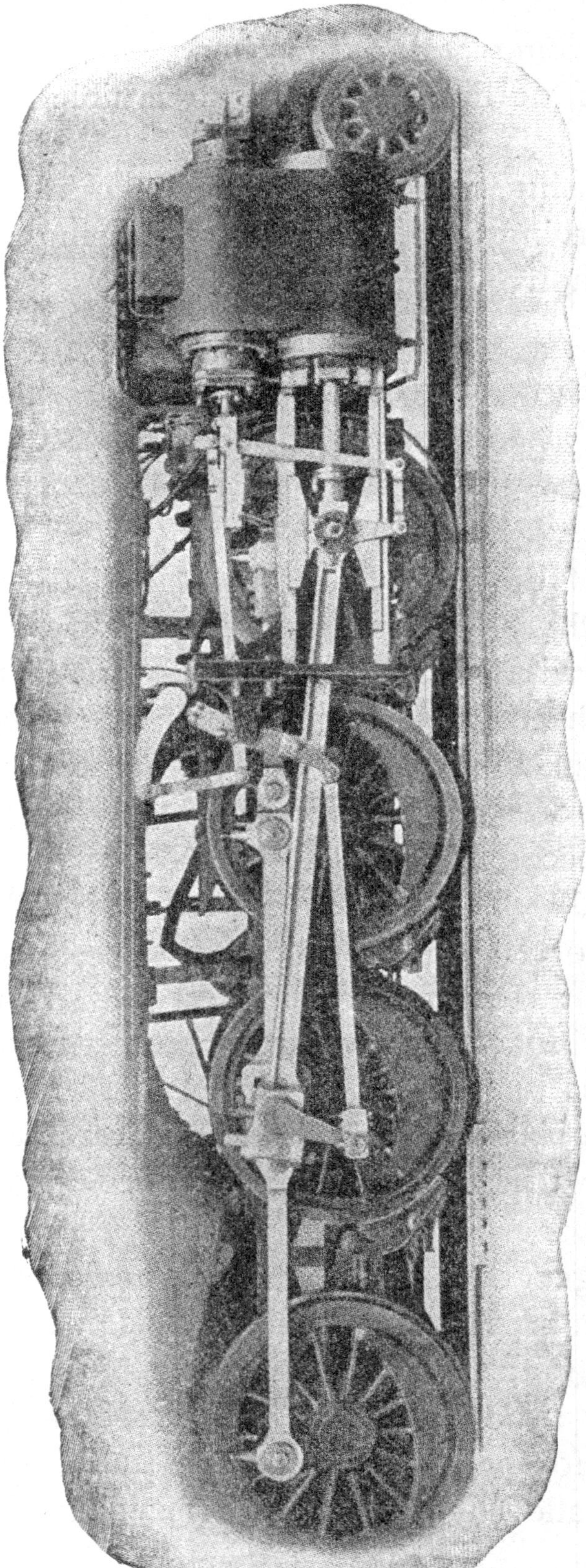

FIG. 311. WALSHAERT'S VALVE GEAR AS APPLIED TO A CONSOLIDATION LOCOMOTIVE FOR THE PENNSYLVANIA RAILROAD

and it is then possible to have an inside bearing for the crank axle. The Walschaert gear, as ordinarily designed, is not symmetrical in a vertical plane, and there is a tendency to lateral bending and unequal wear when so constructed. Mr. de Glehn has taken special care to avoid these objections, and his design shows well-balanced wearing surfaces of ample proportions, which should be quite durable.

The point to which we wish to call particular attention is the great contrast in the weight of the moving parts and the size of the bearings when this Walschaert outside gear is compared with similar parts of a Stephenson link motion driven by eccentrics. A prominent superintendent of motive power, who has made a special study of indicator cards and given much attention to valve gears, in a recent discussion on "Modern Tendencies in Locomotive Design in America," made this statement: "I consider that the increased complication and weight of the valve motion is an exceedingly serious matter in giving distorted steam distribution, due to the destructive effect of the valve motion in causing wear and tear."

The reports on "Weights and Detail Parts of Locomotives," in the proceedings of the Master Mechanics' Association, 1903, page 187, gives the weights of parts of the Stephenson valve gear for large locomotives, in pounds, as follows: Eccentric, 212; eccentric strap, 225; eccentric rod, 125; link, 148; rocker arm, 248; transmission bar, 128; valve rod, 66; valve yoke, 90; valve, 211. These figures indicate that the Stephenson valve gear, including the eccentrics and straps, as found on our modern locomotives, has become a very ponderous affair.

Some attention has been given to the valve pattern in the effort to make it as light as possible, but the same care has not been taken with the moving details connected with it, and which easily become a disturbing factor at high speeds if made too heavy.

The principal load which comes on the eccentrics and straps, causing them to heat, is not the friction of the valve, but it is that due to the inertia of the reciprocating parts of the valve gear whose motion is reversed twice for every revolution. If we include the rocker arm, the weight, as found above, of the moving parts, from valve to eccentric strap for one cylinder is 1,052 pounds, and at high speeds the energy of this moving mass must impose a heavy load on the eccentrics. The eccentrics and straps are the most difficult details in the locomotive machinery to keep properly lubricated, and it requires constant vigilance to prevent them from heating. When they do heat and cut and the straps are taken down, their location inside the frames is the most inconvenient one possible, and with the increasing weight of the machinery this part of the locomotive repairs has become very laborious and expensive. More attention should be given to the reduction of the weight of the moving parts of the Stephenson valve gear, or some other type should be used.

The Walschaert valve gear, located outside the frames, is easily accessible and very convenient for inspection, lubrication and repairs. The main driving bearings are two small pins with bushed bearings, and the contrast with the heavy and cumbersome eccentrics and straps which are their equivalent in a valve gear system is very striking. This gear is simple and light throughout, and it has much to recommend it which would overcome the objectionable features of the shifting link motion driven by eccentrics.

OPINION OF MR. J. C. MUHLFELD.

DELIVERED BEFORE THE INTERNATIONAL RAILWAY CONGRESS.

There are now several newly built locomotives in America which are equipped with the Walschaert valve motion. The chief difference between the Walschaert and the Stephenson valve motions is the constant lead with the former when the valve travel is changed and the combination of the two distinct motions, one derived from a crank arm and the other from the cross-head. The Walschaert valve gear as originally designed was not symmetrical or in a vertical plane; there was a tendency to lateral bending and unequal wear, and when compared with the eccentric-driven valve gear, it had the disadvantage of being affected by the vertical displacement of the axle. With improved designs these undesirable features have been appreciably overcome, and the simplicity, accessibility and operation of the Walschaert motion will insure its receiving due consideration for the steam distribution of modern tonnage and passenger locomotives.

When comparison is made between the weight of the moving parts and the size of the bearings of the Walschaert outside valve gear with similar parts of the Stephenson inside valve gear, it will be found that a simple design of the latter weighs more than double the amount of the former, and is becoming a very cumbersome and heavy suspended reciprocating and revolving arrangement. The increased complication and weight of the Stephenson valve gear has resulted in a distorted steam distribution, to say nothing of the destructive effect of the heavy parts and the increased cost for maintenance on account of excessive wear and breakage. The eccentrics and straps have not only to carry the increased friction of the larger sizes of valves, but the reversing twice for every revolution and the inertia of the reciprocating parts of the gear has contributed the principal load, and they have become one of the troublesome details of locomotive machinery, requiring constant inspection and maintenance attention to prevent liability of heating or failure. Furthermore, the inaccessibility of the entire arrangement, due to its location within the frames, makes proper attention almost impracticable in connection with the modern locomotive displacement conditions.

A valve motion placed outside of the frames certainly has the advantage of accessibility and convenience for inspection, lubrication, repairs and cleaning, and the Walschaert type provides for this and gives an opportunity for better diagonal and cross-bracing which is so necessary between the main frames and the frames and boilers, to maintain alignment and stability and reduce frame failures.

The Walschaert valve gear has long been popular in France, Germany, Belgium, and other foreign countries, and it is used extensively on compound and simple types of locomotives for high and low speeds. It can be made of comparatively light parts, and having no angular advance and no lead that can be given other than by the crank arm, a valve movement can be produced that is equivalent to that given by an eccentric having angular advance.

Unless improvement can be effected in the design and application of the Stephenson valve motion, it will be necessary to inaugurate the use of the Walschaert or some similar type of valve gear for the modern locomotive construction in this country to insure the best results from high pressure with a single valve having large port openings and rapid movement for admission and exhaust, delayed opening of exhaust port for longer expansion and initial expansion, and effective action on crank when the angularity of the main rod and piston pressure is most favorable to produce the greatest degree of efficiency as well as economy at high and low speeds.

WALSCHAERT VALVE GEAR.

The Walschaert valve gear differs chiefly from the Stephenson link motion in that it requires for each cylinder but one eccentric or its equivalent, to insure the movement of the valve, and the proper distribution of the steam for both forward and backward motion. The elimination of the heavy eccentrics and their connections relieves the axle of an appreciable portion of its dead weight. It also makes it possible to place the gear outside the driving wheels in a more accessible position, where the parts can readily be oiled, inspected and repaired.

While it may not be possible to adjust the valve as readily with the Walschaert valve gear as with the Stephenson motion, for the reason that the parts and connections are not as suspectible to change, it is not as liable to become disarranged, and if correctly designed and fitted up will give accurate results with less difficulty in application and greater economy in construction.

Equal cut-offs in both ends of the cylinder are more easily secured than with the Stephenson valve motion, and the play of the engine on its springs has practically no influence on the steam distribution.

In the Walschaert valve gear the operating eccentric is secured to the driving axle either directly or by a return crank from one of the crank pins.

The position of this eccentric or crank is such as to give the proper valve travel, the throw corresponding with the movement of the valve irrespective of its lap and lead; the angular advance of the eccentric being 0°.

The link is of any convenient form and is usually pivoted to a support on the engine frame or suspended from the guide bearer. The trunnion is rigid and there is no chance for twisting strains. The link is actuated by the eccentric rod which is commonly attached to its lower extremity. The sliding block in the link is secured to one end of the radius rod. The raising or lowering of this rod by means of the reversing shaft, shifts the block from one end of the link to the other above or below the pivotal connection; this reverses the movement of the valve with relation to that of the eccentric.

In many cases the gear can be so designed that the motion is transmitted from the eccentric to the valve stem in one vertical plane so that practically all of the pins can be put in double shear, and all tendency to twist the valve motion is avoided. In some cases, as for

FIG. 312. TEN-WHEEL LOCOMOTIVE—CHICAGO, ROCK ISLAND & PACIFIC RAILWAY CO.

example, the Chicago, Rock Island and Pacific locomotive, it is necessary to have the eccentric and the valve stem in different vertical planes. In such a case it is not practicable to have all the pins in double shear, but the valve motion can be made of a high degree of simplicity and solidity. It is sometimes urged against the Walschaert motion that it requires more moving parts than the Stephenson. This is, however, offset by the better opportunities for solid construction and ready inspection and adjustment.

The end of the radius rod opposite the link is attached to a combining lever, the function of which is to give the required lap and lead to the valve. The lower end of this lever is connected to and travels with the crosshead, while to the upper end is secured both the valve rod and the radius rod, one being placed above the other. The point at which the radius rod is attached to the combining lever becomes a fulcrum. The relative movement of the two ends of the lever must be such that the full movement of the crosshead imparted to the lower end of the lever will give a movement of the upper end equivalent to the required lap plus the lead.

Under ordinary conditions with steam chest valves having outside admission, the connection or fulcrum between the radius rod and the combining lever is placed below the valve rod connection. With valves having inside admission, this fulcrum is usually placed above the valve rod connection.

The link should have a radius equal to the length of the radius rod. If this is so it will be seen that when the engine is on the dead center the link block can be moved from end to end of the link without altering the position of the valve with relation to the ports and the lead will be constant.

As any variation in the length or relative position of the link, the radius rod, and the combining lever or its connections, will necessarily change the resulting movement of the valve, it is absolutely essential first, that the motion shall be correctly designed and plotted, and second, that the detail parts shall be accurately constructed according to the diagram. With these two points assured the adjustment of the gear on the locomotive is quite simple. The dead center marks on the rim of the driving wheel and the port tram marks on the valve stem are found in the usual manner. After connecting the gear any slight variation which may occur between the forward and backward position of the valve can be adjusted by lengthening or shortening the eccentric rod.

American practice has, until recently, adhered, with few exceptions, to the Stephenson link motion, actuated by eccentrics. Experiments were made with the Joy valve gear, following the English practice, but this design has not been perpetuated in America. In the construction of locomotives for export, there has been a more or less continuous demand for the Walschaert gear. Its first use in the practice of the Baldwin Locomotive Works was in the year 1878, upon locomotives for the Mexico, Toluca and Cuantitlan Railroad. This gear valve has been growing in favor abroad and owing to its accessibility and the other advantages set forth in these pages, has lately attracted wide attention from American engineers.

For large locomotives where the driving axles are of

such diameter as to greatly increase the diameter of the eccentrics, the Walschaert valve gear is particularly useful. It is also found of essential advantage on locomotives having relatively small wheels and engaged in freight service, necessitating a high rotative speed.

The size and arrangement of parts in a modern locomotive make it difficult for an engineer to properly examine the eccentrics and link motion when the engine is on the road, and breakdowns are more frequent on this acount. The conditions of service also tend to make it more and more difficult for the engine men to give the close inspection and care which is demanded in other branches of engineering service with high speed machines. With the Walschaert valve motion only a single eccentric or its equivalent is necessary for each valve. As usually constructed it is found more convenient to substitute a return crank, thus reducing the pin bearings to the smallest possible diameter, so that they may be readily lubricated, and, owing to the small amount of work they have to do, give satisfactory service and absolute freedom from heating.

So far the distribution of steam in the cylinders is concerned, the constant lead, which is a feature of this motion is not considered objectionable, and it has some distinct advantages. Under such conditions it is possible to determine upon the amount of lead the engine should have at the most economical point of cut-off. This point determined, and so designed, it cannot be altered by anyone in the shops or roundhouses. Another advantage is that it prevents valve setters from attempting to produce results by moving the eccentrics into improper relations one to another.

WALSCHAERT VALVE GEAR

The constant lead of the Walschaert motion prevents the sealing of the cylinders by the piston valve when the piston is at the end of its travel or approaching it. Whereas with the link motion, either by derangement or excessive wear, the valve laps the ports at the end of the stroke, thus causing excessive compression and many other troubles.

Another feature of the motion which appeals to the engineer, is the ease of handling the reverse lever when the locomotive is running at a high rate of speed.

GENERAL INSTRUCTIONS.

In setting the Walschaert valve gear it must be borne in mind that two distinct motions are in combination: the motion due to the crosshead travel, and the motion due to the eccentric throw.

The crosshead motion controls the lead, by moving the valve sufficiently to overcome its lap, by the amount of lead in both front and back positions. The eccentric throw controls the travel and reversing operations. It will be seen that the movement due to the eccentric, without the crosshead motion, would place the valve centrally over the ports when the piston is at the extreme end of the stroke. The combined effect of these two motions, when the parts are properly designed, gives the required movement of the valve, similar to that obtained by the use of a stationary link. To reverse the engine, the link block is moved from end to end of the link, instead of moving the link on the block. This operation is accomplished by means of a reversing shaft connected with a reversing lever in the cab.

Walschaert gears should be correctly laid out and constructed from a diagram, as the proportions cannot be tampered with by experimental changes without seriously affecting the correct working of the device.

The only part capable of variation in length is the eccentric rod, which connects the eccentric with the link. This rod may be slightly lengthened or shortened, to correct errors in location of the link center, from center of driving axle which carries the eccentric.

The eccentric usually assumes the form of a return crank on one of the crank pins, and its center is at right angles to the plane of motion, at ninety degrees to a line drawn from the point on the link at which the eccentric rod is attached, through the center of the driving axle. This eliminates the angular advance of the eccentric, and allows the use of a single eccentric for both forward and backward motion. The throw as specified must be correctly obtained, and great care taken that the position shown in the design be adhered to. The crank representing the eccentric is permanently fixed to the pin, and the slightest variation will be detrimental.

When the engine is assembled, the throw of the eccentric should be checked up by the specifications, and any error should be at once reported in order that the mistake may be rectified by either correcting the position of the eccentric, or by a change in the design of the other parts to compensate for the error.

In case of accident, if any of the rods or connections are broken, it is advisable if possible to disconnect the eccentric rod. The combining lever should be uncoupled from the crosshead and securely fastened in forward position. If for any reason the eccentric rod cannot be taken down, the radius rod must be removed in order that no motion may be imparted to the valve. The valve can then be placed in central position and held either by suitable blocking or by clamping the valve rod. This seals both steam ports and cuts out the cylinder on the damaged side.

SPECIAL INSTRUCTIONS FOR ERECTING AND SETTING VALVES.

1. Check carefully the dimensions of the following parts, rejecting any that are not exactly to drawing:

a—Valve
b—Valve stem
c—Valve crosshead or slide
d—Combining lever
e—Crosshead link
f—Link radius rod
g—Reverse link
h—Location of combining lever on crosshead
k—Length of eccentric crank

2. Check eccentric throw to see that it is exactly as specified.

3. Be sure that guide bearer is correctly located from cylinder, as the reverse link is usually attached to it, and variations in the location of the link cannot be allowed. If the link is attached to separate crosstie, similar precautions must be taken to insure its correct location.

4. Exercise great care in the location of the link so that the trunnion center is exactly to dimensions from the center of cylinder.

5. See that the reverse shaft center is correctly located to dimensions given, and that the lifting arm and link are of the exact lengths as specified.

6. Connect crosshead gear to valve, and radius rod to link, without connecting eccentric rod to link.

7. Hook up radius rod to exact center of link, and then revolve driving wheels, seeing that crosshead gear gives correct lead as specified for both front and back admission ports.

8. Connect link to return crank by eccentric rod, and obtain full travel front and back, and in both forward and backward motions, correcting any errors by lengthening or shortening eccentric rod as noted under general instructions.

The valves may now be considered as definitely set, and may be tested to any cut-off points in the usual manner.

A simple additional check should be made as follows:

Set one side of the engine so that the piston is at its extreme forward position in cylinder and check lead on admission port.

In this position it should be possible to move the link block through its entire travel in the link, without in any way disturbing the movement of the valve.

This operation should then be reversed, and the other side of the engine similarly tried with the piston located at its extreme backward position in the cylinder.

CHAPTER XII

SIGHT FEED LOCOMOTIVE OIL PUMP.

The pump shown in Fig. 313 is an eight feed outfit arranged to run first feed on the right to air end of air pump. Oil for this feed is contained in a cavity drilled in the pump body, access to which is had by removing dummy valve stem seen standing out by itself on the right. Oil for all other feeds is contained in the large tank back of main pump body, which is filled through a large hole under a removable threaded cap seen on top to the left. The tank cover may be turned end for end if desired for convenience in filling.

This motor is known as right hand and is here shown set vertically, but may be shifted at any angle to suit surroundings, if found necessary.

The gauge glass and heater stud are here shown on the left, but may be interchanged with the draining plug on the other end of the tank if desired.

The above style of check, shown in Fig. 314, is furnished for each feed running to main cylinders or valves.

The short length of perforated ¼ inch pipe shown extending down is used on line to steam chest on slide valve engines and should extend down close to balance plate and be capped on the end as shown.

On piston valve engines omit this quill and screw direct into casting at desired points, being careful to locate oil hole so it will never be uncovered to the exhaust.

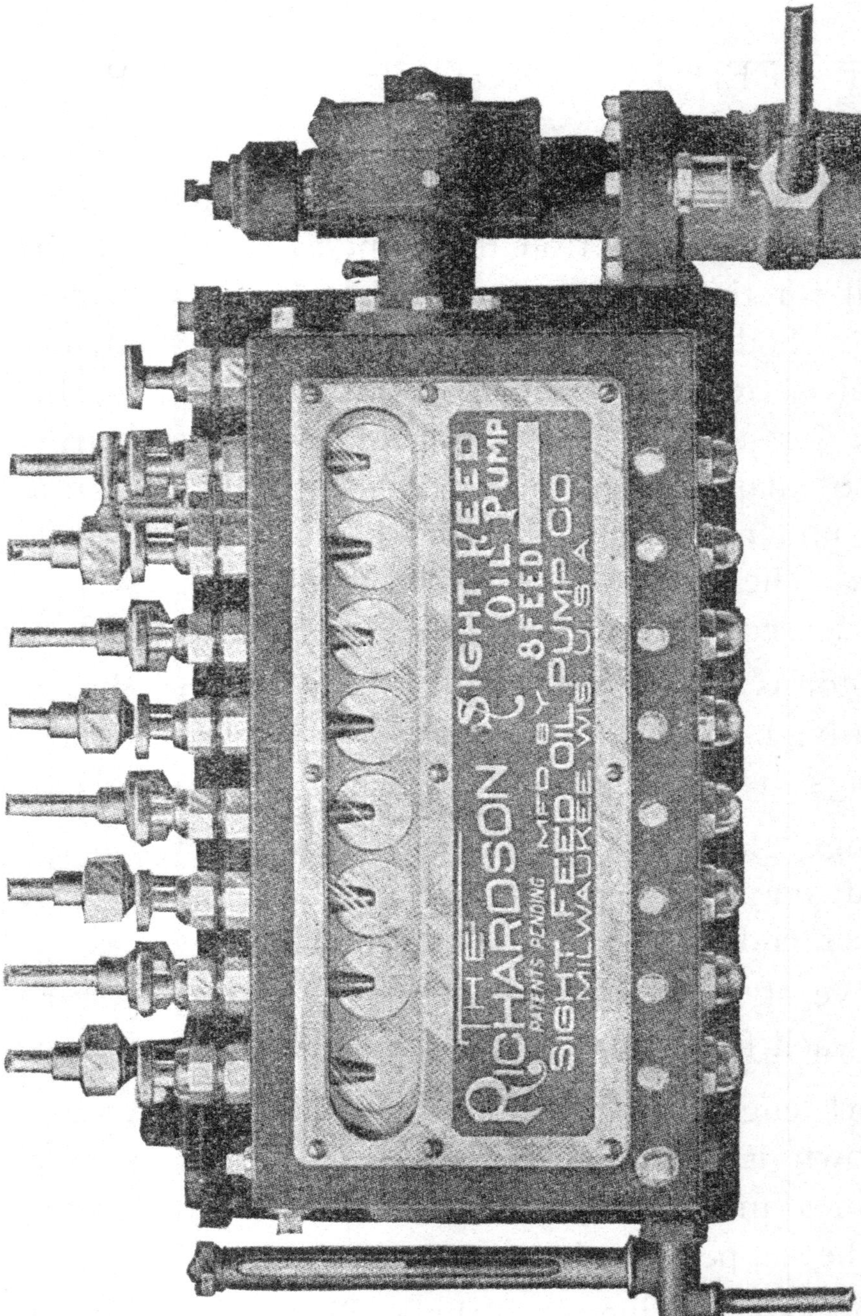

FIG. 313. SIGHT FEED LOCOMOTIVE OIL PUMP

FIG. 314. NO. 7 CHECK VALVE

INSTALLATION.

Place cap screws in flange on housing, inserting cap screw with extra long head in back hole next to tank, and tighten all firmly. Locate pump in cab as low as possible, snug up to the boiler, the sight feeds in plain view of engineer.

Bracket for supporting pump to be made of ¾ or 1x5 inch iron, and rigidly fastened to the boiler. Connection to bottom of oil tank is made by inserting three ¾ inch 10 thread studs, of suitable length to project through this iron bracket and allow for nuts on the under side. Bottom of tank is tapped out to receive these studs.

Pipe the air supply to motor from engineer's brake valve, placing the valve in this line as near as possible to the brake valve for convenience to the engineer in opening and closing.

Run steam connection also to the motor, placing a T in the line close to motor. Between this T and motor place a valve. Run side outlet of T through a valve into heater connection on tank, noting carefully direction of arrows cast on this fitting. This arrangement keeps water drained out of steam line to motor at all times.

Motor should, by all means, be piped for both air and steam, using steam only whenever air supply fails, or motor becomes clogged with old oil, in which case a little steam will warm it up and work out the gummy accumulation.

Drain from heater and exhaust from motor may either be piped into drip cup under gauge cocks or down through the cab floor and no valves should be placed in these lines.

All piping should be made up of ⅛ inch pipe size copper tubing with liberal allowance in all cases for expansion and contraction. In no case use larger size tubing or piping than specified.

All oil lines should be made up of same tubing as above mentioned, with joints at unions on pump and the check valves brazed tight.

This piping may be placed same as ordinary lubricator piping.

If the pump is installed on an engine requiring additional feed lines to those already in place, same may be enclosed in hand rail or run along running board.

Replace old oil cup in air end of air cylinder with a check valve (see Fig. 315) provided for this purpose.

Check valve for oil line to steam end of air pump may be inserted in air pump steam piping in cab or elsewhere in this line if more convenient.

Check valves at main engine cylinders must be screwed into cylinder casting, and not into a piece of pipe extending out of lagging or false cover.

Where old tallow pipe connection is in place, a set of parts to form a check valve, with directions to apply, will be sent, same to be placed in this old fitting.

In piping up motor, if found more desirable, cylinder on same may be shifted as desired, by removing cap screws and turning around, care being taken not to remove cylinder from piston rod.

After the pump is all piped up, stencil plainly on front of body in blank space above name plate, lettering to show point to which each feed is piped.

OPERATION.

Fill tank nearly full with clean oil, and screw down filling cap lightly. There is no pressure at any time in oil tank.

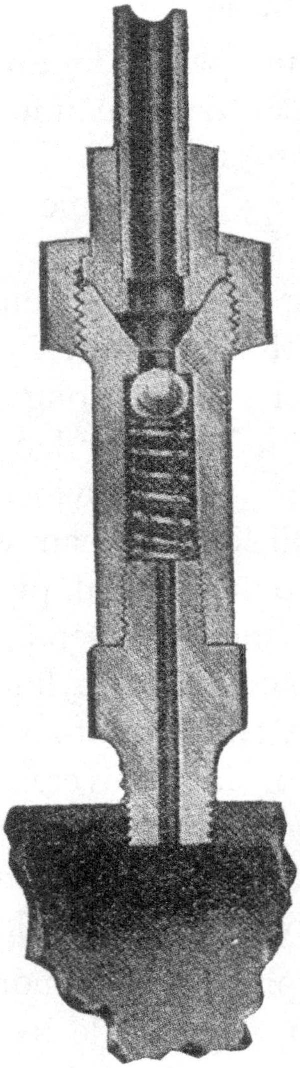

FIG. 315. NO. 6 CHECK VALVE FOR AIR PUMP LINES

Open the valve in steam line to heater, a very little, and leave open. Turn air into motor, adjust valve to run the motor at about 75 strokes per minute. This valve being placed close to engineer, should be closed when stopping long at stations, as the first stroke of motor always delivers oil at all points.

Oiler on cylinder should be filled when leaving round house, and requires no further attention. This oiler must be kept partially filled with curled hair or waste, and strainer left in bottom. Cup may be blown out and cleaned by turning on the air to motor when plug is removed from oiler.

Adjust all feeds with small increase feed valve, handle pointed away from oil tank. Set feed to air end of air pump to drip very slowly, as too much oil will gum up the valves in various parts of air brake apparatus.

To secure increased flow in one discharge line, adjust valve for this feed. To secure increased flow in all lines, as when starting, turn increase feed valve one quarter turn or less as circumstances may require. All feeds will resume their normal flow when this valve is returned to open position.

In case circulating plunger refuses to work, oil may be poured into cavity below drip nozzles from time to time, allowing motor to keep on running.

Powdered graphite mixed with oil may be fed to cylinder and valves in same way, through this opening.

Do not use flake graphite or put any such mixture in tank, as it may choke up the passageways when allowed to settle when the engine is laid up in round house.

To pack motor piston rod, remove cap screws in cylinder (Fig. 316) and lower gland No. 307, insert two rings one quarter inch square spiral packing and replace cylinder. Adjust packing by loosening nut 330 and

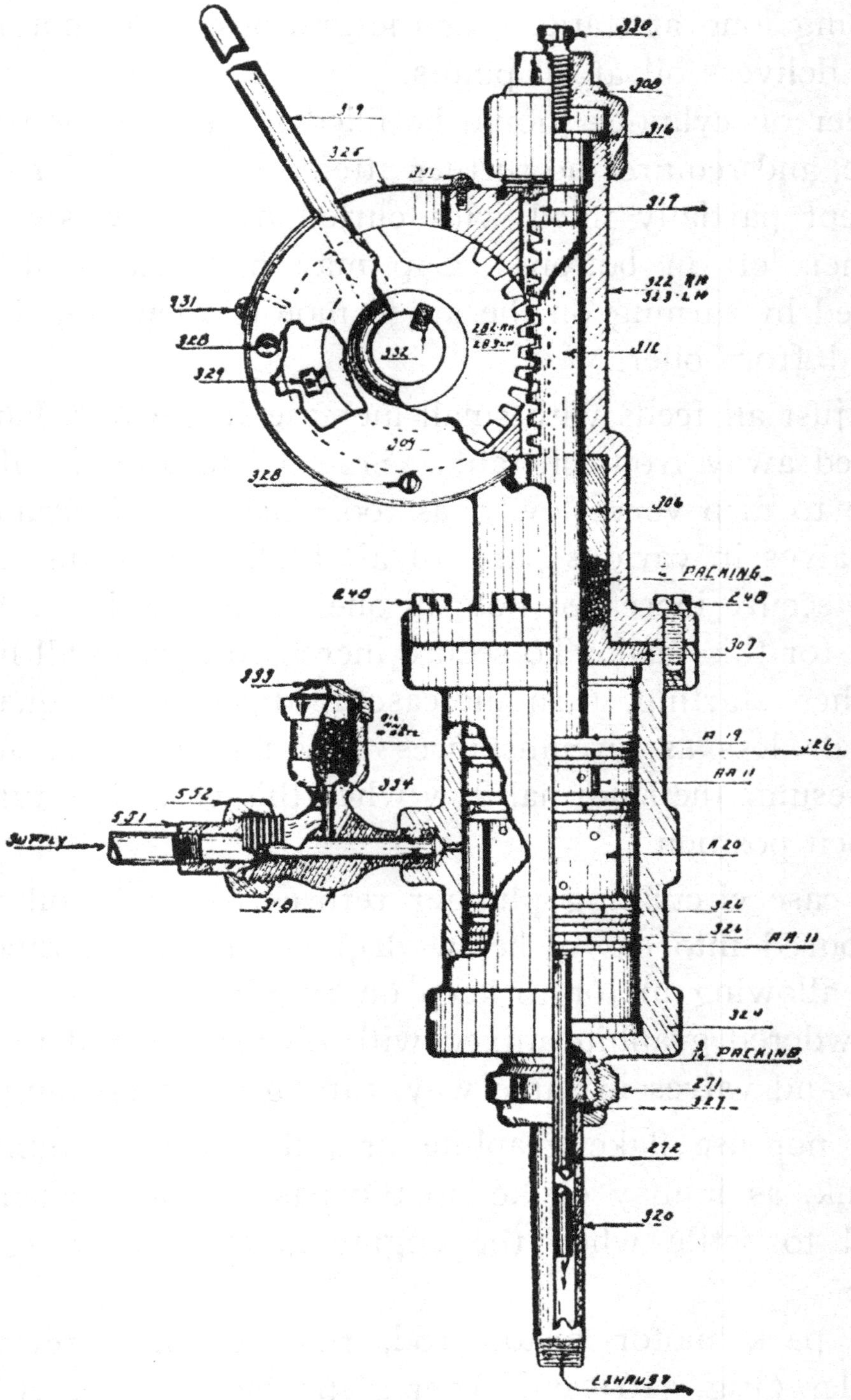

FIG. 316. CROSS SECTION, SHOWING MOTOR PARTS

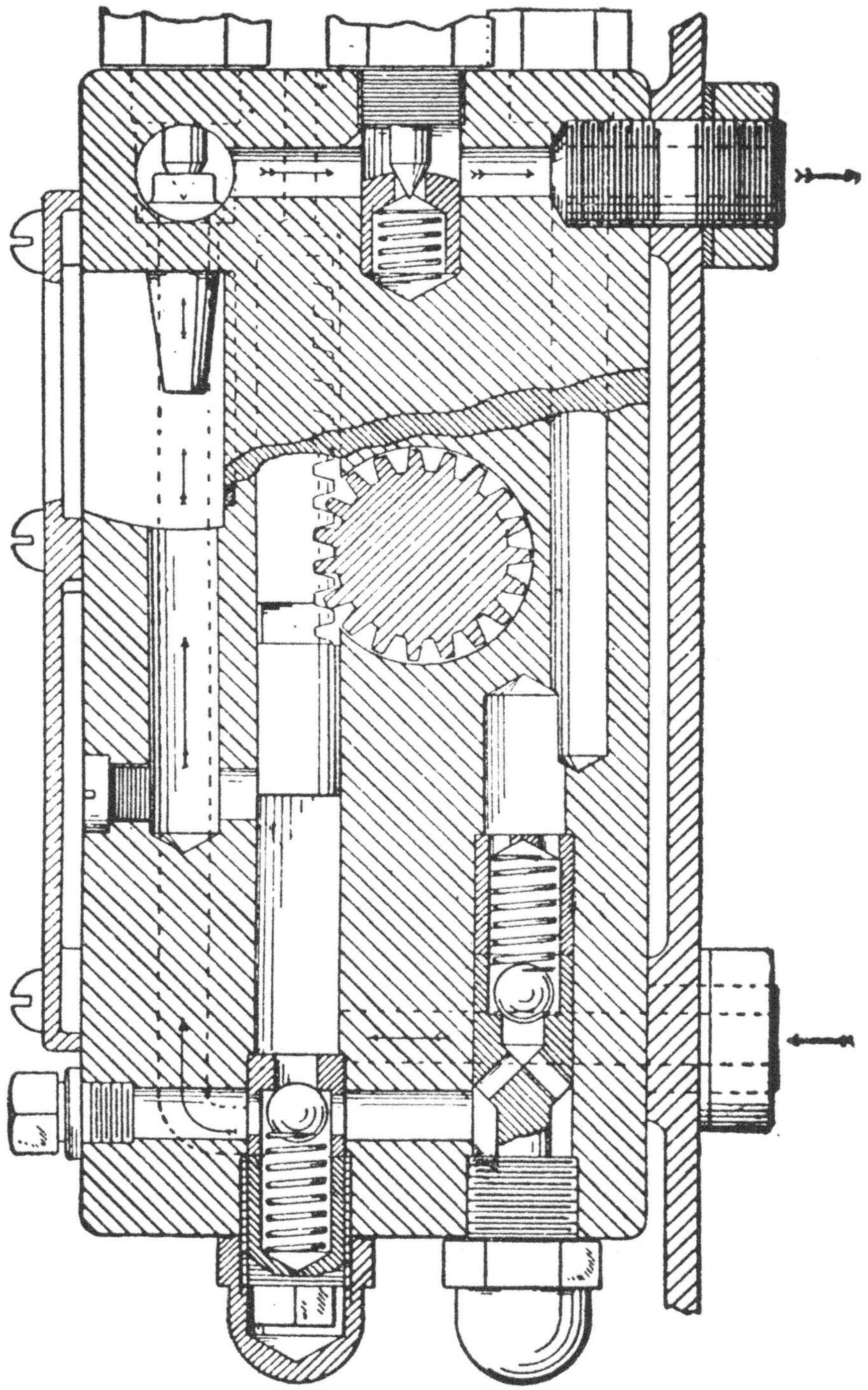

FIG. 317. SECTION OF PUMPING MECHANISM
Showing forcing plunger, valves and overflow dam or increase feed valve

screwing down cap 308, then retighten 330 as a lock nut. If motor refuses to work, remove piston and thoroughly blow out all parts with steam, and see that same are clean and piston rings working freely.

If pump refuses to work, first see that oil is hot. If a feed refuses to take oil, or blows back remove blind nuts for this feed on bottom of pump body and take out valve screws, springs, balls and ball cases, being careful not to lose any parts and noting carefully their position in pump. (Fig. 317.) Clean and replace same.

Do not crowd too much packing in the various packing glands.

In cleaning sight feeds or glass in front of them, raise the cover and wipe with rag. Do not use waste, as small pieces may fall into holes below the drips and get into pump valves. A piece of an old signal flag will answer.

Let all necessary repairs be made by a mechanic who thoroughly understands the assembly and operation of pump and motor.

A FEW SUGGESTIONS.

Keep the valve stems on top of the pump well packed and screwed down fairly snug as the feed can be more easily adjusted and will not rattle open or shut.

Make an examination all around occasionally to see that all union joints are kept up tight.

If the motor slows down a little some cold night don't imagine something is out of order. Cold oil pumps harder and it has probably chilled in the pipes where they run from under the jacket to the cylinder. Just give the motor a little more air or steam.

SIGHT FEED OIL PUMP

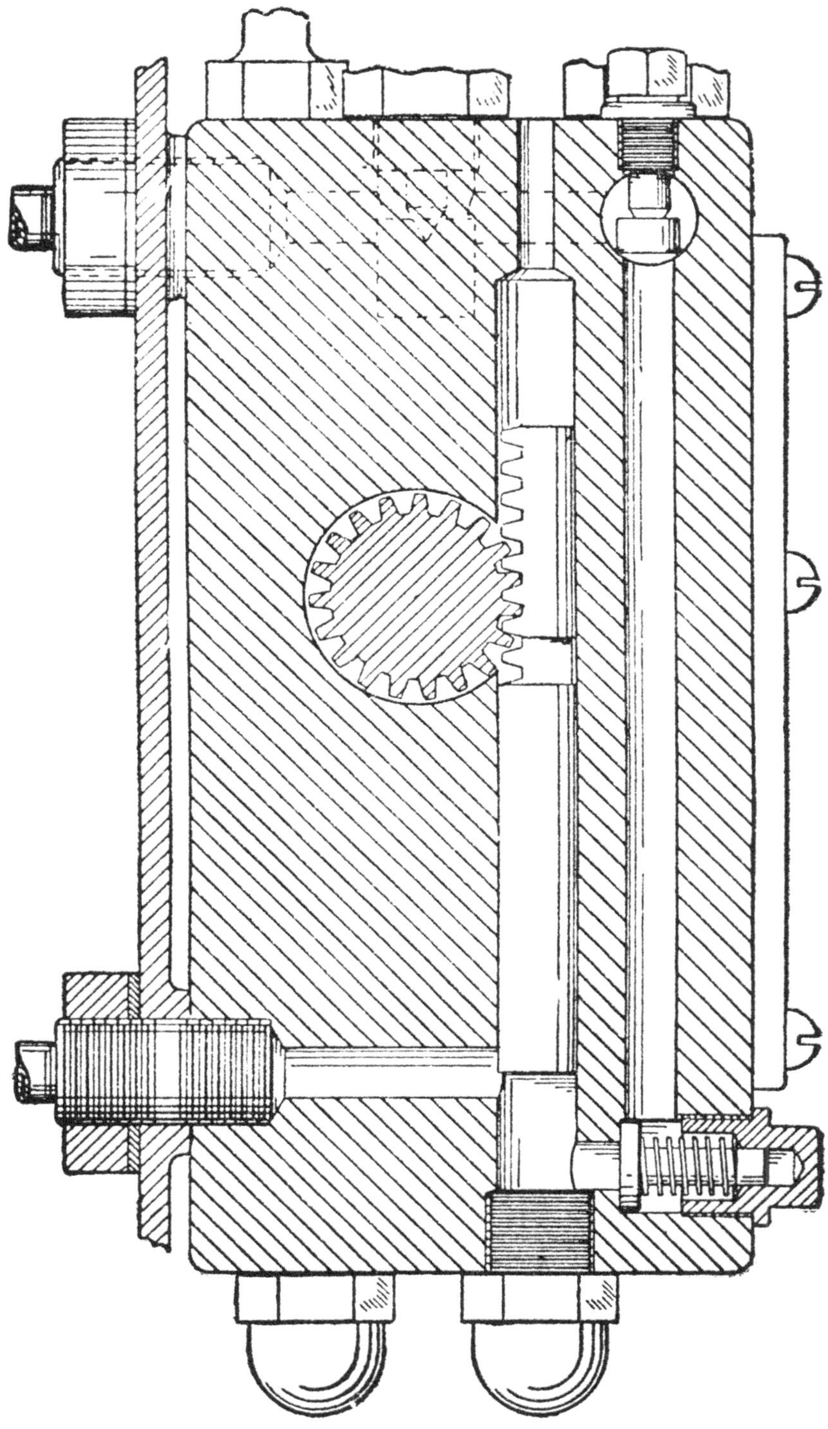

FIG. 318. SECTION OF PUMP

Showing plunger that circulates the oil from tank past the feed nozzles, throwing excess back to tank through the increase feed valve, as shown by upper arrows in Figure 317

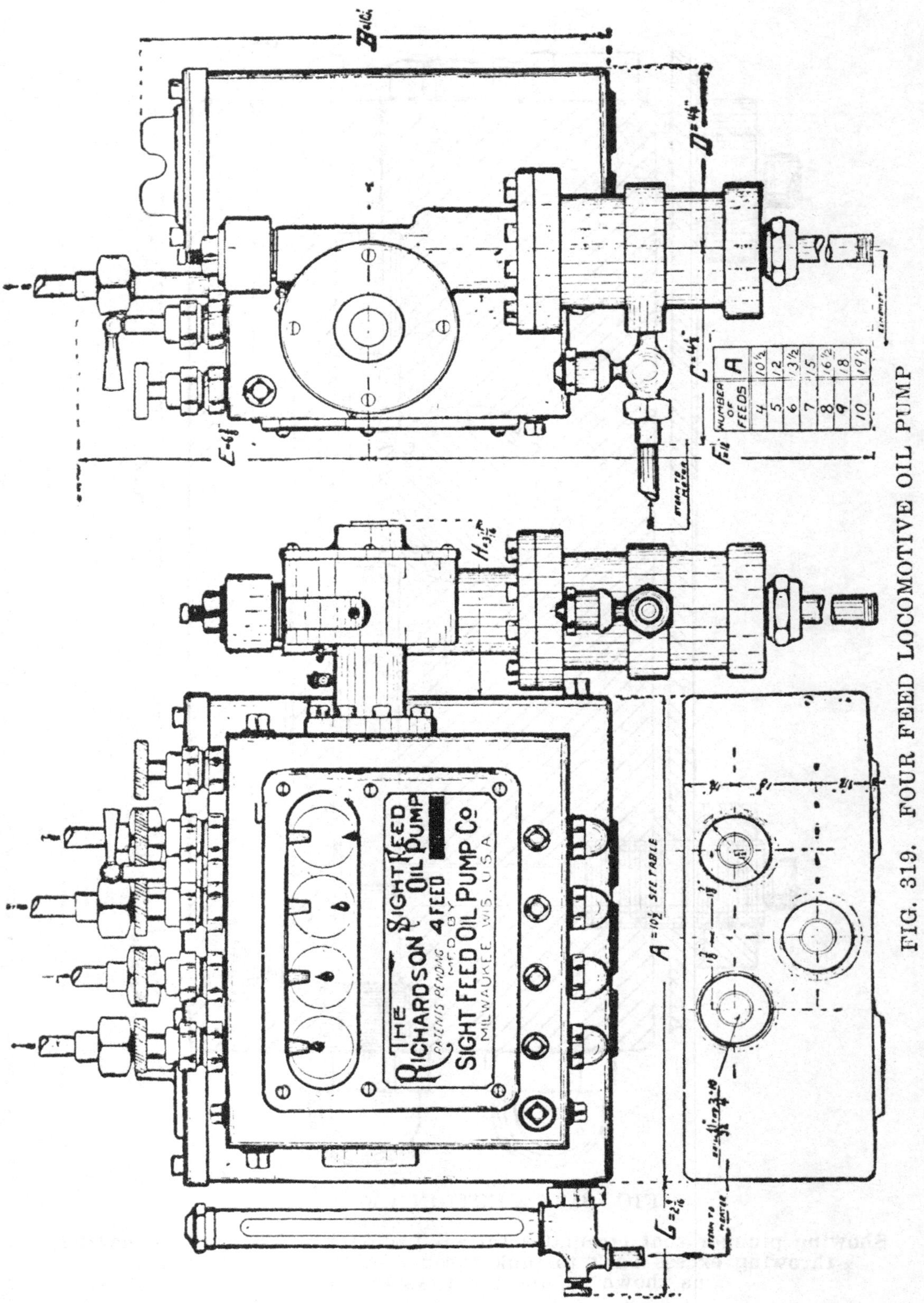

FIG. 319. FOUR FEED LOCOMOTIVE OIL PUMP

SIGHT FEED OIL PUMP

An occasional thorough blowing out of the oil tank with steam will avoid the trouble that dirt under the pump valves will surely cause.

If oil blows back at the feed, remove the cylinder check valves and test them for leaks—then make sure the valves in the pump are seating properly.

No usual amount of ordinary dust will cause this and a good flushing will generally remove the trouble which may be done while floating or shut off for a minute—speed up the motor and turn the increase feed valve around to run the oil in a stream to the particular feed that is giving trouble.

Keep the oil in the tank hot.

Remember the size of the drops this pump feeds are smaller than those fed by a lubricator and it will therefore take a greater number per minute to feed a given quantity of oil in a given time.

Don't put graphite or mica mixture in the tank unless you want to stir the oil up by hand now and then. Feed it to the plungers direct through the sight feed recess in front.

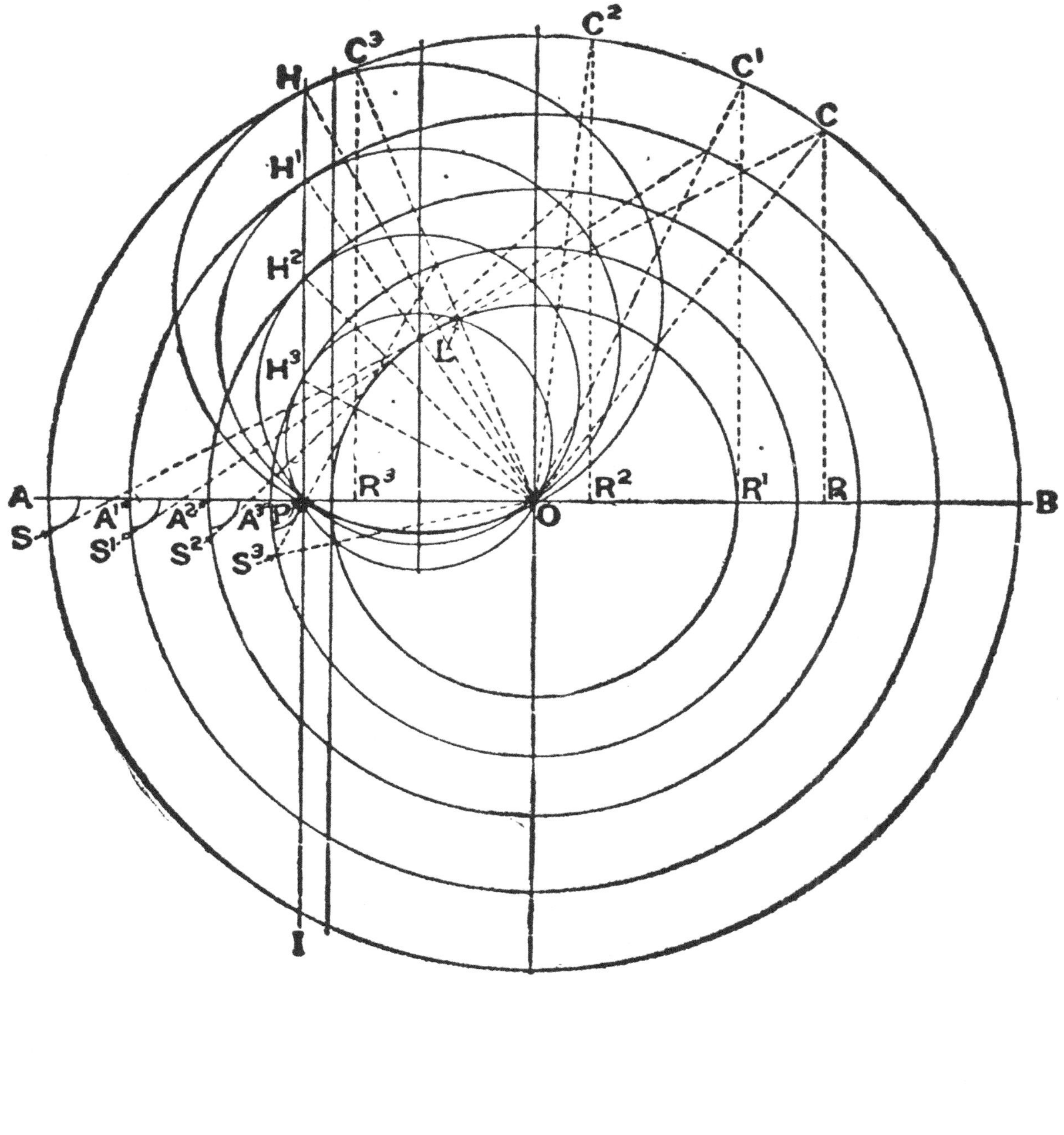

C²
C³
H
C¹
C
H¹
H²
L
H³
A
R³
R²
R¹
R
B
S
A¹
A²
A³
P
O
S¹
S²
S³
I

CHAPTER XIII

FOUR-CYLINDER BALANCED COMPOUND LOCOMOTIVES.

BY F. J. COLE, MECHANICAL ENGINEER.

(By special permission of American Locomotive Co.)

First, let us consider, briefly, the reasons which led up to the use of the four-cylinder balanced compound at home and abroad. Generally speaking, all other locomotives may be characterized as having two main rods connecting the pistons and cross heads to the driving wheels. This construction necessitates the use of excess counterbalance in the wheels opposite the crank pins, in order to make the engine run smoothly. While the horizontal moving or reciprocating parts are balanced, a vertical component is produced by the centrifugal action of these excess weights, producing alternately an increase and decrease in the static driving wheel weights. This is the cause of the so-called "hammer-blow," which is really a wavelike increase and decrease in the normal wheel loads, reaching its maximum when the counterweight is down next to the rail, and gradually decreasing during one-half of a revolution until the minimum is reached, with the counterweight on top of the wheel. For an ordinary ten-wheel locomotive, weighing 163,000 pounds in working order, weight 123,000 pounds on drivers, cylinders 20 by 26 in., 69-in. driving wheels, the static and dynamic weights on driving wheels are:

Speed in Miles per Hour.	Static.	Dynamic.	Increase, per pair of wheels.
50	123,000	140,500	5,840
60	123,000	148,000	8,340
70	123,000	157,200	11,400

For a ten-wheel locomotive, 164,000 pounds total weight, 129,000 pounds on drivers, 21 by 26 in. cylinders, 63-in. wheels, the static and dynamic weights on driving wheels are:

Speed in Miles per Hour.	Static.	Dynamic.	Increase, per pair of wheels.
45	129,000	146,000	5,700
55	129,000	154,400	8,480
65	129,000	164,700	11,900

In an engine having two pairs of drivers, the excess weights per wheel are greater, as they are divided between two pairs of drivers instead of three pairs, but this is somewhat offset by the fact that such engines usually have larger drivers, but where the same size drivers are used, the dynamic weights would be increased in direct proportion to the decrease in number of wheels. A practical demonstration of this is to run soft steel or iron wires between the driving and carrying wheels of a locomotive on a testing plant and ascertain the flattening of the wire at various speeds along a length equal to the circumference of the driver. At the Purdue University testing plant, Professor Goss has shown that wires used under these conditions at very high speeds were not flattened in certain parts at all.

As the actual limitation of weight on the drivers of a locomotive which the rails and roadbed will safely sustain is the dynamic and not the static load, it follows that if no excess weights were used in the driving wheels considerable more weight may be placed upon the drivers, and consequently a more powerful locomotive could be built to come within a given limitation of weight.

Again, the turning moment is more uniform in a four-cylinder locomotive. This arises from two causes:

1.—The angularity of the main rods of a two-cylinder locomtive produces a greater turning moment when both cranks are ahead and at 45° each above and below the horizontal centre line. For the same reason the pressure is less when the cranks are behind. In the four-cylinder engine this condition does not obtain. as one rod neutralizes the other, so that the maximum and minimum turning moments, occurring every 90°, or four times during each revolution, are of equal length. A diagram illustrating this will be shown later on.

2.—The relatively larger cut-off in the compound produces a more uniform crank pin pressure.

These facts are now so well known that it is unnecessary in this paper to enlarge upon them and discuss them further.

It is an old idea to build a balanced locomotive having four sets of pistons, crossheads and rods arranged in pairs, in which the outside horizontal moving parts connected directly to the wheels are balanced by similar inside parts connected to a crank axle at 180° to each other and moving in opposite directions, the other pair working at 90° to the first.

We have now reached a point where the demand for high speed heavy trains is such that advantage must be taken of everything which will materially assist in greater power, produce less injury to roadbed and to improve fuel economy. These advantages we find in the four-cylinder balanced compound, and the time is at hand when locomotives of this description are in demand and can be used to good advantage on American railroads. The principal objection raised in the past, and perhaps occasionally now, is to the use of crank axles, but the answer to this is that such axles can be made at the present time of such quality and design as to give but little or no trouble. The stresses are better understood and can be provided for, and it is probable that the form of axle which will be largely used is one built up and balanced within itself. Briefly stated, some of the advantages of four-cylinder balanced engines are:

(*a*) Balancing of reciprocating parts by similar horizontal moving parts, one outside piston and its attachments moving forward while the one inside is moving backward. These exactly balance one another without the use of unbalanced weights in the wheels.

(*b*) The increase of weight permissible on the driving wheels when considered dynamically. In the ordinary engine, at 60 miles per hour, with drivers 78 ins. in diameter, the increase or decrease at each revolution of the static weights on the rail is about 23 per cent. This is due to the centrifugal effect of the excess weights used to balance the reciprocating parts.

(*c*) Increase of from 25 per cent to 33 per cent in sustained horsepower at moderate and high speeds without any material change in size or style of boiler.

(*d*) Economy in the use of fuel, water and steam.

(*e*) The subdivision of power on two axles and four cylinders. Reduction of bending stresses on the crank axle due to the fact that only half the turning moment is transmitted through each axle. The advantages from light moving parts, such as cross-heads, main rods, piston rods, etc. The lightness of these parts permits them to be easily handled, and as they have to transmit only half the usual amount of power, the probability is that the wear and repairs will be greatly decreased.

We are indebted largely to the French for the design and practical development of this type of engine. Some fifteen years ago the Northern Railway of France built the experimental four-cylinder compound No. 701, and from this engine the number has gradually increased until over 1,500 are now running in Europe. In these engines the idea has been steadfastly adhered to of subdividing the power and driving two axles independently, the forward or crank axle being connected to the low pressure cylinders, and the rear axle to the high pressure cylinders.

In America, owing to the use of bar frames and heavier engines, there is not room enough between the frames for the low pressure cylinders. It is therefore necessary to transpose this arrangement and place them outside. Theoretically, this is not correct, as the larger and heavier parts should be in the center of the engine. The counterbalance for the disturbing portion can be better arranged when placed in this manner. In practice, putting the high pressure cylinders inside is the best compromise which can be made, and as a matter of fact, but little difference is required in the arrangement of the counterweights, and little or no disturbance can be noticed in the working of the engine.

Abroad, the utmost refinements have been used for the improvement of fuel and steam economy. Independent valve motions for each high and low pressure cylinder, arranged so that the cut-offs can be changed independently of each other, are in general use. Separate exhaust valves and adjustable exhaust nozzles are used to improve the starting and to vary the intensity of the blast.

All these appliances, while they make a more efficient machine, increase its complication, and, except where handled by men who are on the alert to save fuel on every occasion, may result in increased cost of repairs which will more than outweigh the saving of fuel. Furthermore, the demands of service on American roads are more exacting and take up a larger proportion of a locomotive engineer's time than abroad, and render it desirable to limit to a minimum the number of handles, levers and other mechanical appliances requiring adjustment and alteration. In the introduction of this type of engine to American roads and to suit the conditions of to-day, it is the consensus of opinion that simplicity should be the keynote, and all devices which are not absolutely necessary to the successful working of the locomotive should be omitted, the only exception being in localities where the price of fuel would warrant the extra cost and maintenance of all the attachments for fuel-saving in use abroad.

For these reasons the balanced locomotives built in this country have generally used only one valve motion to distribute the steam on either side to the high and low pressure cylinders. As few changes as possible have been made over simple engines of similar type other than the application of the crank axle and cylinder arrangement, the ordinary Stephenson valve motion with indirect rocker being used. There seems to be no doubt of some saving in fuel with independently adjustable valve motions which can be changed to suit the varying conditions of railroad service. But it is a question whether we are ready, in this country, at the present time, to sacrifice our ideas of simplicity and directness of means to the end and adopt these ultra refinements which have been perfected and are in successful use abroad. And except for purposes of test or unusual cases when there may be some demand for these additional complications, I think it is safe to say that for some time to come the greatest number of these locomotives will be built with the ordinary single valve motion and with the extra parts reduced to a minimum.

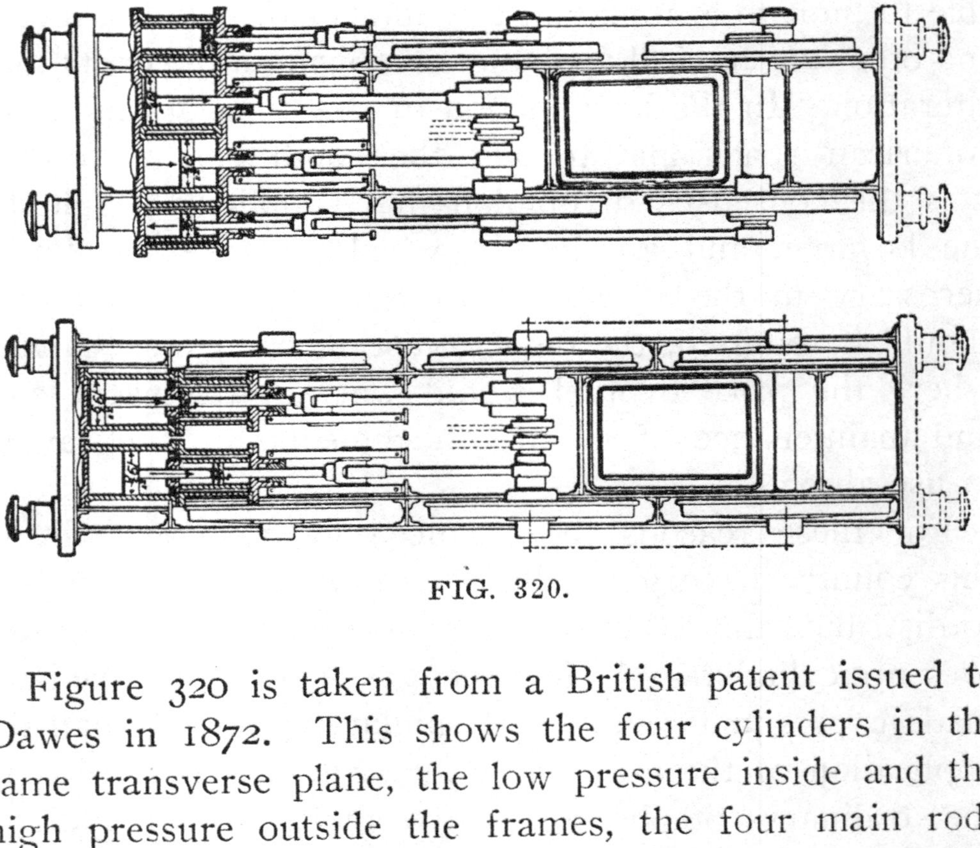

FIG. 320.

Figure 320 is taken from a British patent issued to Dawes in 1872. This shows the four cylinders in the same transverse plane, the low pressure inside and the high pressure outside the frames, the four main rods being connected to the crank axle. The idea of balancing by similar horizontal moving parts moving in opposite directions is plainly shown, the low pressure piston, cross heads and main rods connected to the crank axle balancing the high pressure main rods, etc., connected to the crank pin on the outside of the driving wheel and balancing each other on one side without the necessity for excess weights in the drivers.

Figure 321 illustrates a United States patent granted in 1877 to A. M. Cummings. It is very similar in principle to Dawes', except that it shows simple slide valve cylinders, and is evidently better adapted to American design. The cylinders are on the same transverse plane and the main rods are connected to the crank axle.

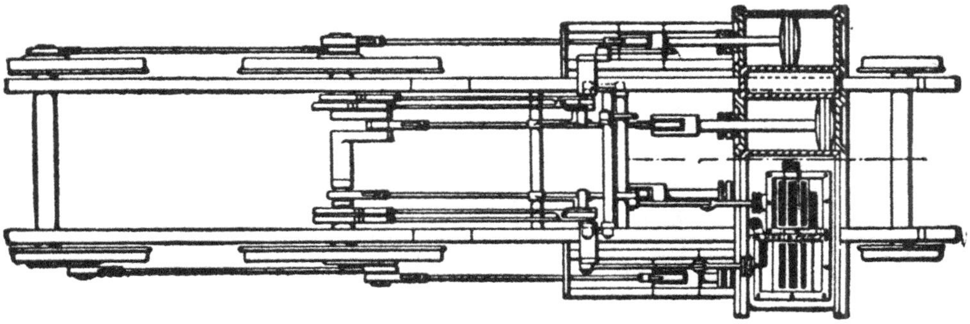

FIG. 321.

An end view and details of the cylinders and arrangement of link motion are given in Fig. 322. Some ingenuity is displayed in this respect, avoiding the use of independent valve motion for each cylinder by the use of long rocker shafts extending across the frames. A peculiarity of this engine is the four simple cylinders and the cranks set at 90° in place of 180°. Of course, in such an arrangement the balancing is imperfect, as the moving parts are too far apart.

Strong's compound locomotive is shown in Fig. 323. It was patented in 1891. Here, as in Dawes' engine, the cylinders are compound and the cranks set at 180°, inside and outside the frames, thus making a balanced construction. A year later Strong patented an arrangement of high and low pressure cylinders in one-half saddle casting, as in Fig. 324. It will be noticed that one piston valve controls the distribution of steam in both cylinders.

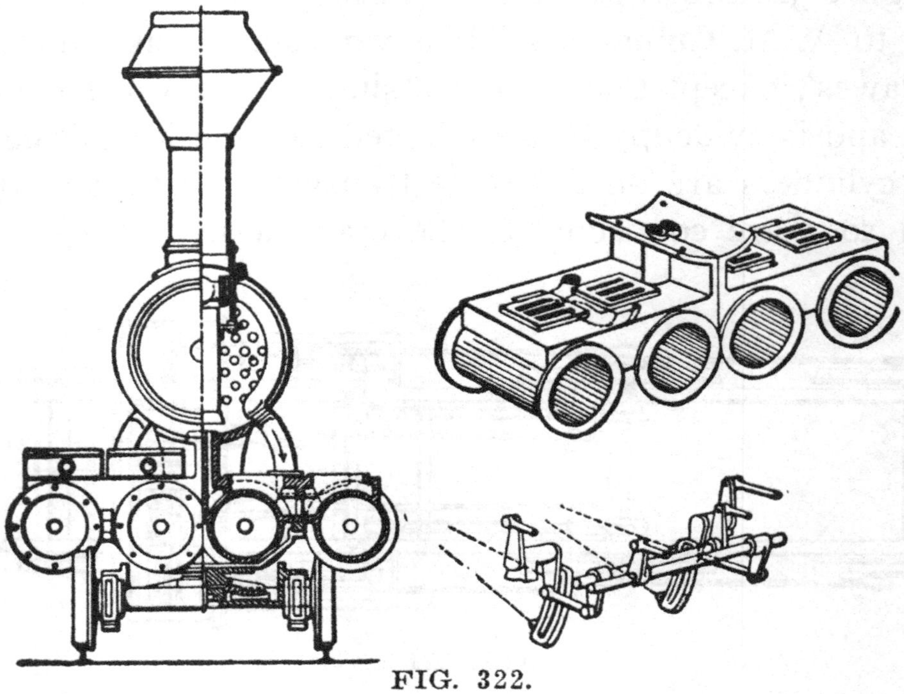

FIG. 322.

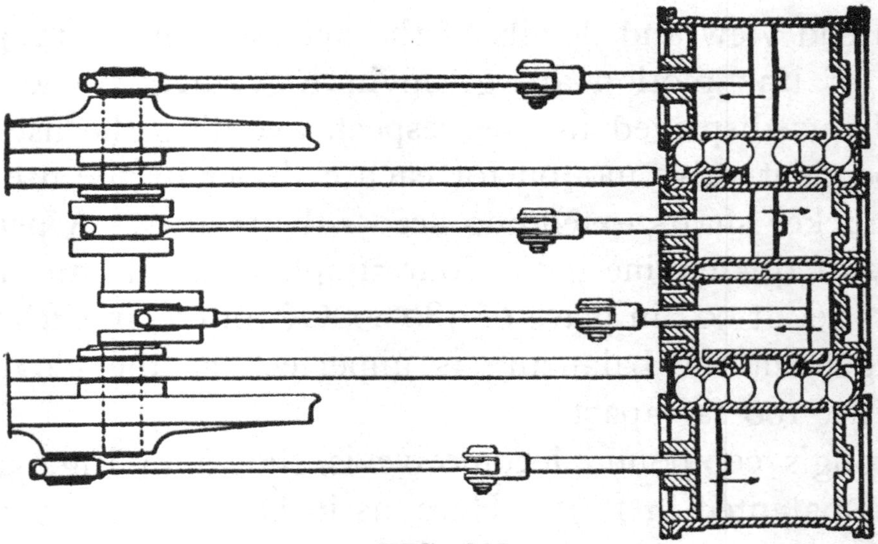

FIG. 323.

In 1895 M. B. Bulla patented in the United States a locomotive having four cylinders in the same transverse plane, arranged in two pairs, Fig. 325, with the cranks set at 180°, half saddle cylinders with one piston valve

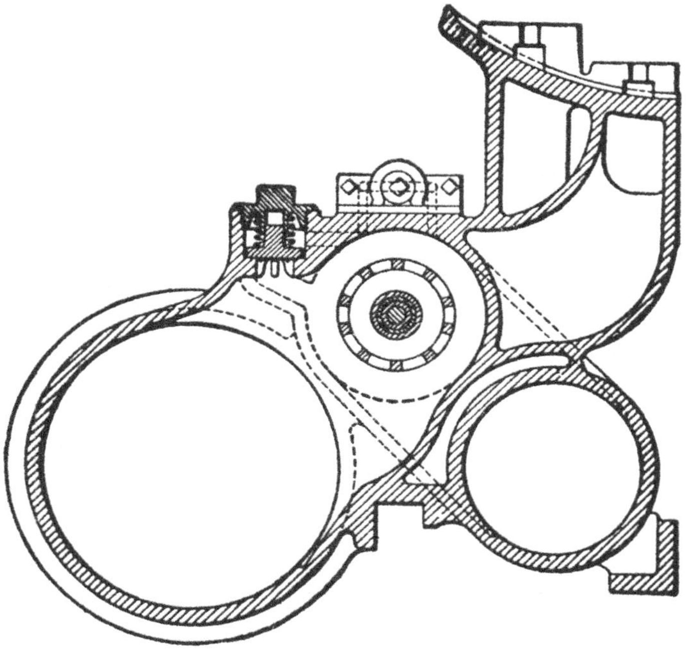

FIG. 324.

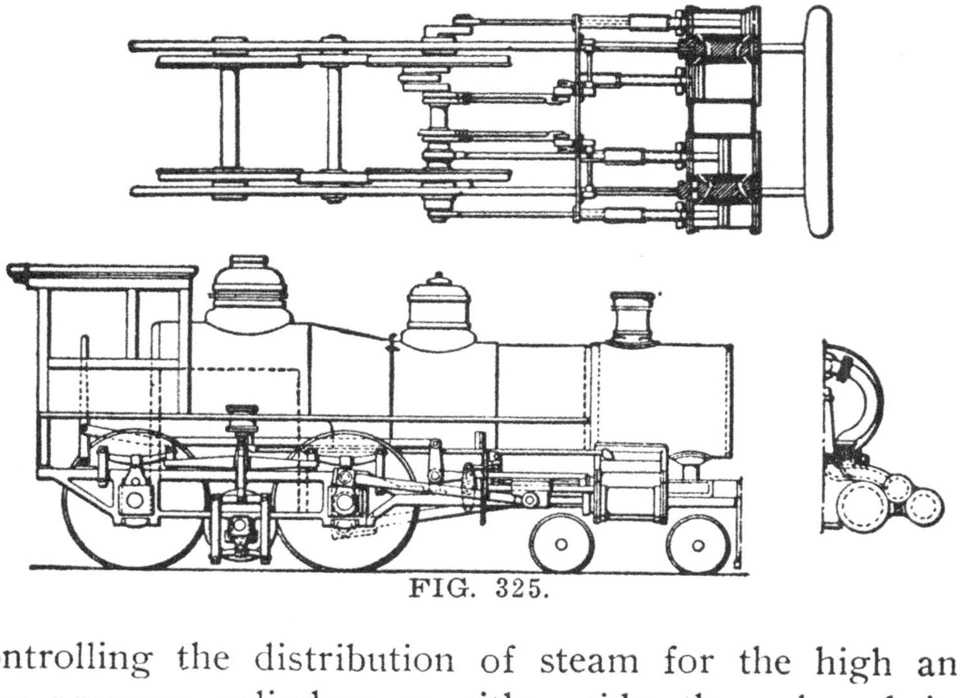

FIG. 325.

controlling the distribution of steam for the high and low pressure cylinders on either side, the valves being operated by the usual Stephenson link motion. But in this locomotive the idea of the balancing is evidently secondary in the inventor's mind, as the small wheels for increasing traction, by throwing more weight on the drivers, are deemed of the most importance and each claim involves their use.

The next picture, Fig. 326, is taken from a Swiss patent granted in 1897 to Von Borries, for a balanced four-cylinder compound locomotive. In this design the cylinders are in the same transverse plane, but independent piston valves are used to control the distribution of steam to the high and low pressure cylinders. The main rods are all connected to the same axle.

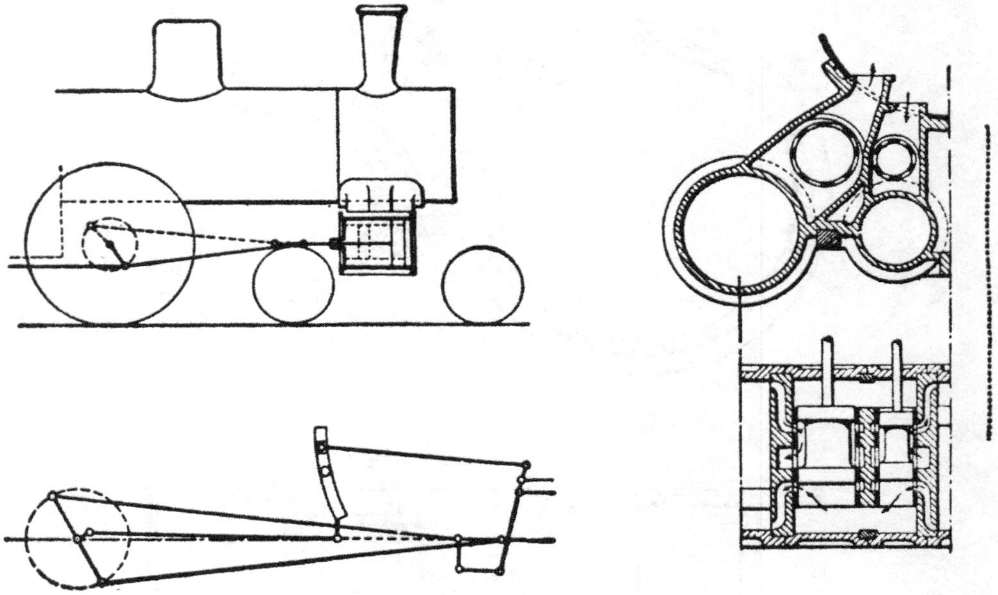

FIG. 326.

After investigating the various forms of crank axles used in Europe, the American Locomotive Company decided upon the Z-shape, or the design having a diagonal member.

Figure 327 shows this form of axle, which was applied to the New York Central locomotive No. 3,000, exhibited at the World's Fair at St. Louis. The weight of this axle finished is 3,200 lbs. It was forged and finished completely by the Krupp Works, at Essen, Germany.

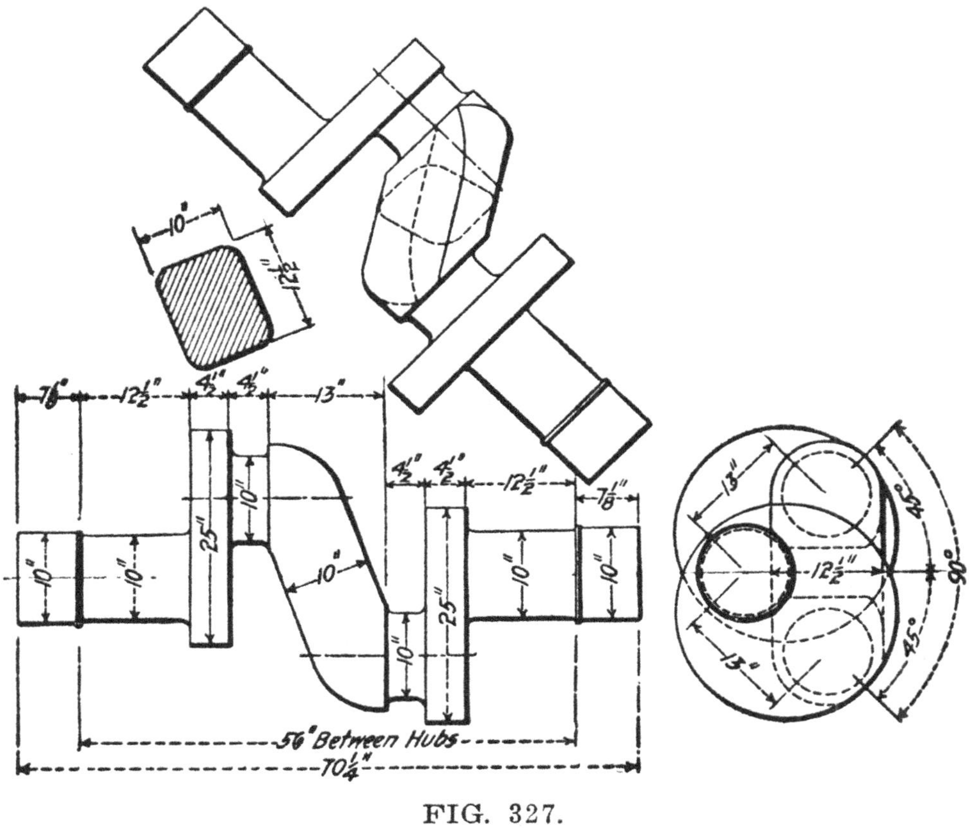

FIG. 327.

Figure 328 shows two forms of crank axles used in Europe, the illustration and description in part being taken from Mr. A. Herdner's paper entitled "The Compound Locomotive and Its Development in France," published by the American Engineer and Railroad Journal, December, 1902.

Mr. Herdner's remarks are as follows:

"Indeed, until 1890, the northern and western railways of France were the only roads that made use of inside connected cylinders for their high speed locomotives. The other roads have shown themselves more reserved. On the Southern, particularly, to which the writer be-

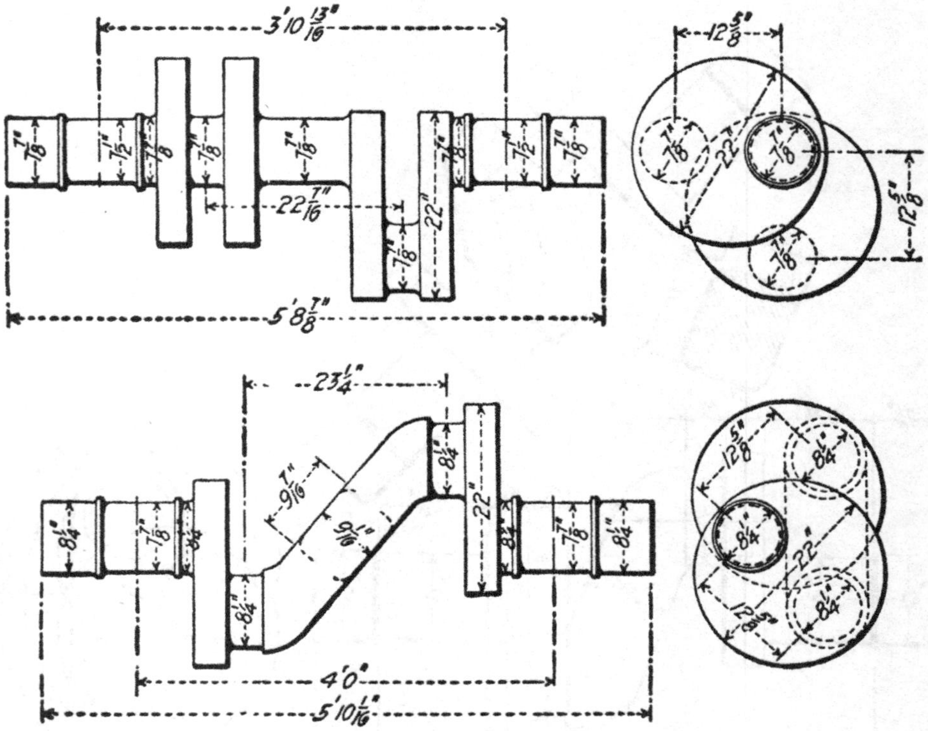

FIG. 328.

longs, the crank axle was absolutely banished from locomotive construction, and when, three years later, in consequence of the constant increase in weight of our express trains, we were led to put in service fourteen four-cylinder compound locomotives similar to those of the Northern Railway of France, we were naturally very much concerned as to what service the crank axles driven by the low pressure cylinder pistons would render in practice. Those axles were of the Worsdell type with circular plates, and are represented by the upper view in Figure 328.

"The successive appearance of cracks on three of the axles after a mileage of 37,500, 58,000 and 133,000 miles respectively, seemed to justify the fears we had conceived in that respect, and yet a much longer experience leads us to-day to believe those fears to have been exaggerated. Of the fourteen crank axles put in service in 1894 and 1896 there remain eleven totally free of any fracture or crack, and their mileage varies from 212,100 to 293,900 miles.

"In the more recent four and six-wheel connected locomotives, the Worsdell axle has been replaced by the Z-shape axle, shown in the lower view in Fig. 328. The design of this axle is obtained from the Worsdell axle by substituting in place of the two inner circular disks and the portion of the axle between a rectangular bar connecting the two journals. Sixty-three of these axles have been put in service between the years 1896 and 1901, inclusive, and none of them have shown the slightest trace of a possible fracture, the oldest of these axles having already made a mileage of 206,200 to 242,000 miles. These results lead us to believe the axle of the Z-shape is more advantageous than the one with the two circular plates, and indeed we give it preference to the former design except when the central portion of the axle has got to be utilized for carrying the eccentrics, as is the case in our new consolidation type engine, about which I shall say a few words at the end of this article (at any rate the matter of viewing this question will be still strengthened at longer experience)."

Owing to the expense of making solid crank axles, the danger of internal strains and of cracking, with the necessity for scrapping the whole axle when cracked, the built-up axle is gradually being substituted and coming into use more and more every year. In the built-up axle the danger from internal strains is very much reduced, as the pieces are either cylindrical or flat, straight, plain ordinary pieces, without abrupt changes of shape. The cracking at the corners is therefore confidently expected to be very much reduced, and if such cracking does occur, the part failing can be replaced without discarding the entire axle.

One contributory cause to the overstraining and consequent decrease in life of the ordinary solid or built-up crank axle is due to the bending moment caused by the centrifugal force of its own unbalanced weight. Calculations show that this is a considerable force which must be provided for, amounting to 5,300 pounds fibre stress per square inch at 320 revolutions per minute in the ordinary design. This is due entirely to the centrifugal action of the unbalanced weight producing a bending moment, the magnitude of which varies according to the speed, the amount the axle is out of balance and the distance between the centres of frames. This naturally directs attention to a design having the flanges or checks prolonged in the opposite direction, so as to balance it within its own structure. Such an axle is shown in Fig. 329. It was made by Mr. Drummond, of the London & Southwestern Railway, and is composed of nine pieces. At first sight it seems complicated and unwieldy, but upon investigation the construction is very simple, and the design is well worth considering. Engineering (London) makes the following comments:

"It is built up, and the axle and webs are of mild steel, the latter having the ends prolonged to form balance weights. All the revolving and part of the reciprocating masses are balanced in the axle, and, as these are bal-

FIG. 329.

anced in the planes in which the disturbances are created, the axle is relieved of the destructive stresses caused by the balancing in the wheels."

FIG. 330.

By this many objectionable features disappear, the bending stresses due to the axle being out of balance are done away with, and the cracking which is characteristic of crank axles will doubtless be greatly reduced. The prolongation of the sides, while it makes an easy way of balancing, does not make use of the metal in the most efficient way. By combining the balanced weight in the center portion, at an angle of 45° with each of the cranks, the excess metal may be located to much better advantage.

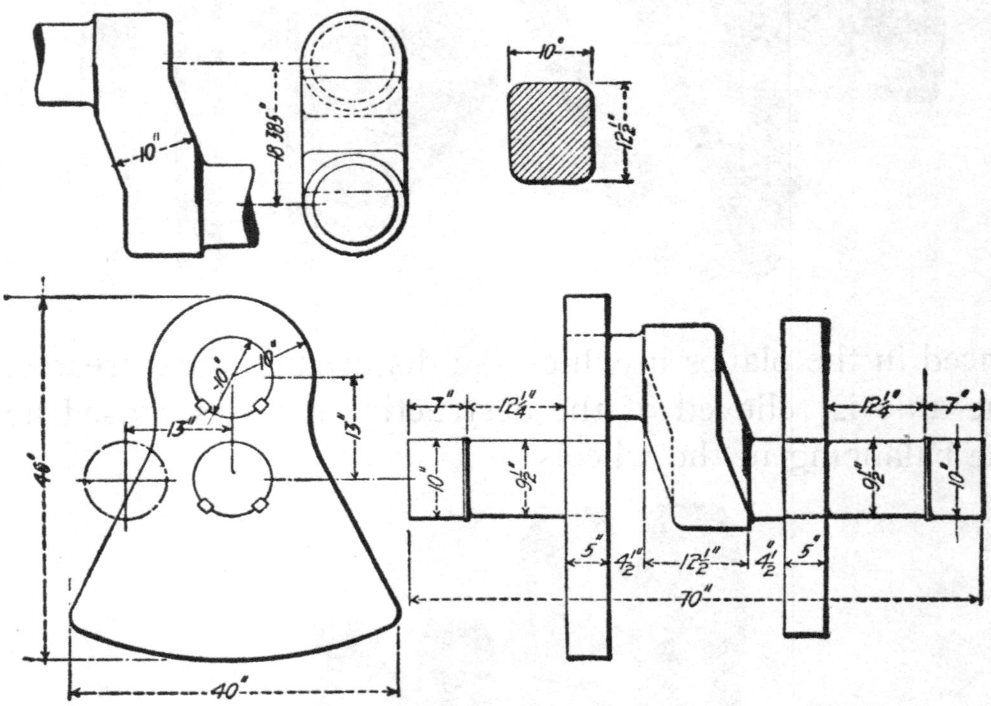

FIG. 331.

Such a design is shown in Fig. 330. The axle is built of seven pieces and the centre portion made with its overhanging weight so that the axle is completely balanced and free from bending strains caused by unbalanced weight. Another form of built-up axle is shown in Fig. 331. This consists of five parts, the diagonal connecting piece being used.

Figure 332 illustrates a four-cylinder balanced compound locomotive, equipped with Pielock superheater and

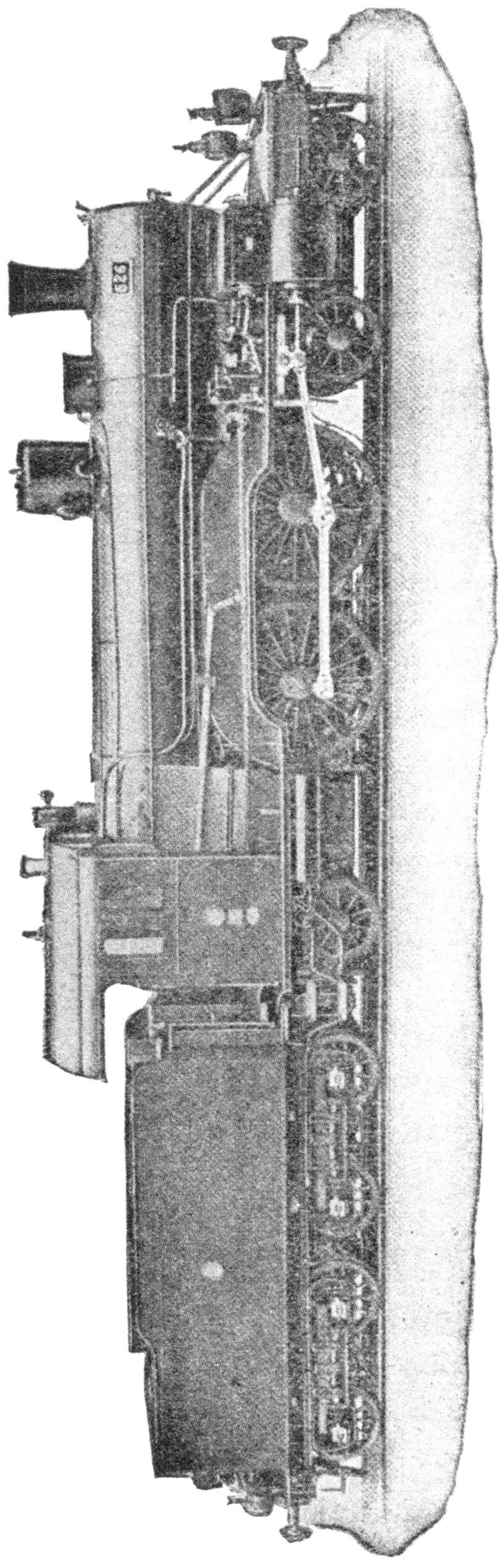

FIG. 332.

built at Hanover, Germany, in the locomotive works founded by Geo. Egestorff. It was exhibited at the World's Fair, St. Louis, and tested on the Pennsylvania Railroad testing plant. The type is Atlantic 442 C 133 with Von Borries valve gear. The principal dimensions are:

Size of cylinders	14 & 24 x 24 in.
Driving wheels	66 "
Wheel base, driving	6 ft. 10¾ "
" " total	29 ft. 3⅜ "
Steam pressure	199 lbs.
Heating surface, tubes	1,504 sq. ft.
" " superheater	310 " "
" " firebox	108 " "
" " total	1,922 " "
Grate area	29 " "
No. of tubes	241
Diameter of tubes outside	2 in.
Length of tubes	14 ft. 7 "
Weight on drivers	67,000 lbs.
Total weight in working order	132,700 "
Weight, empty	119,500 "

The New York Central four-cylinder balanced compound, No. 3000, Class 442 C 200, is illustraed in Fig. 333. The principal dimensions of this engine are as follows:

Cylinders	15½ & 26 x 26 in.
Drivers	79 "
Wheel base, driving	7 ft.
Weight on truck	50,000 lbs.
" " drivers	110,000 "
" " trailer	40,000 "
" total	200,000 "

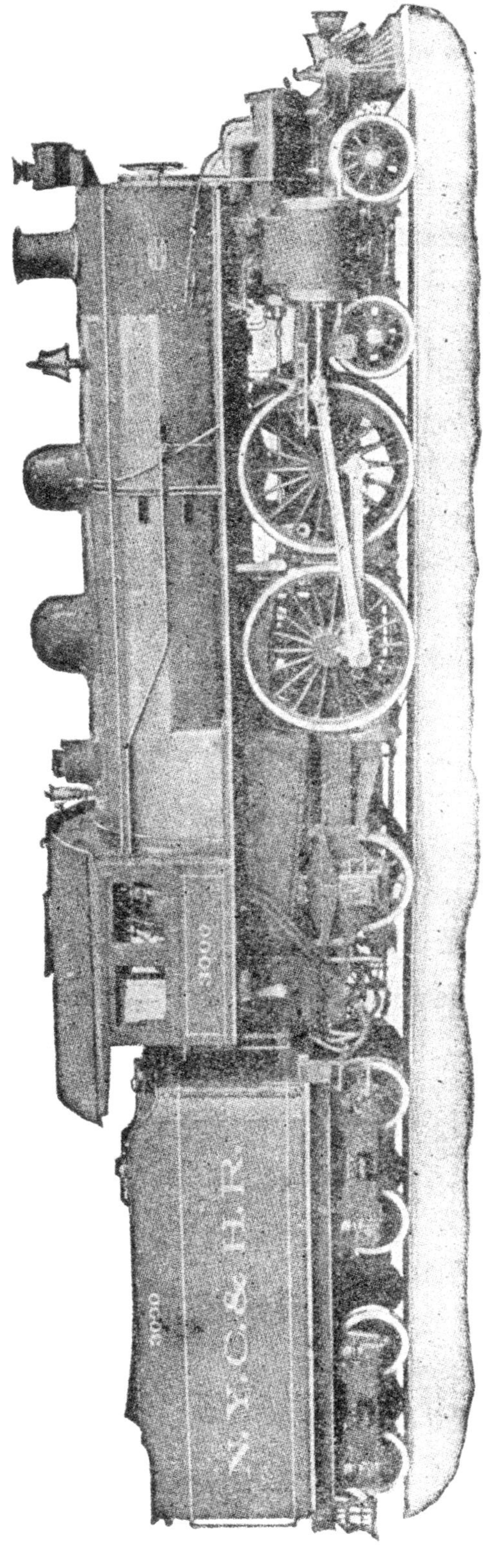

FIG. 333.

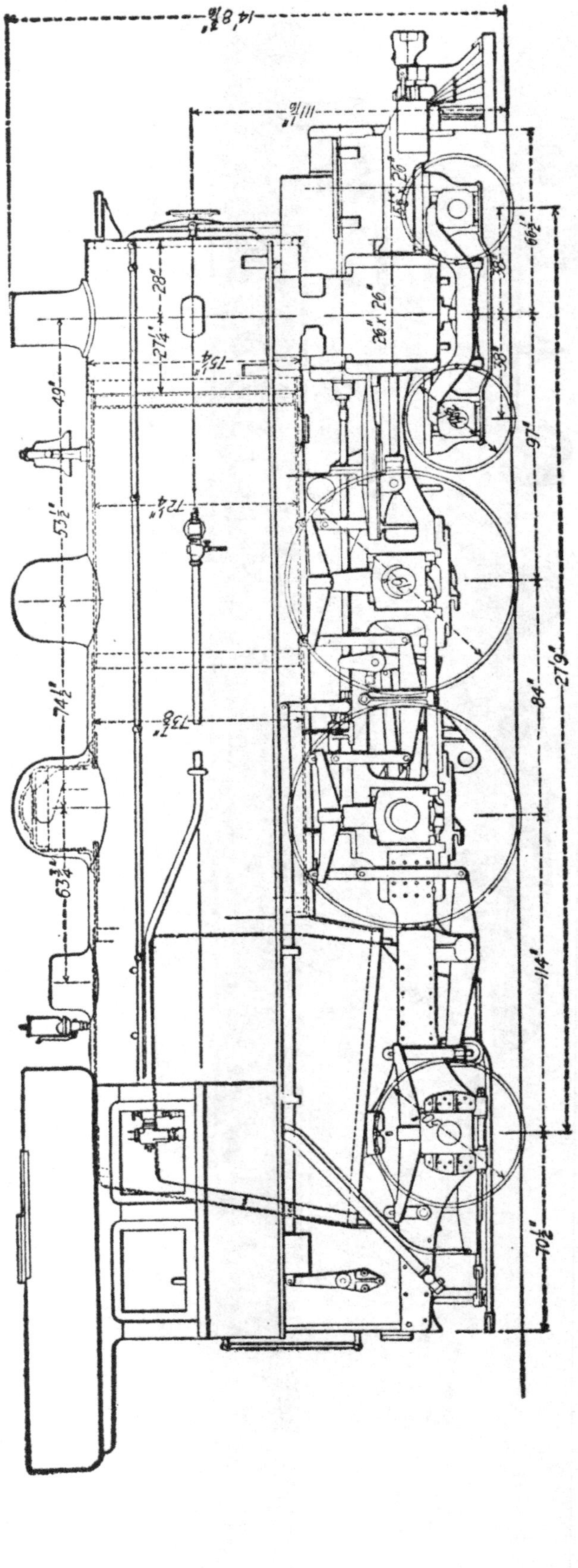

FIG. 334.

Heating surface, firebox		175	sq. ft.
" "	firebrick tubes	23	" "
" "	tubes	3,248	" "
" "	total	3,446	" "
Grate area		50.23	" "

The outline elevation of the engine is shown in Fig. 334. You will notice the ordinary Stephenson link motion is used with the conventional type of reversing rocker, except that it is somewhat modified in order to pass behind the front driving wheels. The valve stem is exactly over the centre of frame and passes through the spring hangers and spring saddle. Apart from the crank axle and the arrangement of cylinders, the engine is substantially the same as the New York Central standard Atlantic. The weight, however, is increased about 14,000 pounds, owing to the extra pair of cylinders, piston rods, cross-heads, crank axle, etc.

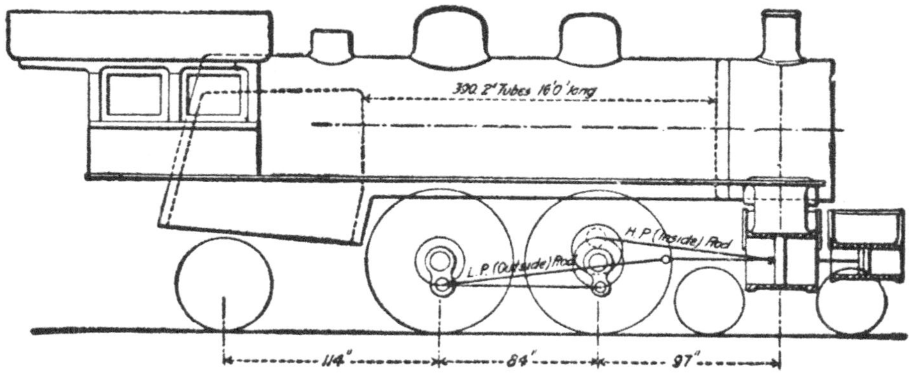

FIG. 335.

Figure 335 shows an outline of the arrangement of the engine, representing particularly the cylinders, main rods and driving wheels. The object in placing the high pressure cylinders ahead is to avoid an unnecessary length of boiler and wheel base and to obtain a good length of high pressure main rods. In this particular engine but few changes were made over the regular New York Central Atlantic type engine. The flues are maintained at 16 feet, the distance from centre of cylinders to forward drivers is only 3 inches more than the regular engine. This was in order to obtain necessary clearance for the low pressure cylinders and engine truck wheels. Any other design of Atlantic engine, having cylinders in the same transverse plane, necessitates lengthening the boiler from 30 inches to 3 feet.

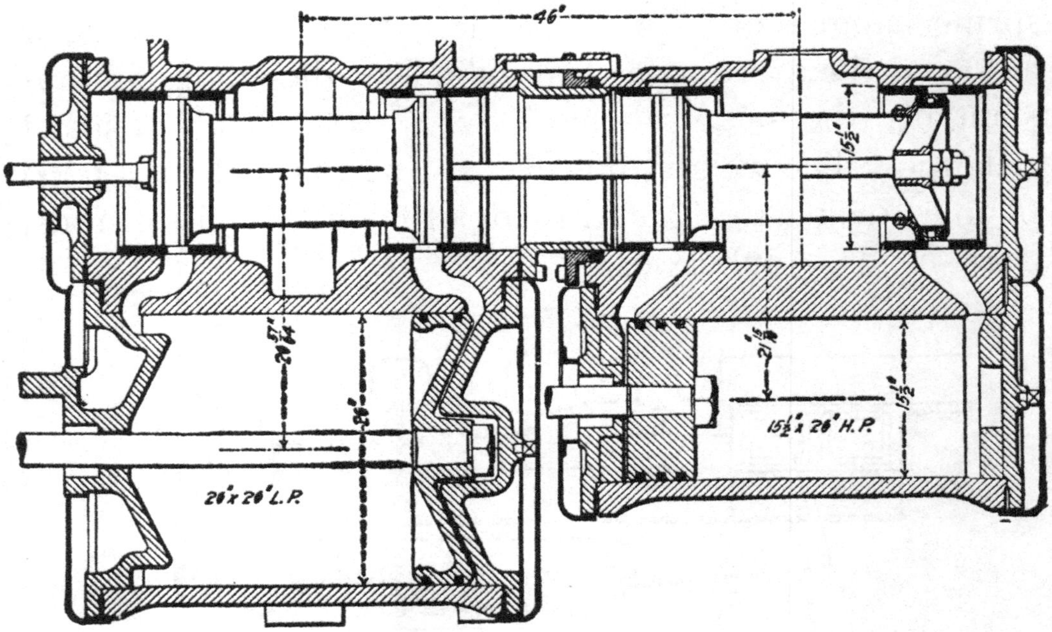

FIG. 336.

A longitudinal section view through the valve chamber and the high and low pressure cylinders is shown in Fig. 336. For convenience, the three sections are assumed to be on the same plane. The live steam from the boiler enters at the top and centre of the inside admission high pressure valve, and is exhausted at the ends of the valve into the receiver. That from the forward port passes through the valve to the outside admission low pressure valve. This arrangement of inside and outside admission valves corrects the effect of the crank axle and distributes the steam in relation to the cylinders so that both valves can be driven from one valve stem. The rocker, links and eccentrics are also shown. These, it will be noticed, are substantially the same as used for ordinary single expansion engines. So that, apart from the two valves, no additional complications are introduced.

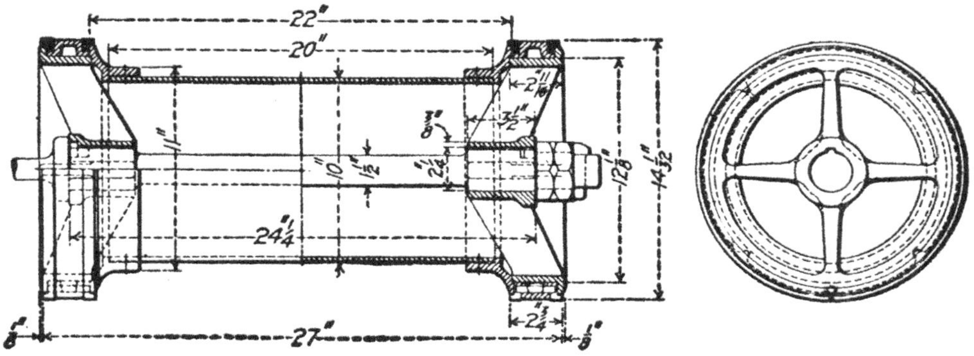

FIG. 337.

A detail of the valve is shown in Fig. 337. It is of built-up construction, with cast iron bull ring and L-shaped rings. These rings are inserted in the groove without springing, and therefore are not distorted, but fit the valve chamber, and are much tighter to prevent the loss of steam. The body is of wrought iron pipe for strength and lightness, the spiders of cast steel, the several parts being accurately turned to fit one another. The valve stem passes through the centre and is arranged with double nuts and a flat key, so that the different parts are securely held in place. In reality, this arrangement is lighter and better adapted to the purpose than a single valve cast in one piece. Only eight packing rings are used, while the usual arrangement of single valves is in reality three valves in one instead of two valves. These single valves require twelve packing rings.

Cross sections of the engine are shown in the next view, Fig. 338. An enlarged cross sectional view through the cylinders is given, showing the relation of the high pressure cross-heads to the engine truck and low pressure cylinders. Also the live and exhaust steam passages in the lower pressure cylinder.

Figure 339 is a reproduction from the working drawing of the low pressure cylinders. The next, Fig. 340, is a similar view of the high pressure cylinders. This engine was built in April, 1904, and was run on the New York Central for about six weeks, operating the fastest and heaviest trains, such as the Lake Shore Limited, Fast Mail, etc. It was afterward exhibited at the World's Fair at St. Louis, and tested in November on the Pennsylvania Railroad testing plant. Very careful calculations were made of the counterbalancing, and it was adjusted so as to produce a minimum amount of vertical effect on the track and nosing of the locomotive. During the test at St. Louis, 1,630 indicated horse-power, 1,470 dynamometer horse-power and 9,796 pounds draw-bar pull was developed continuously at 57 miles per hour, 50 per cent. cut-off; 1,641 indicated horse-power, 1,476 dynamometer horse-power and 9,896 pounds draw-bar pull at 57 miles per hour, 55 per cent. cut-off, was developed continuously for two hours; 1,369 indicated horse-power, 1,189 dynamometer horse-power and 6,788 pounds draw-bar pull at 66 miles per hour, 40 per cent. cuff-off, was developed continuously for one hour and a half; 1,336 indicated horse-power, 1,045 dynamometer horse-power and 5,224 pounds draw-bar pull at 75 miles per hour, at 40 per cent. cut-off, was developed continu-

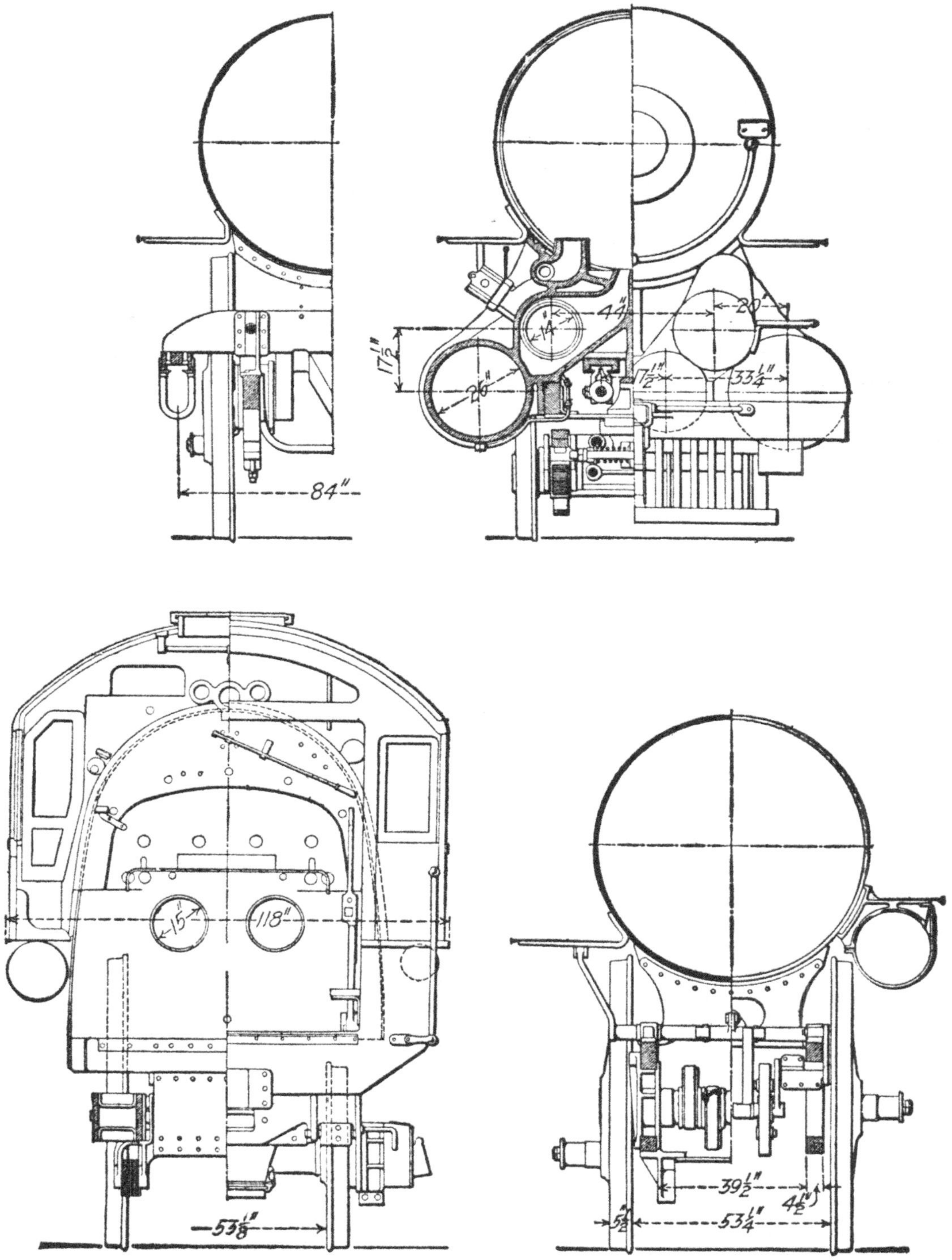

FIG. 338.

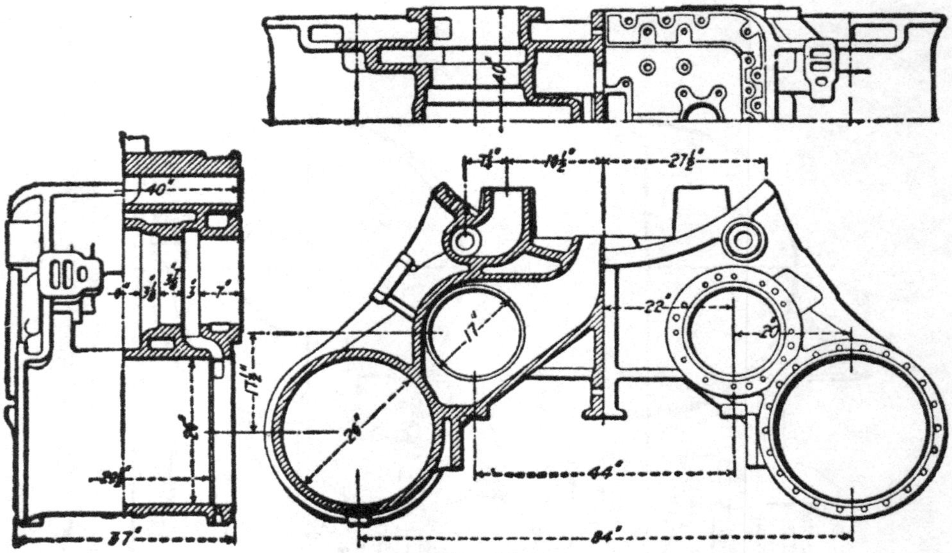

FIG. 339.

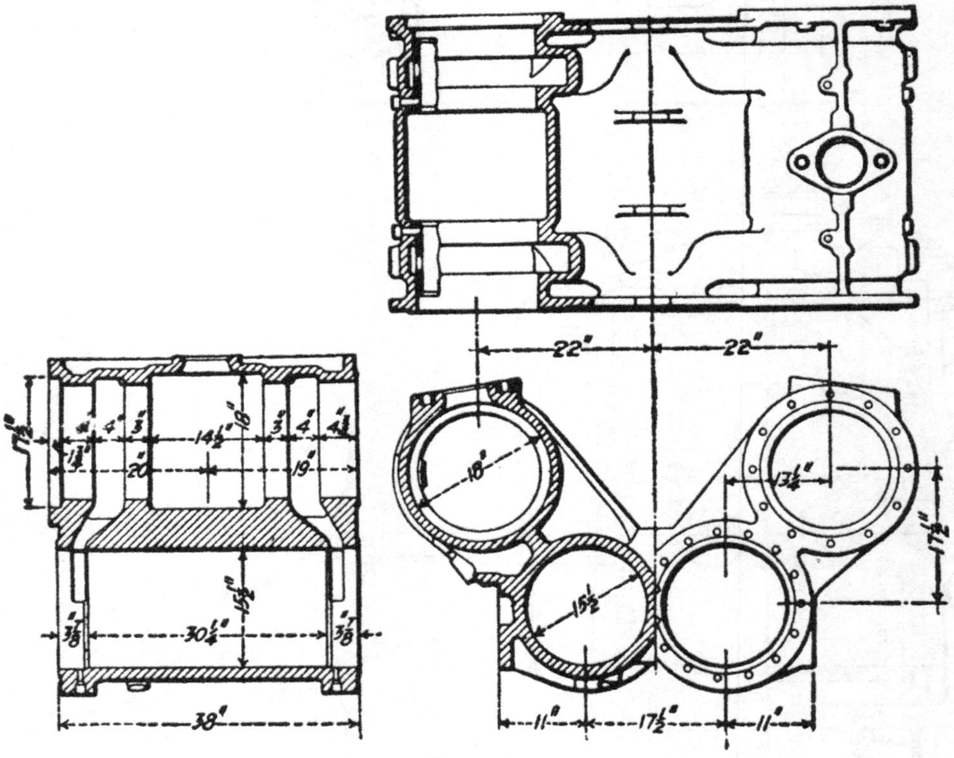

FIG. 340.

ously for one hour. This later test was run without any evidence of heating, and could have been readily sustained for a much longer period if required.

The next illustration, Fig. 341, shows the de Glehn four-cylinder balanced Atlantic passenger engine, Class 442 C 160, built for the Paris-Orleans Railway, similar to that purchased by the Pennsylvania Railroad and imported into this country by them. The Pennsylvania engine has been run on different parts of the Pennsylvania Railroad in regular service, was exhibited at the World's Fair in St. Louis, and tested on the Pennsylvania Railroad testing plant. It represents the latest development of the de Glehn engine and was not modified or changed in any essential features, with the exception of the tender and arrangement of air brake couplings, and a slight increase in the length of the cab. Otherwise it is exactly similar to the engines running on the Northern Railway of France. The principal dimensions are:

Cylinders	14 3-16 & 23⅝ x 25 3-16 in.
Drivers	80 3-16 "
Wheel base, driving	7 ft. ½ "
" " engine	28 ft. 6½ "
Weight on drivers	79,500 lbs.
" " truck	46,500 "
" " trailers	34,000 "
Total weight in working order	160,000 "
Heating surface, firebox	181 sq. ft.
" " tubes	2,435 " "
" " total	2,616 " "
Grate area	33.9 " "
Serve ribbed tubes	inside diam, 2 9-16 in., length 14 ft. 5¼ in.
Boiler pressure	227 lbs.
Maximum tractive power simple	18,270 "
" " " compound	19,527 "

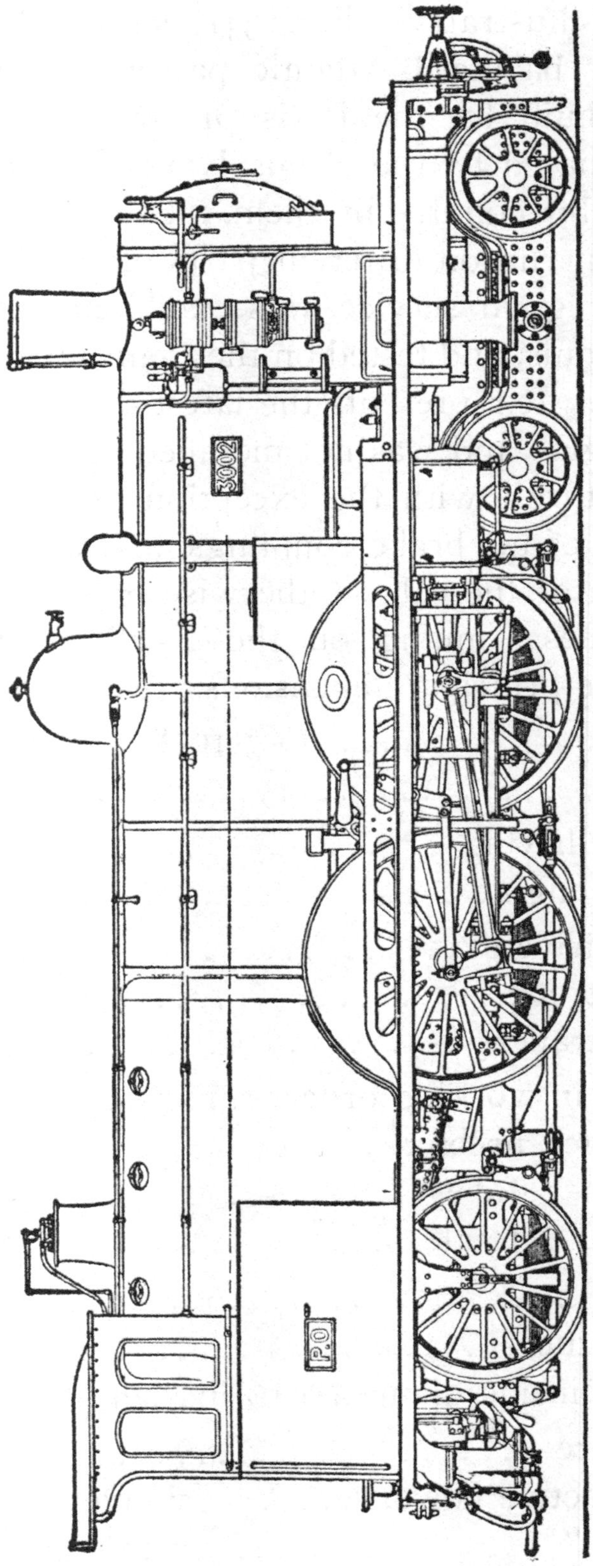

FIG. 341.

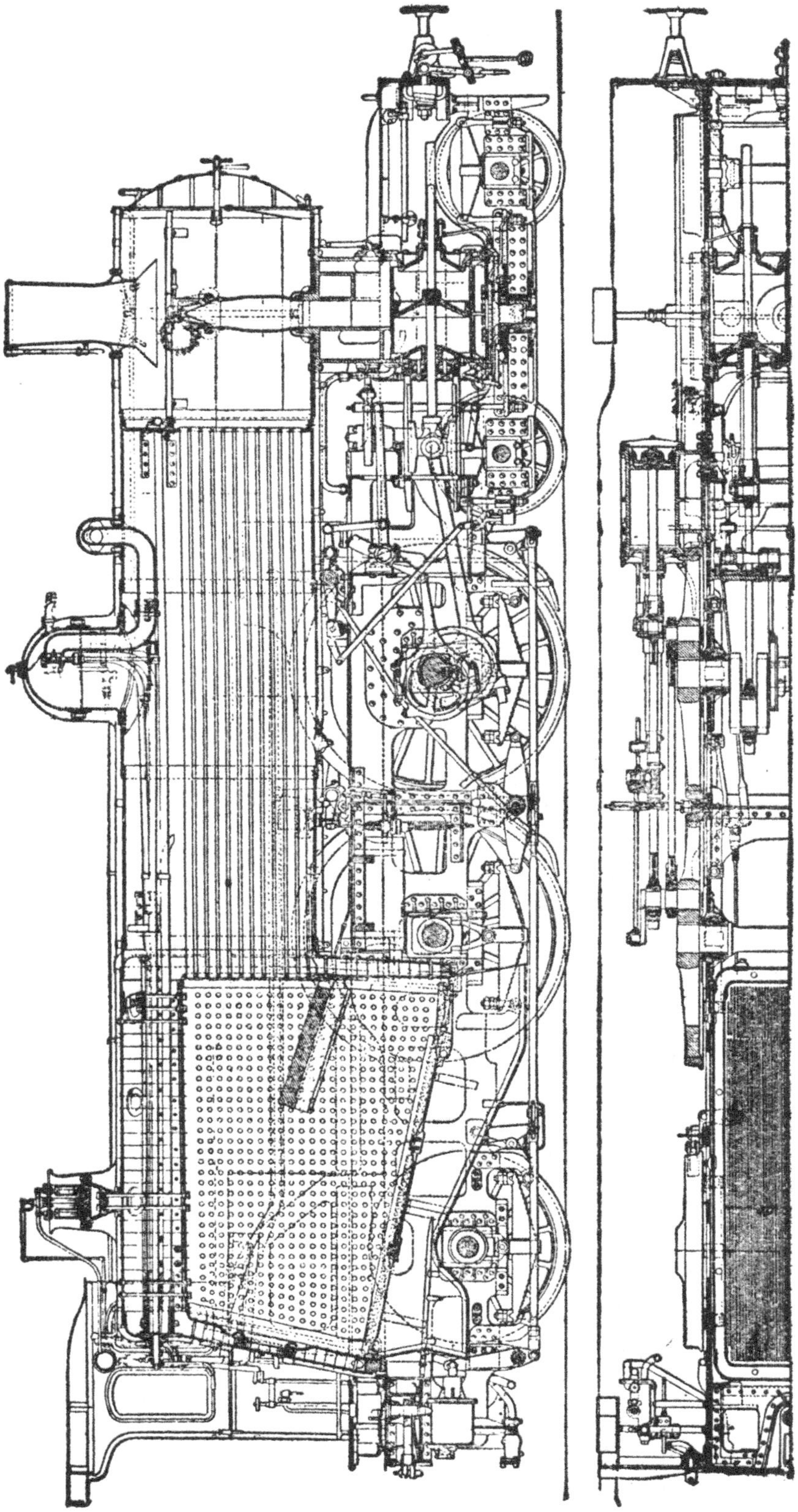

FIG. 342.

This is a well designed and finely finished locomotive, representing the latest and best French practice. Four separate sets of valve motion are used, one for each cylinder. The cut-off can be adjusted independently for both high and low pressure cylinders, the adjustment being made by a screw reverse gear which can be readily changed to alter the relative positions of the high and

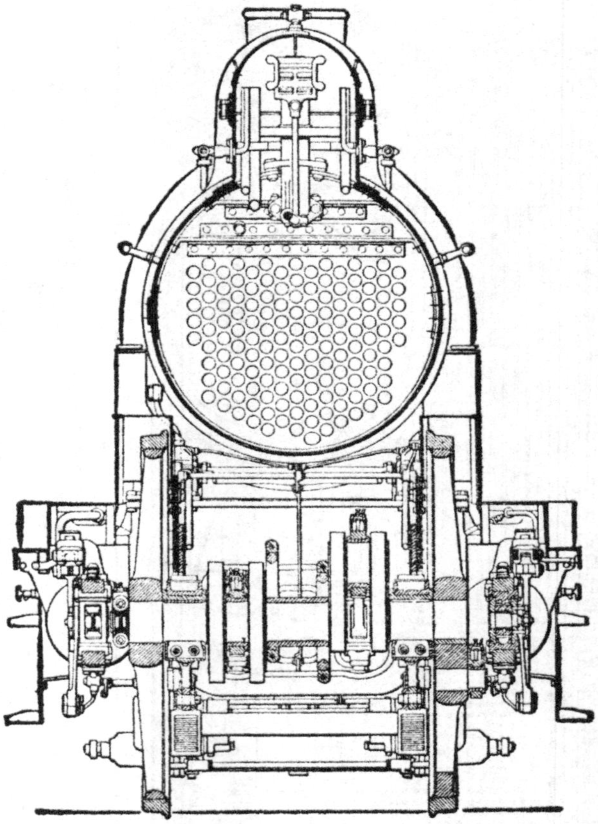

FIG. 343.

low pressure cut-off. A separate exhaust for starting and adjustable exhaust nozzle are used. The engine contains all the refinements for fuel-saving and efficient working, with the exception of the super-heater, which have been developed at the present time.

Attention is called to the Walschaert valve gear. Apart from the four-wheel leading truck there are but few American features on this engine. The usual European construction of plate frame, small cab, sand box in driving wheel covers, etc., are adhered to.

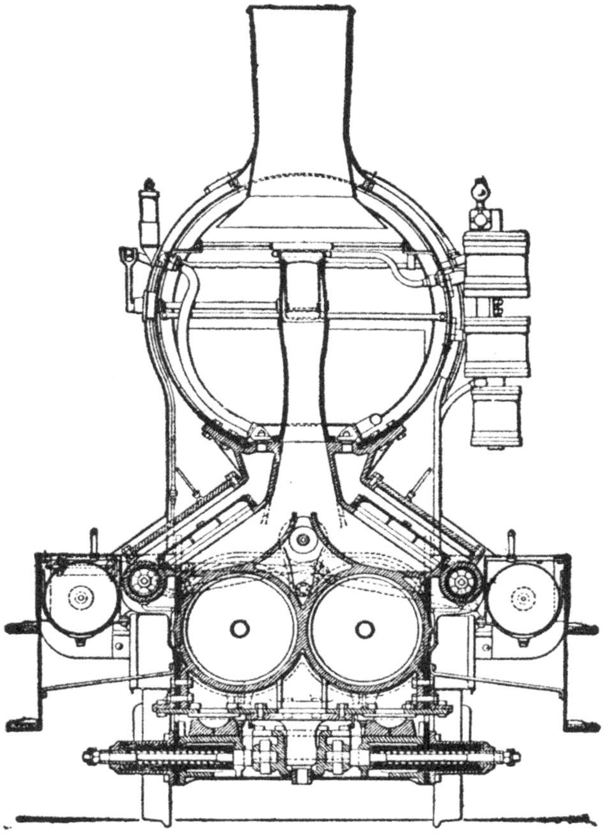

FIG. 344.

The next view, Fig. 342, is a longitudinal section and half plan. This shows clearly the arrangement of the cylinders and method of coupling the low pressure cylinders, which are inside the frames, and the forward driving axle. The high pressure cylinders are outside and placed close to the front driving wheels, with rods connected to the rear drivers. Mr. de Glehn lays particular stress on the advantages of subdividing the power and driving the two axles independently. The cross sectional arrangement of the cylinders is shown in Fig. 343.

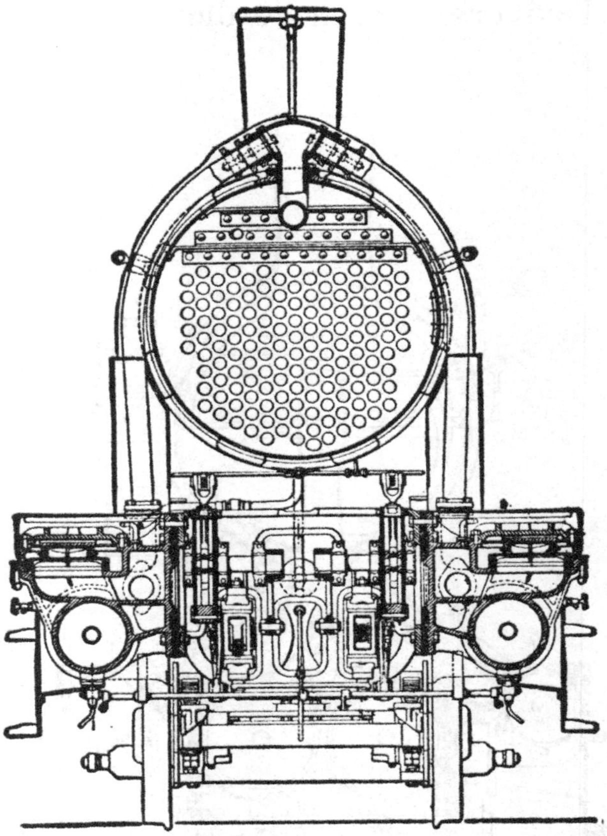

FIG. 345.

Next, Fig. 344, is a sectional view just ahead of the front driver and shows the crank axle and plate frames, and Fig. 345 a sectional view through the high pressure cylinders showing arrangement of steam pipes with high pressure pipe from the boiler. The illustrations of this engine are taken from the American Engineer and Railroad Journal.

The next illustration, Fig. 346, is taken from the Railway Age, and shows four different arrangements of four-cylinder balanced compound Atlantic engines. The one at the bottom is that of Von Borries, in which the four cylinders are in the same transverse plane, each cylinder having an independent valve, but connected in pairs. Walschaert gear is used, with an arrangement of rockers imparting motion to high and low pressure cylinders on

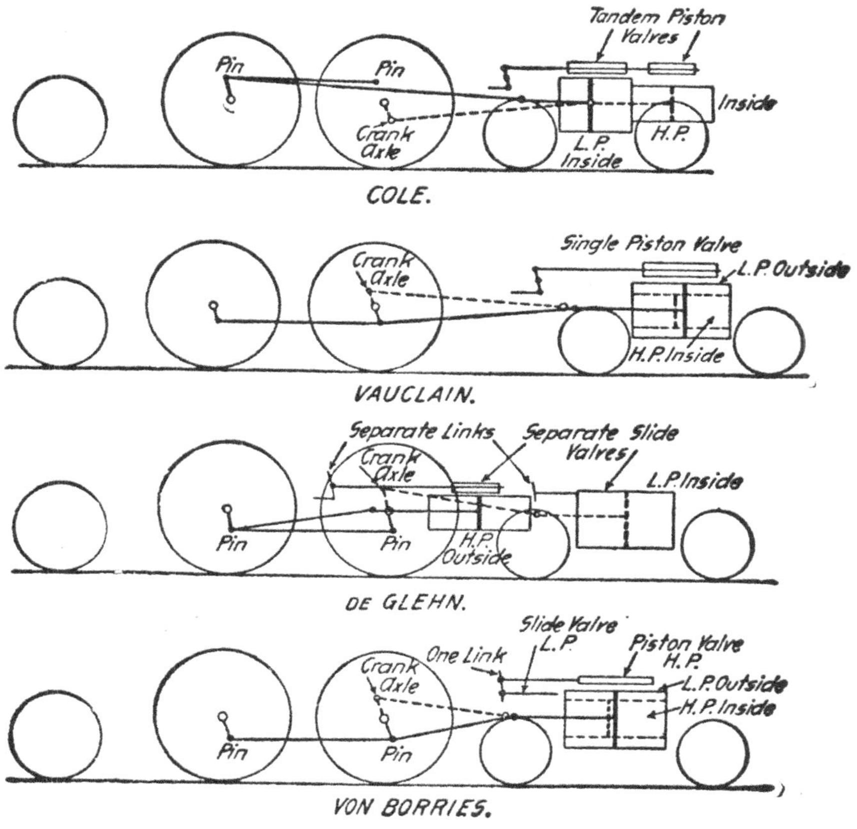

FIG. 346.

one side, giving a different cut-off to each cylinder, which is fixed and cannot be altered by the engineer. The high and low pressure rods are all connected to the first pair of driving wheels.

The next is the de Glehn system, with independent valves and valve gear, capable of adjustment by the engineer. The low pressure rods drive the forward drivers and the high pressure rods the rear drivers, the parallel rods being used merely to keep the wheels in proper relation with each other.

The next is the Baldwin arrangement of cylinders in the same transverse plane, one piston valve operating the low pressure cylinders on each side, driven by the ordinary Stephenson link motion, the high and low pressure rods being both connected to the crank axle. While this valve is in one casting, yet in reality it is three valves, twelve packing rings being used.

The next is the American Locomotive Company's design in which the high pressure cylinders are placed inside, the single valve motion and valve chamber being used to operate both cylinders on one side. While there are two distinct valves on one stem, yet only eight packing rings are used. The high pressure rods are connected to the forward drivers and the low pressure to the rear, as in the de Glehn.

The next view, Fig. 347, is an eight-coupled four-cylinder compound freight locomotive, Class 280 C 157, built for the Southern railways in France.

Cylinders	15⅜ & 23 19-32 x 25 19-32 in.
Drivers	55⅛ "
Total wheel base	23 ft. 9-16 "
Weight on drivers	142,120 lbs.
Total weight	157,520 "
Heating surface, tubes	2,586 sq. ft.
" " firebox	170 " "
" " total	2,756 " "
Grate surface	31 " "

The next, Fig. 348, is a longitudinal section of the same engine, showing the arrangement of guides for the inclined high pressure cylinder. The cylinders are in the

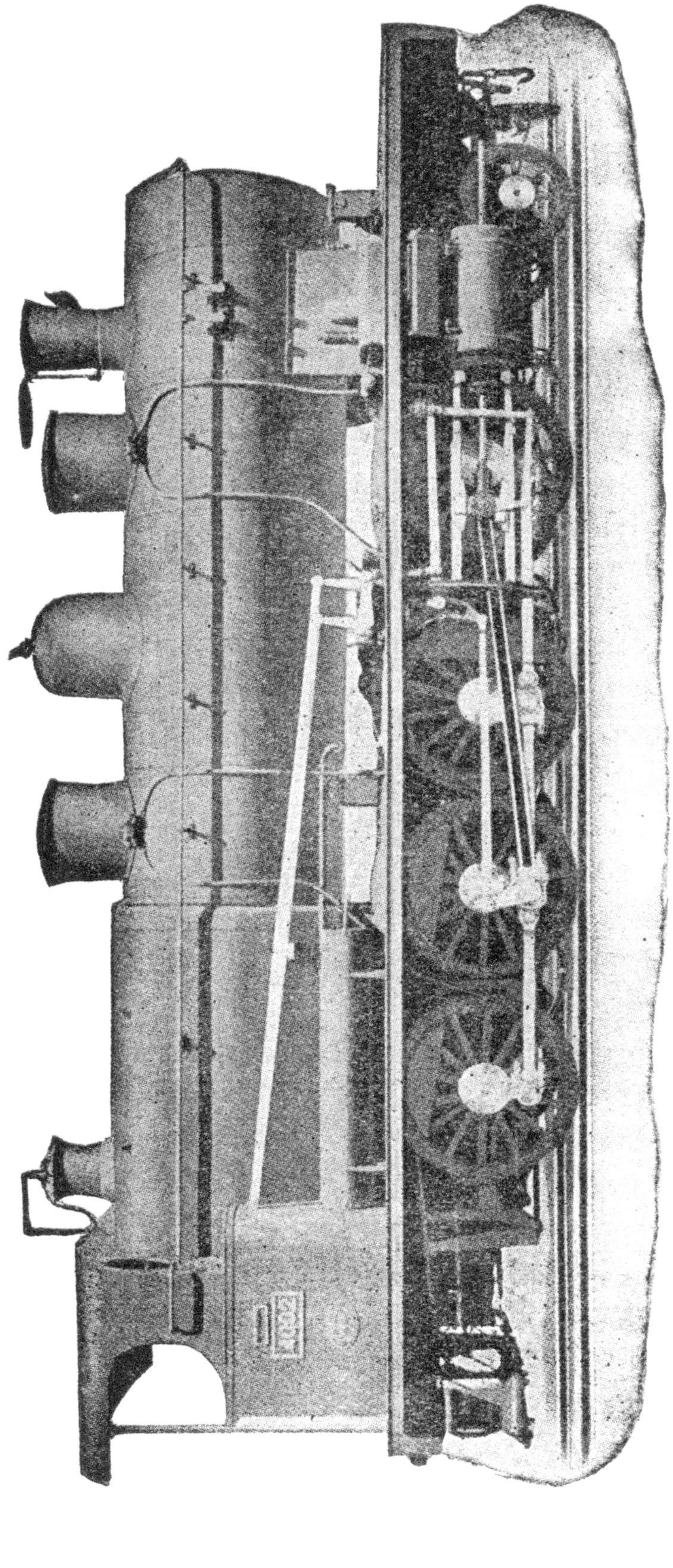

FIG. 347.

same transverse plane, the high pressure being inclined sufficiently so that the guides and rods clear the front driving axle, the second pair of driving wheels being arranged with crank axle. The low pressure cylinders are

FIG. 348.

outside and connected with the third pair of drivers. The Walschaert valve gear is used, as is customary on at least 80 per cent of European engines.

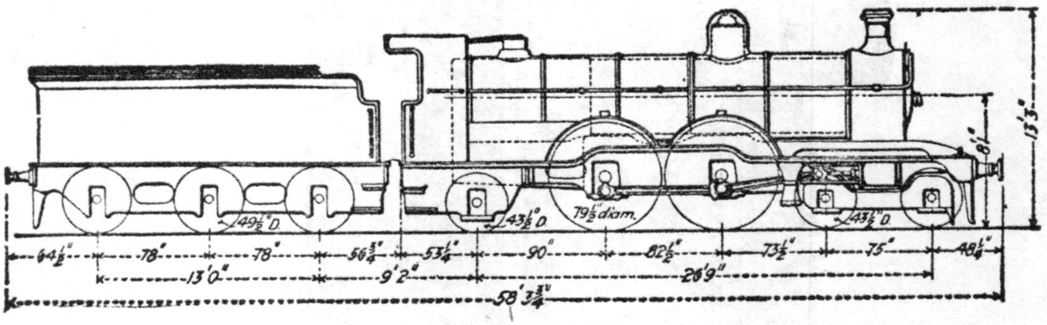

FIG. 349.

Figure 349 shows a four-cylinder simple locomotive, Class 442-134, built by the Great Northern Railway of England, and the principal peculiarity of this engine is the four simple cylinders. Designers have usually considered that four-cylinder engines of the balanced type are essentially a compound proposition and regard the use of simple cylinders as very questionable practice. The fact that very few engines of this description have been built would indicate that in comparison with the compound cylinders it has not proved as successful as the builders might have anticipated. The principal dimensions are:

Cylinder	15 x 20 in.
Driving wheels	79½ "
Driving wheel base	82½ "
Total wheel base of engine	26 ft. 9 "
Weight in working order	134,400 lbs.
Heating surface, firebox	140 sq. ft.
" " tubes	1,160 " "
" " total	1,302 " "

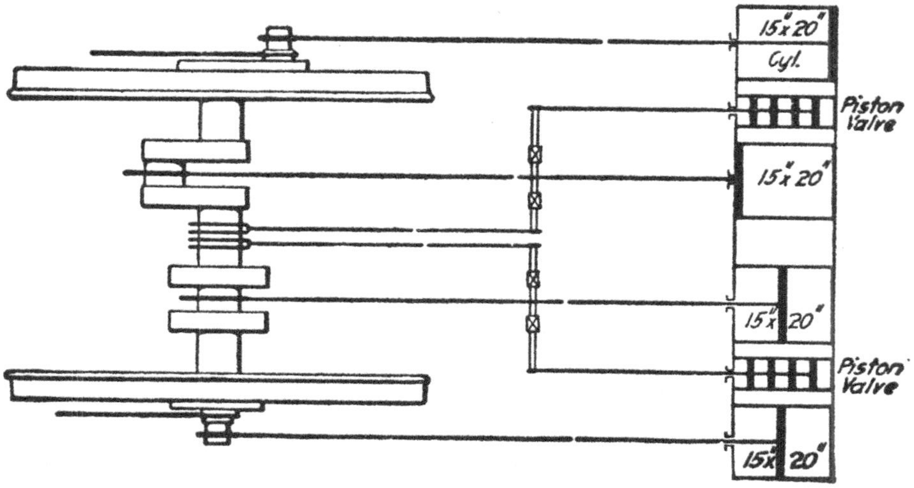

FIG. 350.

The illustrations, Figs. 349 and 350, and descriptions are taken from the Railway Age.

Figure 351 is a four-cylinder compound Atlantic passenger, Class 442 C 146, built by Mr. de Glehn for the Great Western Railway of England. It has been illustrated widely in the technical papers, and is interesting

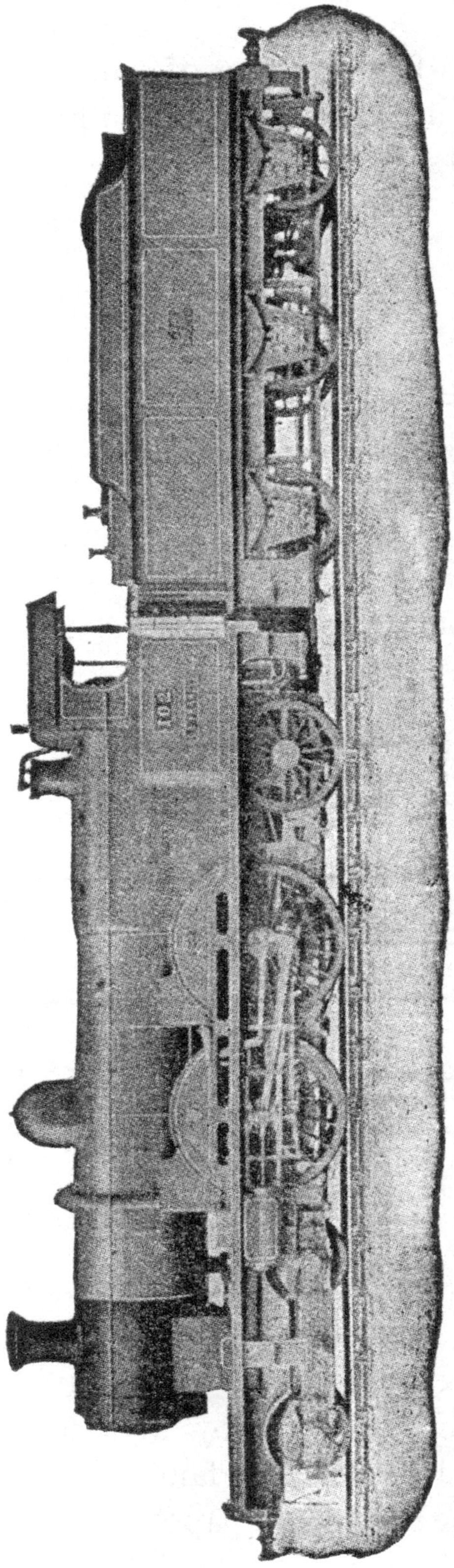

FIG. 351.

particularly from the fact that owing to the excellent results obtained in France from these engines a prominent English railroad is willing and anxious to try a similar engine in direct comparison with its own build. The principal dimensions are as follows:

Cylinders	13⅜ & 22 1-16 x 25 3-16 in.
Drivers	80½ "
Total wheel base of engine	27 ft. 10⅝ "
Weight on drivers	34½ tons
Total weight in working order	65 "
Heating surface, firebox	167 sq. ft.
" " tubes	2,158 " "
Total heating surface	2,325 " "
Boiler tubes	126 of the Serve pattern
" " internal diameter	2 9-16 in.

You will notice all the European characteristics, such as sand boxes in the wheel covers, plate frames and Walschaert valve gear. The outside journal boxes on the engine trucks are rather an unusual design for a locomotive, but present some advantages from accessibility for oiling the bearings and repacking. I have not been able to obtain any accounts of tests of this engine in comparison with English engines.

To apply the four-cylinder balanced principle to 442 Atlantic, 462 Pacific, 460 ten-wheeler, 480 twelve-wheeler types of American engines requires either the high pressure cylinders to be located ahead of the low, with only a slight increase in wheel base or length of boiler, or an increase of 30 to 36 in. in the normal length of engine between the front driver and the cylinder centre. This also involves increasing the boiler and flues a like amount.

When applied to a locomotive having a two-wheeled leading truck, three different methods are available:

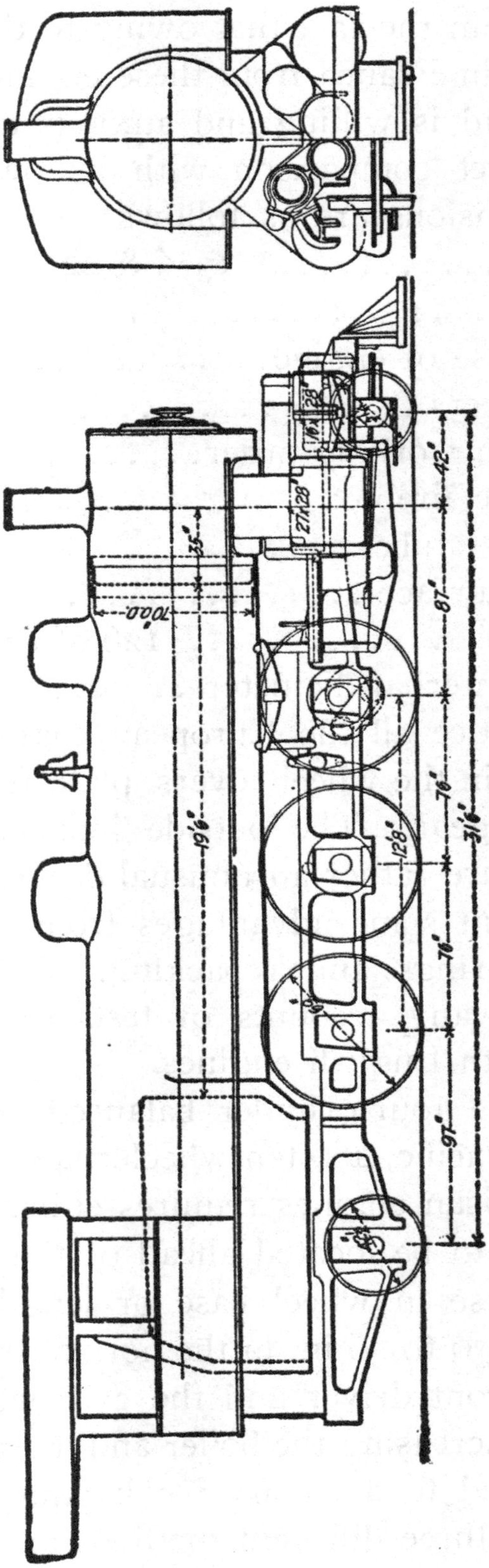

FIG. 352.

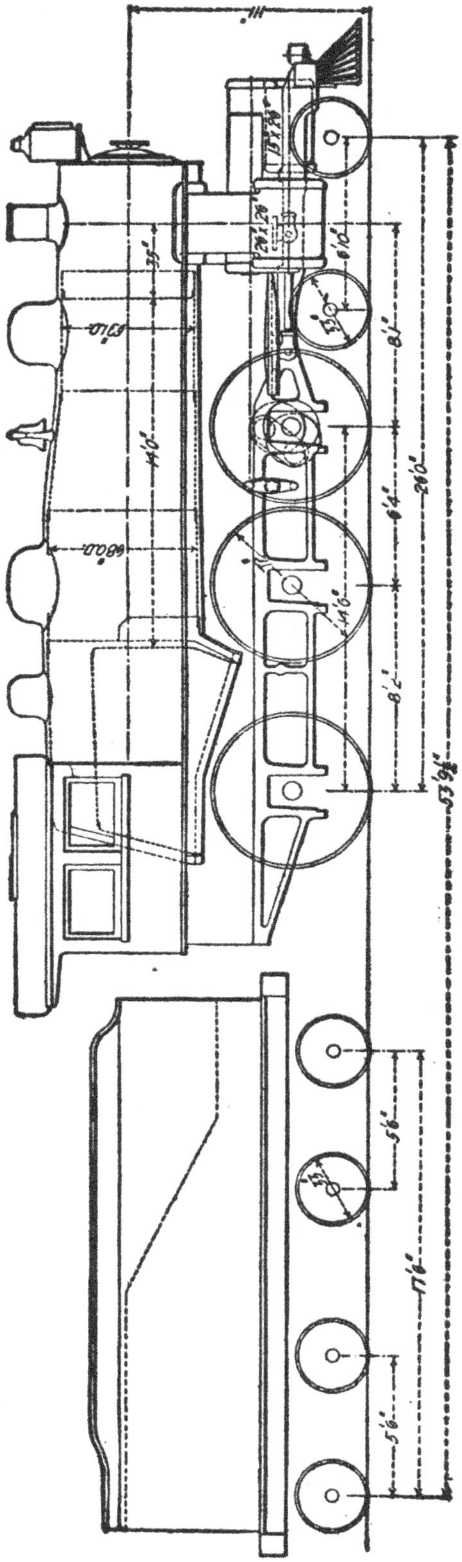

FIG. 353.

(*a*) Locating the high pressure cylinders ahead and increasing the distance between front driver and centre of cylinders about 24 in.

(*b*) Inclining the high pressure cylinders to clear the front driving axle.

(*c*) Looping the high pressure main rods so as to encircle the front driving axle.

These arrangements are named in the order of their practical availability. In considering the advantages of balanced engines, it is well to bear in mind that their use is by no means confined to high speed passenger service. The destructive effect on the track of excess weights in driving wheels is a matter of revolutions and not necessarily of speed. The dynamic effect of a medium-sized wheel running at a high speed in revolutions, while making a moderate number of miles per hour, is just as great as a large wheel, at the same number of revolutions, making a much higher speed in miles per hour. Furthermore, the necessity for subdivision of power in a very heavy fast freight engine is more urgent. The probable decrease in the repairs of the lighter parts will go far toward paying for the first cost and maintenance of the extra number of moving parts.

Figure 352 shows a preliminary design of Pacific engine, 462 type, arranged as a balanced compound with the crank axle at the forward drivers. The high pressure cylinders are ahead of the low. No increase has been made in the length of the boiler or flues.

A similar arrangement of balanced compound is shown in Fig. 353, applied to a ten-wheel passenger engine, 460 type.

Figure 354 shows the principle applied to a twelve-wheel freight engine, 480 type, the forward drivers being

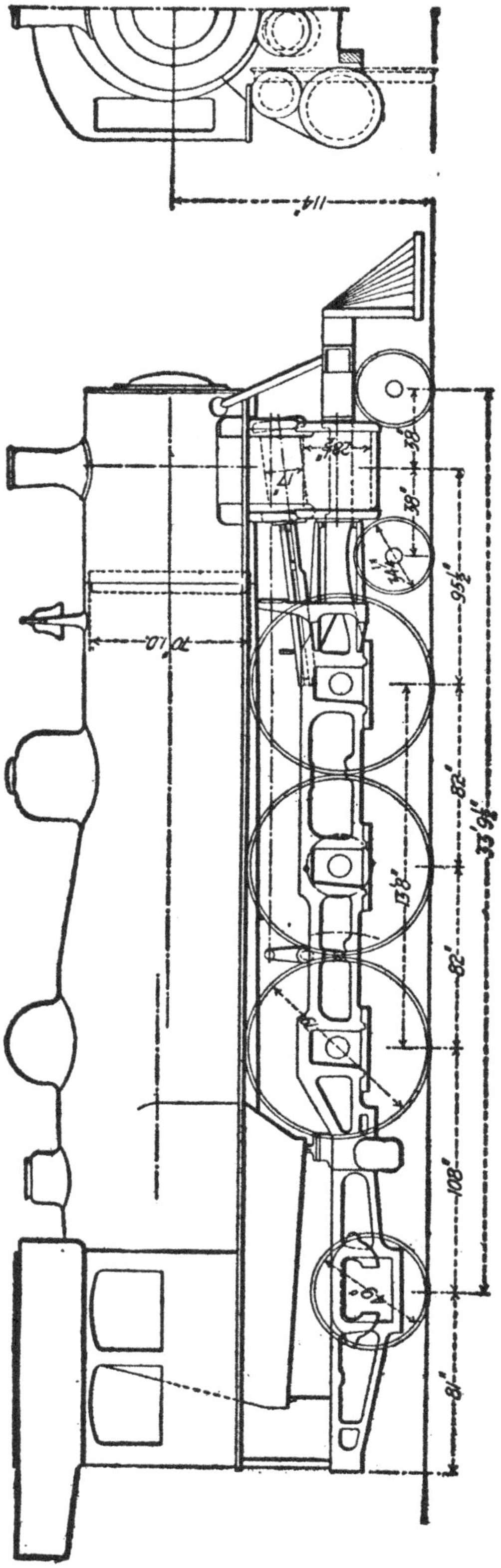

FIG. 354.

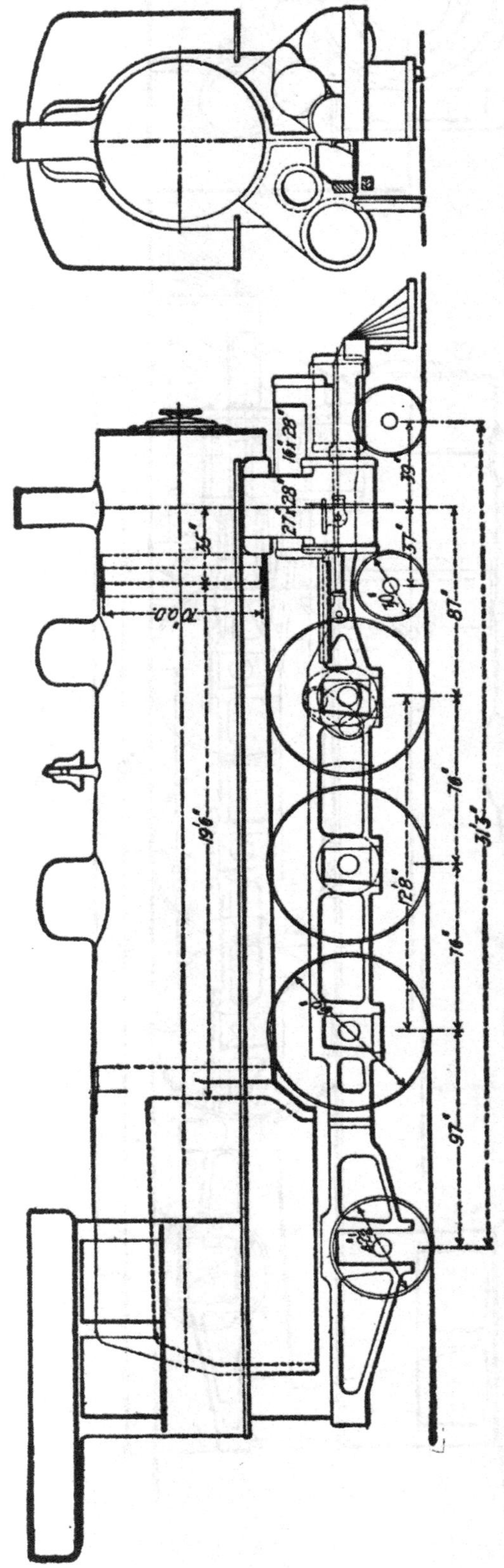

FIG. 355.

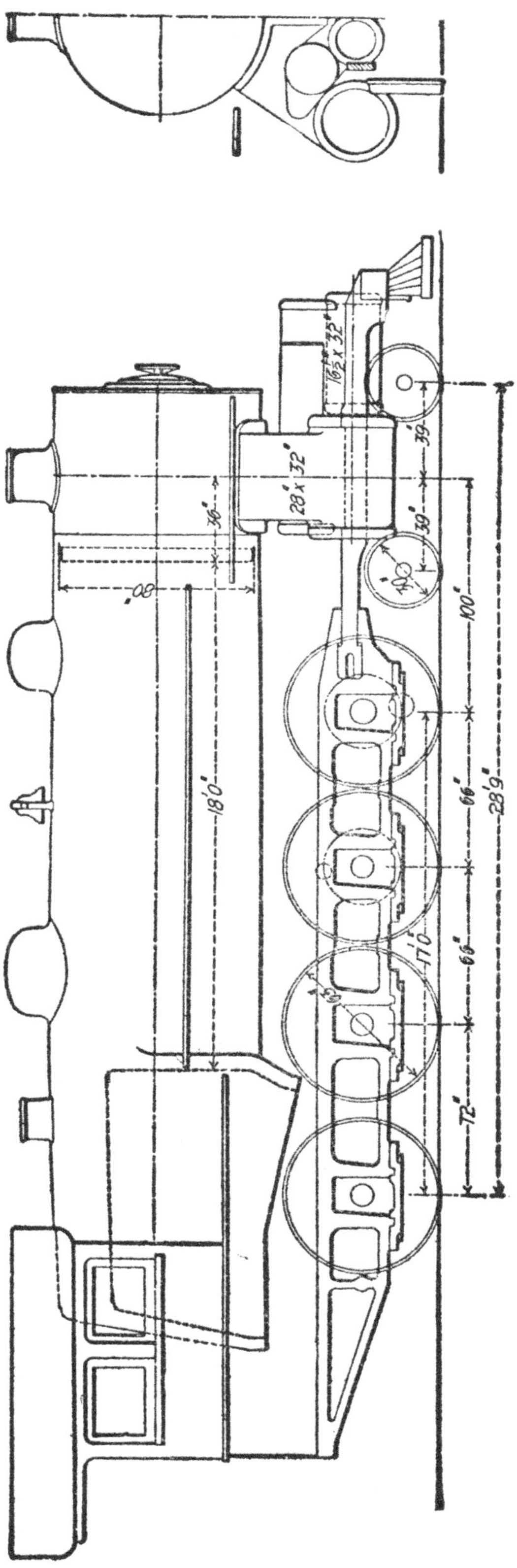

FIG. 356.

provided with crank axle and the high pressure cylinders ahead of the low, as in previous designs.

Figure 355 shows the application to a Prairie passenger engine, 262 type. In this design it is necessary to use a forward truck with outside bearings, so as not to interfere with the high pressure cylinders. The distance from the front driver to the centre of cylinders is increased about 23 in. over the normal distance required for simple cylinders.

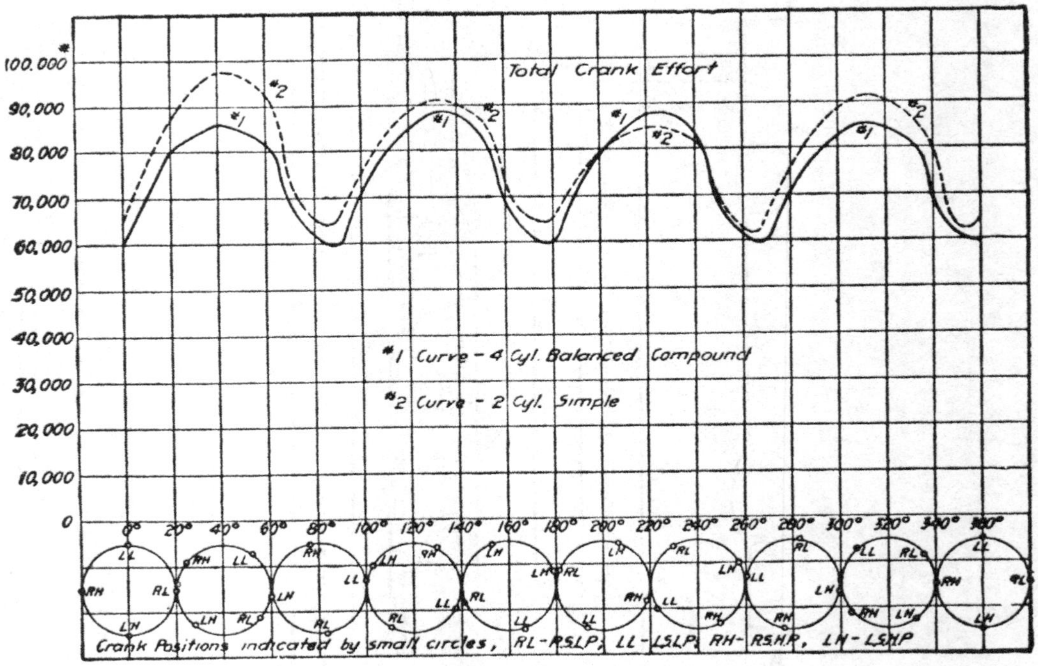

FIG. 357.

Figure 356 shows a Pacific 462 type, passenger engine, in which the high and low pressure cylinders are in the same transverse plane, the inside high pressure cylinders being inclined so that the main rod and guides are over the forward axle. In this design the middle or main wheels are arranged with a crank axle and the high and low pressure rods connected on this pair of drivers. The eccentrics are placed on the rear driving axle with an ordinary reversing rocker between the main and rear pairs of drivers, the connection to the valve being made by a long valve stem.

The foregoing illustrations show some of the applications of four-cylinder balanced compounds to different types of locomotives, illustrating its wide application to a variety of service. Fig. 357 shows the crank effort diagram comparing the turning moments at the crank pin of a four-cylinder balanced compound Atlantic locomotive, with a simple engine of similar tractive power. Curve No. 1, of solid lines, show the great uniformity of the turning moment of the balanced compound. Curve No. 2, of dotted lines, shows the two-cylinder simple engine. It will be noticed from this the greatest turning moment and consequently tendency to slip is when both cranks are forward of the centre of the wheel at an angle of 45° from the horizontal.

THE OHIO LOCOMOTIVE INJECTOR.

Type.—The Ohio Injector is a simple constructed lifting injector. It lifts and forces the water with one set of tubes placed in a center line through the body of the injector.

Description.—As shown in the sectional cut are the following parts:

1—Body (back part).
2—Body (front part).
3—Delivery end connection.
4—Steam valve hub complete.
5—Steam valve and primer complete.
6—Steam nozzle.

7—Lifting tube.
8—Combining tube.
9—Delivery tube.
10—Line check valve.
11—Stop ring.
12—Overflow valve complete.
13—Water valve complete.
14—Starting lever.
15—Overflow nozzle.

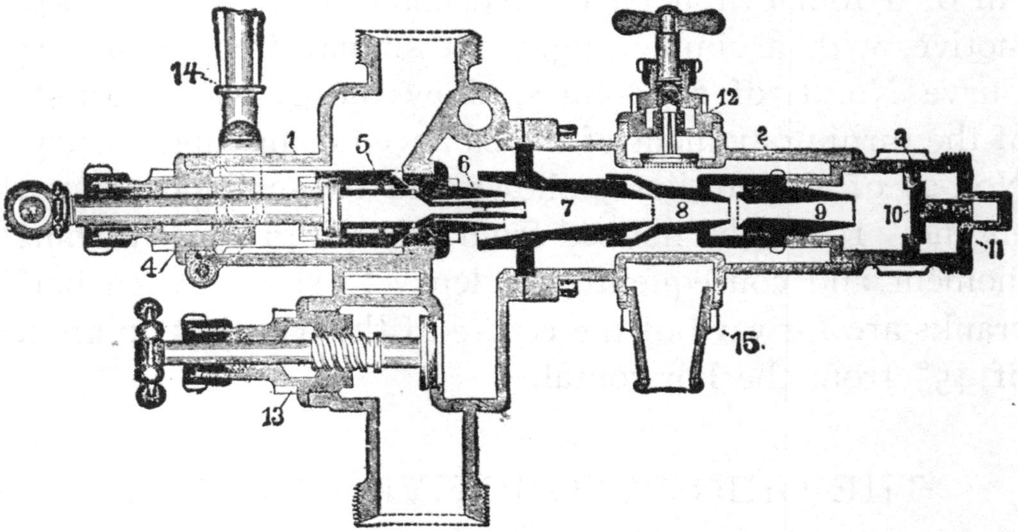

OHIO LOCOMOTIVE INJECTOR.

The steam pipe leading from the dry steam space of the boiler is connected to the top of the injector above steam valve 5. The water or suction pipe is connected to the bottom of the injector below water valve 13. The branch or delivery pipe is coupled to the delivery end connection in front of line check valve 10, and stop ring 11. The overflow nozzle 15 should be fitted with a drain pipe leading to some convenient place outside of the ash pan.

To Operate.—First pull the starting lever 14 back until the resistance of primer is felt. As soon as full volume of water appears at the overflow, pull starting lever back as far as it will go.

Regulate the quantity of water needed by the water valve. When operating the injector the steam pressure should be turned full into the steam pipe, as this governs the working of the injector.

When the steam valve 5 is in priming position, the inside priming nozzle is seated against the steam nozzle 6; this permits the steam to pass into the priming nozzle through the small ports, back of the seat and to pass out through the contracted opening at the point of priming nozzle. As the steam leaves the priming nozzle it passes into the lifting tube 7; from lifting tube 7 the steam passes into combining tube 8, from which it passes through spill holes to the chamber surrounding the tubes. From this chamber the steam pases out through overflow valve 12 and nozzle 15 into the atmosphere.

The rush of steam passing from the point of the priming nozzle into lifting tube 7, causes a partial vacuum to be formed in the chamber in front of water valve 13. The partial vacuum acting in combination with the atmospheric pressure on the water in the tank causes the water to rush up into the injector body and follow the rush of steam out of the overflow to the atmosphere. The starting lever 14 is pulled full back. The full jet of steam traveling at a high velocity mingles with the water and is condensed by it; the resultant mixture then has a velocity and pressure greater than the boiler pressure, which causes the water to be forced into the boiler.

To use as heater; close the overflow valve 12, and pull lever 14 back until the resistance of the primer is felt and let stand at this point. Overflow valve 12 must not be closed except when injector is to be used as a heater.

Important points in applying and operating the Ohio Injector. The pipes should be the specified sizes for each size of injector. Keep the tank valve open wide; keep all joints tight; keep hose strainers clean; keep packing in good condition.

Working pressure and capacity. The Ohio Injector works under steam pressure from 250 lbs. down to 25 pounds. The capacities are regulated according to size.

CHAPTER XIV

MECHANICAL STOKERS.

Several vital conditions have caused the general demand for mechanical locomotive stokers.

During the last ten years the railroads have been obliged to double the capacity of their locomotives to handle the large increase in traffic and meet the demand for faster schedules.

This naturally resulted in the proportionate increase in the labor of the firemen, making it more difficult every year to get good men.

The consumption of coal has also increased enormously, making the fuel account one of the largest operating expenses.

Consequently, the leading officials have sought the most practical and efficient way to reduce the coal consumption, and at the same time make the fireman's position attractive to the right grade of men, not only to secure the best results as firemen, but also to insure the supply of the right class of firemen to become high grade engineers.

The natural and practical way of reducing the expenses and overcoming difficulties due to hand firing is to *use mechanical stokers,* and consequently a large number of men have attempted to produce successful locomotive stokers, but with few exceptions they have all failed in some vital respects.

The Locomotive Stoker is a successful, simple, durable and easily operated, steam driven mechanical device that throws the coal into the fire box of locomotives and spreads it uniformly in small quantities over the entire grate area or where needed when fire burns unevenly.

The Stoker is of the self-contained, horizontal, direct acting, steam-driven, reciprocating plunger type.

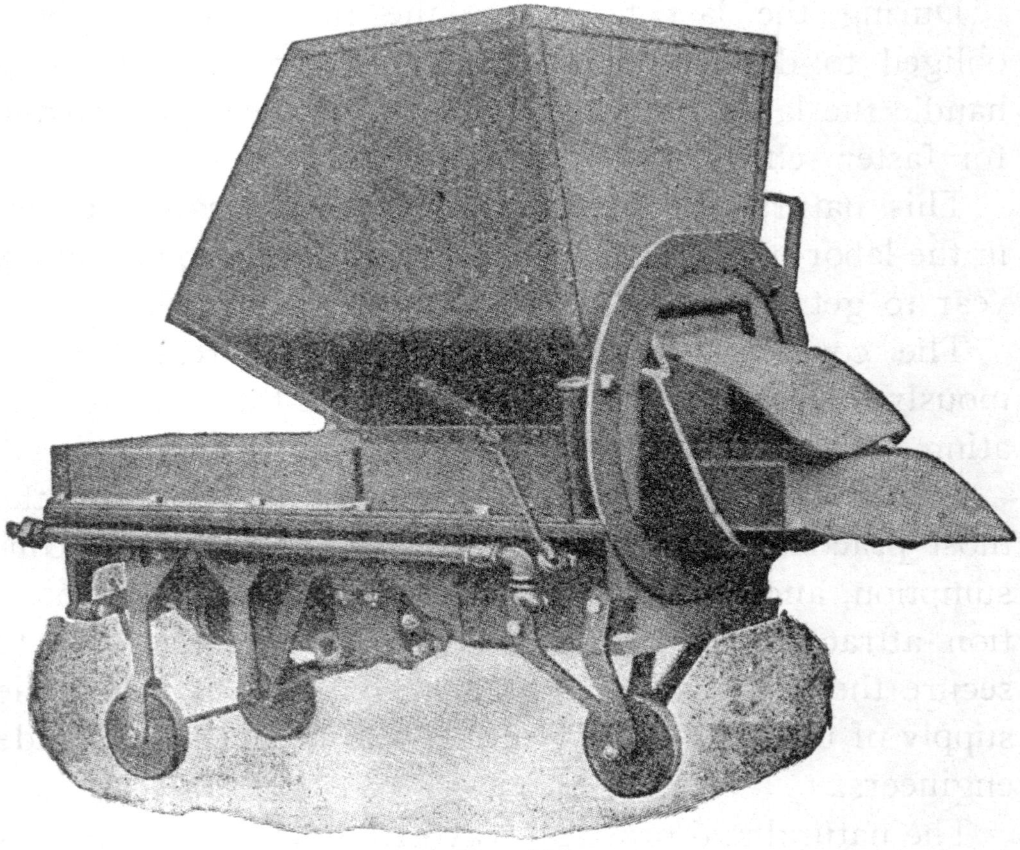

Fig. 358.

PRINCIPAL PARTS—The principal parts of the Stoker are:

The detachable coal hopper for receiving the coal and delivering it in front of the plunger.

The Plunger or spreader for throwing the coal into the fire box and spreading it with the assistance of the special swinging fire door.

The fire door located at fire box opening, in front of the plunger opens inwardly, being hinged at the top to the special fire door ring (furnished with the stoker), which is hung on the regular hinges from which the ordinary fire door has been removed.

OPERATING MECHANISM.

The horizontal steam cylinder, cushion valves, steam chest and valve motion with operating levers complete.

The steam cylinder contains the usual piston secured to the piston rod with the other end of rod secured to the underside of the plunger for operating same.

Main Frame. A rectangular main frame mounted on legs having castors for easy handling.

The Frame contains or supports all parts of the stoker except the special fire door and ring as before mentioned.

The stoker is easily and quickly attached or detached by means of two slotted lugs provided in the fire door ring and two keys for insertion behind recesses provided in the front of the stoker frame.

The Locomotive Stoker is started by simply moving the throttle lever forward and stopped by moving it back.

The steam pressure is easily governed by an adjustable valve in the steam supply pipe.

The length of stroke is controlled by the short hand lever and quadrant located outside of the throttle lever.

To use the stoker the fireman simply fills the hopper, then by the throttle and stroke lever governs the quantity of coal thrown into the fire box and the distribution of same.

Fig. 359.

ADVANTAGES.

1. They are very *simple,* having only a few standard, simple parts.

2. They are very *durable,* being made of cast steel, wrought steel and brass.

3. They are *easily operated* by any experienced locomotive fireman, no previous experience with mechanical stokers being required.

4. They *spread the coal uniformly* over the entire grate surface, *or where needed, if fire burns unevenly.*

5. They *feed in small quantities* of about one-fifth to one scoop per stroke.

6. They *handle any kind or size of coal,* suitable for proper combustion: Bituminous or Anthracite.

7. They *keep the cold air out of the fire box,* saving repairs and producing better combustion.

8. They *keep the fire clean, bright and uniform,* preventing clinkers and allowing much longer runs to be made without pulling the fire.

9. They *keep the steam pressure up and more uniform* than with hand firing, producing all the steam that the engine requires.

10. They do not take the place of the firemen, but *reduce their work and assist them in properly firing the engines under all conditions.*

11. They *save considerable fuel,* and because of the better combustion, also reduce the quantity of smoke produced.

12. They are *easily, cheaply and quickly attached or detached.*

13. They *do not interfere with the easy inspection of the fire.*

14. They *fit nearly every type of engine having single fire door.*

15. They are *self-contained* having no attachments.

16. They *do not cause steam to enter the fire box.*

17. The *stoker fire door swings entirely clear* without removing stoker.

18. No parts of stoker are fixed inside the fire box to warp out of shape.

CHAPTER XV

RAILWAY SIGNALING.

The signal service of railways has progressed in equal ratio with that of other branches, the motive power department not excepted. Indeed, without the progress made in the signaling department the high speeds now attained would have been impracticable, notwithstanding powerful locomotives and improved air brakes.

Perfection in signaling insures to a great extent safety of life and property, and the efforts of railway companies to get this branch as nearly perfect as may be are constant. This is natural. For upon it depends not only the safety of passengers and employes, but also the easy and economical operation of trains.

Formerly each railway company had its own particular signal system, usually quite different from those used by other companies, and to some extent this still obtains, although, like other things are learned, experience is rapidly teaching the advisability of greater uniformity.

THE INTERLOCKING SYSTEM.

In the early days of railroading in England the necessity of concentrating in one cabin the control of a number of switches, and the signals governing them, became apparent and the result was a combination of several levers, each controlling a switch, and its corresponding signal, the whole arrangement to be worked by one or two men, and the movements of trains were as a consequence greatly facilitated.

But this combination of half a dozen or more switch levers in one frame, to be controlled by one man, brought to the front a new danger, viz.: the attendant might carelessly give all clear signals, at the same time to two trains moving in different directions, and this might cause the trains to proceed over routes crossing one another.

The function of all fixed signals, no matter by what name they are called, is to inform the engineer whether or not he may run past the signal, and if he is allowed to pass, on what conditions he may proceed.

The engineer depends entirely upon the signal, and is guided in handling his train by the position of the signal before him.

Wherefore, as before stated, should the signal man through carelessness or otherwise give all clear signals simultaneously to two trains moving in opposite directions, bad consequences would result.

It was to overcome this danger that the interlocking system was devised.

In these days of large terminals, having numerous tracks, concentration of levers for operating switches and signals is a prime requisite, and interlocking is one of the essential factors of concentration.

Interlocked signals are operated from an elevated cabin by rods or wires, supported near the ground at regular intervals by ground rollers.

Two wires are used as a rule, the second or return wire being added in order to make sure of the prompt return of the signal to the normal or stop position, although the counterweight alone should perform this function.

The universal standard signal for switches, crossings, drawbridges, etc., is the well-known semaphore. At some drawbridges and crossings of one railroad with another, where interlocking has not been introduced, various balls, targets, gates, and other old-time devices are still in use.

For slow movements such, for instance, as those to and from side tracks, dwarf signals about four feet high are used.

The arrangement of interlocked switches, and signals is the same no matter whether the yard be simple or complex.

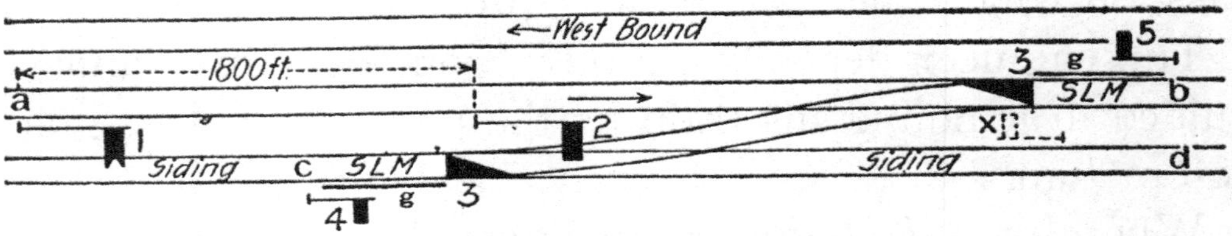

Fig. 42.—Interlocked Switches and Signals

LOCKING SHEET

1 locks 2 reversed.
2 locks 3 normal 5 normal.
3 locks ———
4 locks 3 normal and reversed (when 3 is reversed locks 5 normal)
5 locks 3 normal and reversed.

Fig. 360.

Fig. 360 is a diagram showing the principle of the interlocking system. This cut shows a cross-over track connecting the siding "cd' with the east bound main line "ab."

The switches are in their normal position, allowing trains to pass from *a* to *b* and from *c* to *d;* and the signals are all in their normal positions, each indicating to an approaching engineman that he must stop before he passes the signal. Signals 1 and 2, which are for high speed movements, are on tall posts, while signals 4 and 5, being for low speed movements, are dwarfs. For complete signaling there would be a dwarf signal at x for movements from *d* to *c;* but to simplify the illustration that signal is here omitted. To permit a train to pass from *a* to *b* the signalman pulls ("reverses") lever 2, moving the arm of signal 2 to the all-clear or go-ahead position (60 degrees to 75 degrees downward), and the interlocking of the levers in the cabin prevents him from doing this unless the switch in the main track is in the right position. If the switch has been moved to the other position the lever of 2 is immovable. If signal 5 has been cleared for a movement in the other direction (from *b* to *a*) 2 cannot be cleared. Signal 1, when horizontal, does not stop a train, but simply gives notice to the engineman that he must expect to be stopped at 2. After 2 has been cleared, then 1 can be cleared, thus giving notice to the engineman that he can pass 2 at full speed. If, in the 1,800 feet between these signals, there is a curve, or if there should be a dense fog, shortening the view, this distant signal would thus appreciably economize the time of the train.

Until recently the usual practice has required a signal with two arms at 4 and the same at 5, the upper arm to be cleared for a movement along the straight line, and a lower arm for a movement through the crossover. With the practice as shown here, a single arm indicating a clear track, but not showing which of the diverging tracks is clear, the engineman is required to keep his speed sufficiently low to enable him to discover, from the position of the switch rails, which line he is traversing, in season to avoid any possible collision with a preceding train or car. For high speed movements over diverging routes, the two-arm arrangement is used.

In this drawing the two switches marked 3, and the two locks which lock them, together with the detector bars, which move simultaneously with the switch-and-lock movement, are worked by a single lever, though in many cases a separate lever is used for the lock. The detector bars, the position of which is indicated by the lines *g g*, are to prevent the signalman from inadvertently throwing a switch while a train is passing over it.

A machine consists principally of as many levers placed side by side, on a common shaft, as are necessary to move the switches and signals. Two switches may be moved

Fig. 361.

by a single lever; and two or more signal arms on the same post, or near together, may be so arranged that either one will be moved by the same lever, the position of the switch connections being made to govern the selection of the arm to be moved.

The main function of the interlocking feature of the machine is to prevent the display of conflicting signals.

Fig. 361 shows the Saxby and Farmer interlocking machine and Figs. 362, 363 and 364 illustrate the construction of the main working parts of this machine. Fig. 362 shows a rear view of part of the machine.

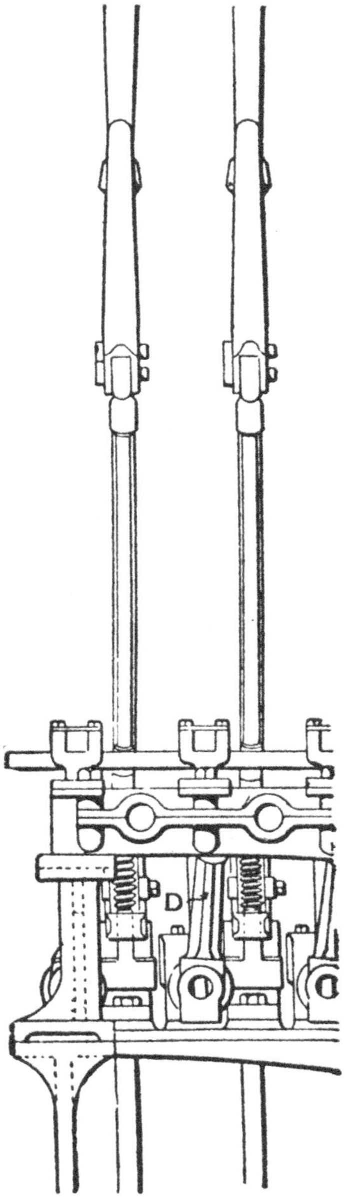

Fig. 362.

Fig. 363 shows a side elevation of one lever, and the interlocking frame partly in elevation and partly in section. Fig. 364 a plan view. The principal parts are the lever A, lever latch L, rocker B, segment C, link D, locking shaft E, and locking bars FF.

Each lever in a machine has a corresponding locking bar F. Other parts of the machine are G driver for bar, H locking dogs fixed to the tops of bars, K cross locks, sliding in grooves in a plane with the locking dogs H and u tappet.

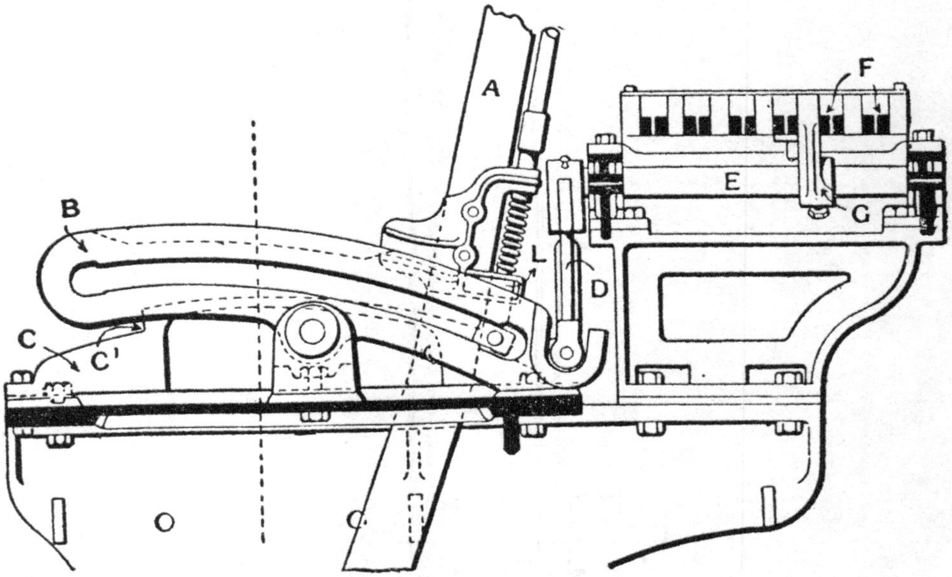

Fig. 363.

The lever, Fig. 363, is shown in its normal position.

Locking dogs are shown also at *k, p, r, s, t, v, w;* and cross locks at *x y* and *z*.

The operator, in grasping a lever preparatory to pulling it, lifts the latch rod, *L,* and in so doing turns the locking shaft *E,* on its axis, the latch acting through the link *D;* this, by means of *G,* moves one of the locking bars *F, F.* lengthwise, and, by means of dogs *H,* and cross locks *K,* obstructs the movement of some other bar. By this operation, called preliminary locking, we may insure that no lever shall be pulled until all conflicting levers are first in a safe position. The latch having thus performed its locking function, and having unlocked the lever *A, A* is pulled over and the latch is allowed to drop. It drops behind the notch C^1, and, by tilting *B*, lifts *D* still further,

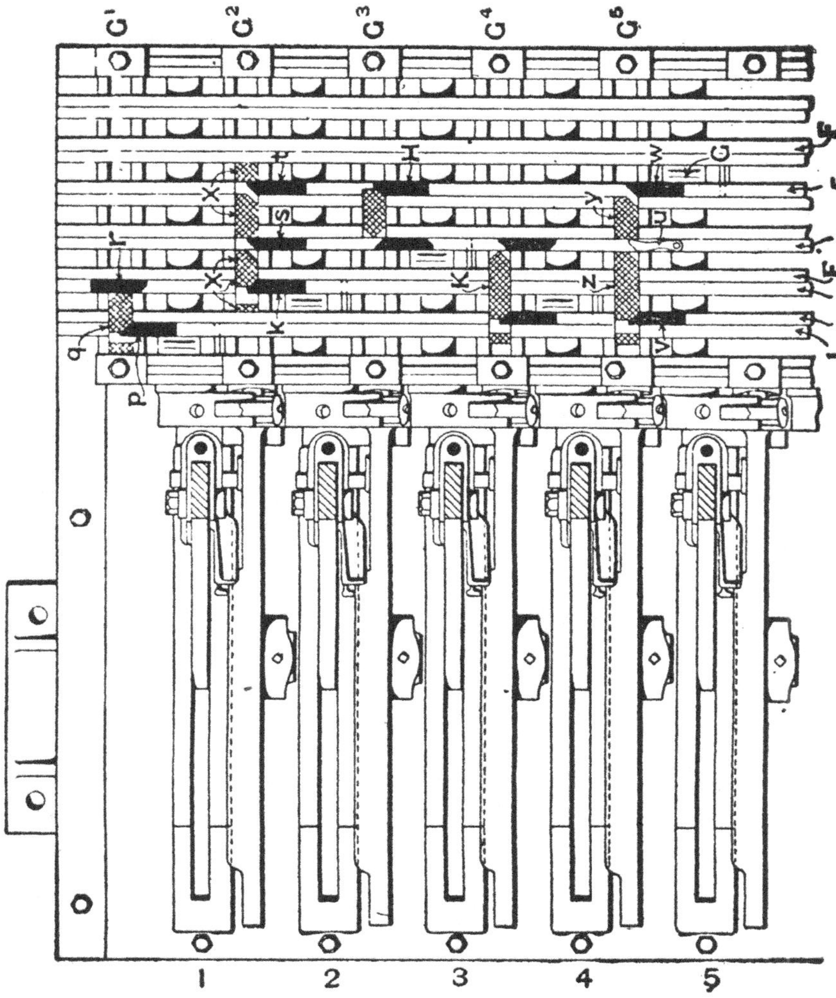

Fig. 364.

pushing the bar *F* further, thereby unlocking such levers as may properly be unlocked after this one is reversed.

The process of locking and unlocking may be followed by taking an example of one or more movements from the locking sheet of Fig. 360. In doing this it is to be borne in mind that each reversal of a lever (or, more exactly, each unlatching and latching again which must accompany the movement of a lever) moves the corresponding locking bar F to the *left* (upward in Fig. 364). The bar carries with it its dog or dogs, the mitered ends of which push the cross locks one way or the other. The cross lock in each groove in Fig 364, as x x, groove G^2, is in a single piece, with notches cut for the dogs, as shown; except that in groove G^5 there are two separate pieces y and z. These do not perform any function except when tappet u lies between their square ends, so that when either is pushed toward the other that other is moved along also.

To reverse lever No. 2, for the purpose of clearing signal 2, lever No. 3 and lever No. 5 must be normal. Now, in pulling lever No. 2 (Fig. 363) locking bar 2 (Fig. 364) is made to slide as just described; and this, by means of dog k, pushes cross lock x to the right, thus holding dog s in its present position; which locks lever 3 by holding locking bar 3 motionless. The movement of 2 also has the same effect on lever 5 by locking t. With the movement of locking bar 2, dog r, by releasing cross lock q unlocks bar 1 by making it possible to push dog p. Dog r does not clear q until the lever 2 has been reversed and its latch dropped.

When the bar 1 is unlocked, the signalman may clear the distant signal 1 by lever 1. This movement pushes p, and thus, by means of q, locks 2 in its new (reversed) position. Thus the signalman cannot throw up home signal arm 2, and try to stop a train after he has allowed the train to pass signal 1.

If, at the beginning, the switch has been set for the side track (lever 3 reversed) signal 2 could not have been cleared, because dog *s* would block dog *k*; but signal 4 or signal 5 could have been cleared. When the switch lever (3) is reversed, tappet *u* is pushed in between cross locks *y* and *z*, so that clearing signal 4, which moves bar 4 and dog *v*, locks signal 5 by blocking dog *w*. If the signalman should clear signal 5 instead of signal 4, the movement of bar 5 (dog *w*) would lock 4 in its normal position, by blocking the path of dog *v*.

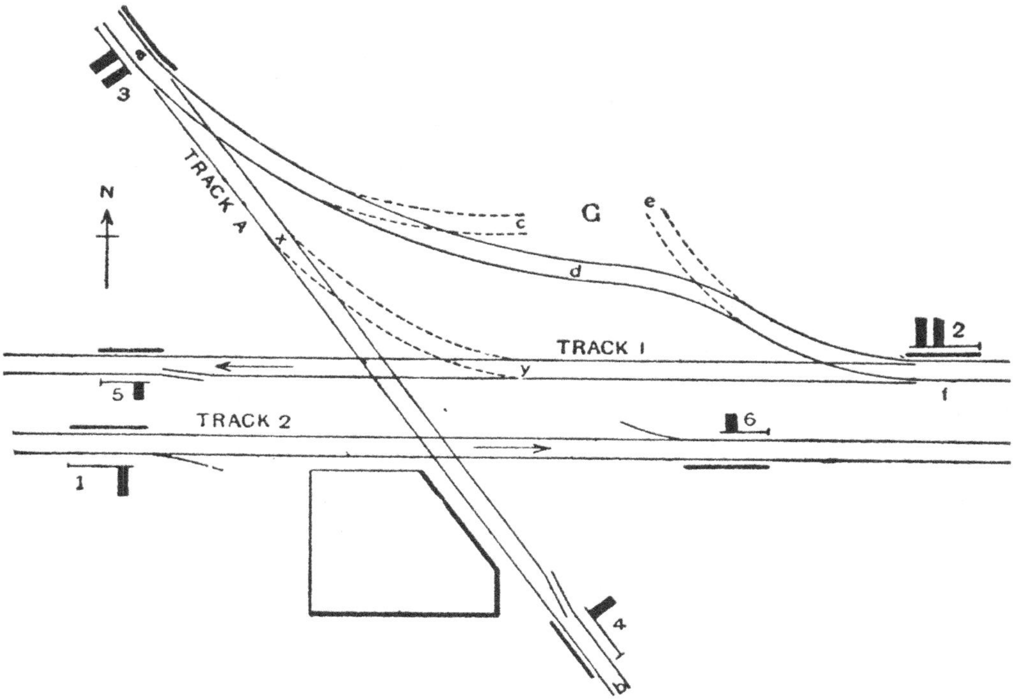

Signals and Derails at a Crossing
Fig. 365.

Where a track diverges from the main track in the high speed direction, as at signal 2, Fig. 365, two arms are provided, as shown; the lower arm being for the side track. Sometimes this lower arm is smaller than the upper one, and on some roads the dwarf signal would be used instead of the second arm. In the most recent practice this one arm is deemed sufficient even if there be two or more diverging routes. In many existing plants the earlier practice still prevails of having the upper arm indicate for the route at the extreme right and the other or others in the order in which the tracks lie, the lower arm indicating for the route to the extreme left.

The arrangement of signals at a crossing of one railroad with another is shown in Fig. 365. A simple crossing as of track A with track 1 would require four single arm signals, placed as shown at 1, 2, 3, 4. To provide against a collision on the crossing in case the engineman neglects to stop at the signal when it is against him, a derailing switch (called "derail") is used. An outline of such a switch is shown at signals 1, 4, 5 and 6. A "derail" is simply half of an ordinary facing point switch, and it is normally left open as at 4 and 5, so that a train disregarding the signal is thrown off the track. Where possible the use of derails should be avoided by putting in a diverging track. Such a track is shown in the drawing, leading from A at *a*; also one from track 1 at *f*. In some cases such a track is made continuous, as *a d f*, so as to be used, when necessary, for transferring trains from one road to the other. If such a connecting track is so short that its use would not be likely to promote safety, "dead ends" may be laid as shown at *c* and *e*. A mound of earth should be made at *G*. Signals 5 and 6 are for "back-up" movements which, according to the usual regulations, must be made at low speed.

Each switch and derailing switch has a detector bar to provide against careless throwing of the switch while a train is upon it. The lower arms of signals 2 and 3 are to admit trains to the side tracks.

Signals 1, 2, 3 and 4 should, of course, have distant signals, suitably interlocked, set back far enough to obviate the necessity of an excessive reduction of speed in approaching a crossing.

After a train has passed a distant signal at "all clear" (the engineman expecting to find the home signal also clear) a hasty signalman might restore the distant to the caution position, then restore the home signal and then clear the conflicting route, and cause a collision. To prevent this the levers of the signals for the conflicting route or routes are sometimes fitted with electric locks. The circuit for these locks runs through the rails of the track between the distant and the home signal, so that the presence of the train locks the levers. For example, signals 3 and 4 being fitted with the locks, the track circuit controlling them would extend from signal No. 1 back to its distant signal; and a train passing this distant signal at all-clear would, by shunting the electric current, lock 3 and 4 and hold them locked until it passed over the crossing. If the distant signal were at caution the train passing it would have no right to foul the crossing, and it would not thus act on levers 3 and 4; for the locks on these levers are never operative except when the lever of the distant signal of signal 1 is pulled to the clear position. When it is pulled it closes a circuit through magnets on levers 3 and 4, withdrawing a lock bar, which, until it was thus withdrawn, prevented the track-circuit lock from acting.

It has been stated that a train, in locking conflicting levers, secures them in the locked position until it has passed entirely over the crossing.

This, however, is not done by the single track circuit before mentioned. That circuit extends only from the distant signal to signal 1. Beyond signal 1 there is another track circuit, performing a similar function; but not controlled or interfered with either by signal 1 or by its distant signal. It locks 3 and 4 whenever a train is on the track between signal 1 and signal 6.

Signals, arranged as here shown, are used to protect trains approaching drawbridges. If track *A* were a stream, with a drawbridge spanning it carrying tracks 1 and 2, the approaches would be signaled as at signals 1 and 2, but with a single arm in each case. The back-up signals are not usually required and, indeed, are not always required at crossings. The derailing switches (and signals) at a bridge must, of course, be far enough back so that a train which is derailed will not run into the river. The levers are interlocked with the lever by which the draw is locked in position, so that the bridge can never be unlocked until the signals are at "stop," and the signals can never be cleared until the bridge is in position and locked.

Switches, signals and locks are connected with the cabin by iron rods (angle iron or gas-pipe) as before stated. The foundations of the supports of these are of wood, cast iron or concrete. Concrete foundations are comparatively recent, but are cheap and durable. For signals (but not for switches) wire connections are generally used, being cheaper than rods. In changing the direction of a line of rodding (pipe) a bell crank is used, but where a wire has to make a sharp turn a piece of chain is inserted and run around a grooved pulley.

To compensate for expansion or contraction of signal rods due to variations in temperature a rocker or a simple bell crank is inserted every 500 ft.; and where bell cranks have to be introduced those can be used as compensators, if they are so situated that the length of rod pushed is approximately equal to the length pulled.

In a straight rod the most common form of compensator

is the "lazy-jack," Fig. 366, a pair of levers, connected, turning on vertical pins in a frame which is fixed in a horizontal position. Pushing *a* pulls *b,* and vice versa. The original direction of the movement can, of course, be restored at the next crank.

With wires, which cannot be pushed, the problem of compensation is more difficult. Numerous devices have been tried and rejected, including a cylinder filled with liquid, in which a piston moved by expanding liquid took up the slackness of the wire. This was found too delicate for the rough usage that such a device is subjected to. The most common type of wire compensator is that which

Compensator for Signal Rods.

Fig. 366.

may be described as a disengaging clutch. The end of the wire which enters the cabin is not connected directly to the lever of the machine, but runs over a pulley to a weight or spring which keeps it always taut; and, a row of suitable notches, teeth or holes being provided in a solid bar (which is virtually a part of the wire) the signal lever, when grasped for a pull, is made to engage with a near or distant tooth according to the distance the toothed bar has been drawn in by the weight (as in hot weather) or let out (as in cold weather).

Fig. 367 shows a compensator, used extensively on the Pennsylvania Railroad, which was invented by Mr. Joseph R. Jones of Philadelphia.

In the figure the lever is "home" or normal and the signal is at "stop." The main features of the compensator are the ratchet wheel *N* with a sprocket hub, the

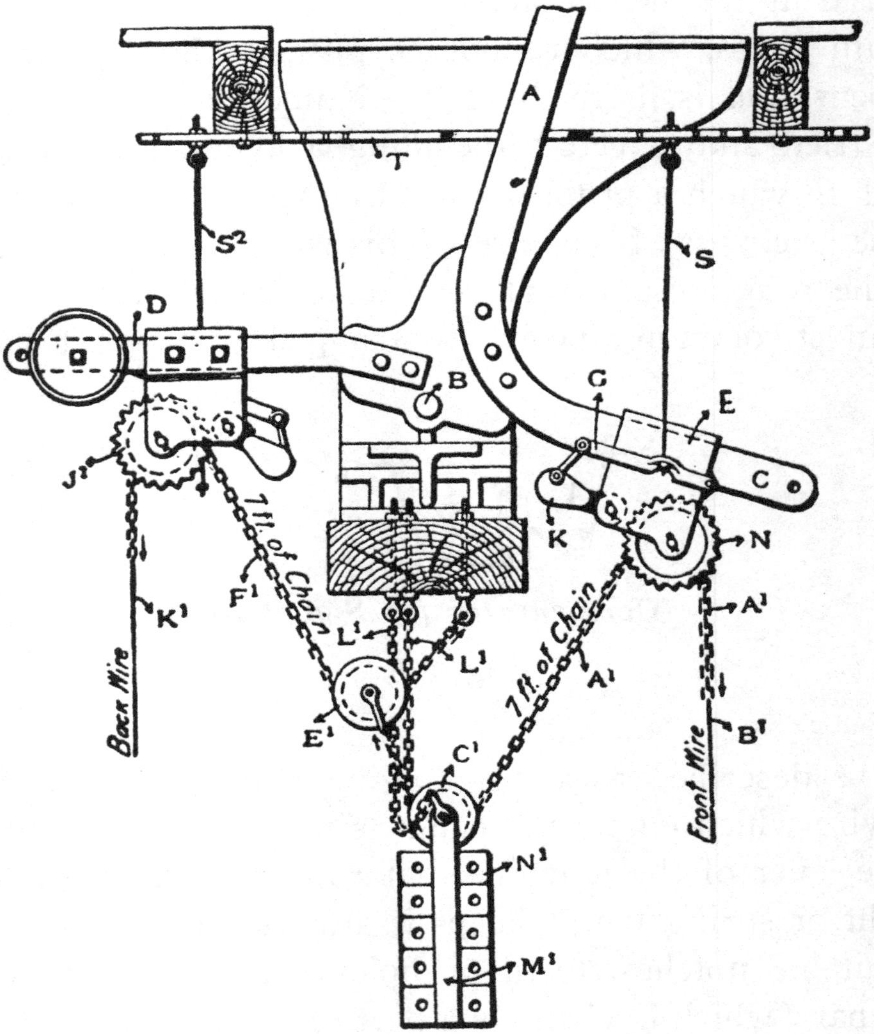

Jones' Compensator

Fig. 367.

weight *N* the pawl *K* and the rod *S*, with their connections. The front wire, used to pull the signal to the all-clear position, is connected, as shown, to the chain A^1, which runs over the sprocket hub of *N* to and around pulley C^1 and to the link attached to the axle of floating pulley E^1. The similar connection for the back wire needs no explanation. The wheels N are supported on the arms of the signal lever by jaws E, in which they are suitably mounted. The chains L^1 support the weight N^1 in case of breakage of either wire. Rods S and S^2 are adjustable at the top, and at the bottom they act on their respective levers by means of adjustable nuts.

In the position of the apparatus here shown, wheel N is free to revolve, and any expansion of the front wire is taken up by the weight N^1. When, for the purpose of pulling down the signal arm, the lever A is pulled to the left, lifting arm C, the lever G (pivoted at the right-hand end to C), being no longer supported by rod S, allows the weighted end K of the pawl to drop; this pawl, having at the end opposite to K a suitable nose, now engages with the teeth of ratchet wheel N and keeps it from revolving; this practically makes the front wire, which is taut, fast to the main lever A, and the movement of the distant signal arm, therefore, corresponds precisely to the movement of the lever. As arm C rises, arm D falls, and the slackening of chain F^1 allows chain A^1 to be pulled up without lifting the weight N^1. Each of these chains is 7 ft. long.

When the lever is in its normal position, as shown in the cut, the back wire K^1 is held fast by the sprocket wheel J^1 the action of the pawl on this side being opposite to that of pawl K. Whenever the signal is pulled off, as described in the preceding paragraph, the depression of arm D causes rod S^2 to engage with the pawl, lifting it out of contact with the ratchet wheel and allowing the latter to revolve. Every time, therefore, the signal lever is pulled, any undue tension or slackness in the back wire is relieved or taken up, as the case may be, by the weight N^1 acting through pulley E^1. The pawls on both sides have springs by which, combined with the gravity of their heavy ends, they are made to quickly engage with the ratchet wheels when not controlled by the vertical rods.

The ceiling bar, *T,* supporting the rods *S* and S^2 *need* not be continuous; short bars extending out from each timber, answer just as well.

The weight N^1 is adjustable, the several pieces being cast with necks so shaped that they can be readily put into or taken out from the yoke M^1. Under normal conditions wires 500 ft. long require a weight of 75 lbs.; wires 1,000 to 1,500 ft. long, 100 lbs.; 2,000 ft., 125 lbs.; 2,500 ft., 150 lbs.; 3,000 ft., 175 lbs. Approximately one-fourth of the weight acts on the back wire and one-half on the front wire, the remaining fourth being carried by the fixed support. It is found desirable to divide the weight between the two wires in about this proportion, the tension on the back wire being, of course, less than that on the other. In the event of the breakage of either wire the weight drops until it hangs on chain L^1, and the tension on the other wire is released.

QUESTIONS.

Ques. 1. What are the functions of all fixed signals?

Ques. 2. Upon what does the engineer depend for guidance in the handling of his train?

Ques. 3. What danger is there in the concentration of a half dozen or more switch levers in one cabin to be controlled by one man?

Ques. 4. How may this danger be avoided, to a large extent?

Ques. 5. How are interlocked signals operated?

Ques. 6. What is the universal standard signal for switches, crossings, drawbridges, etc.?

Ques. 7. What is the normal position of a semaphore signal?

Ques. 8. What kind of signals are generally used for slow movements?

Ques. 9. What type of signal is used for high speed movements over diverging routes?

Ques. 10. What is the purpose of detector bars in connection with interlocking signals?

Ques. 11. Of what does an interlocking machine consist?

Ques. 12. How many switches may be moved by a single lever?

Ques. 13. What is the main function of the interlocking feature?

Ques. 14. What is the purpose of a "derail"?

Ques. 15. What is a "derail" and what is its normal position?

Ques. 16. What is a better arrangement than a derail, when it is possible to put it in?

Ques. 17. If such a track is too short to promote safety, what may be done?

Ques. 18. Assuming that a train has passed a distant signal at "all clear" (the engineer expecting to find the home signal also clear) should a hasty signalman restore the distant signal to the "caution" position, then restore the home signal, and clear the conflicting route, what would be the probable consequence?

Ques. 19. What kind of a preventive is sometimes used with success in such cases?

Ques. 20. In such a device how is the electric current conducted between the distant and the home signal?

Ques. 21. Does a train in locking conflicting levers secure them in that position until it passes entirely over the crossing?

Ques. 22. How far back from a drawbridge should the derail switches and signals be placed?

Ques. 23. Can the bridge be unlocked and opened independent of switches and signals?

Ques. 24. How are the switches, signals and locks connected with the switchman's or signalman's cabin?

Ques. 25. What kind of connections are generally used for operating signals (not switches)?

Ques. 26. What kind of device is used to overcome the expansion and contraction of the rods, caused by variations in temperature?

Ques. 27. What is a "lazy jack"?

Ques. 28. What kind of a compensator is most commonly used with wires?

THE WESTINGHOUSE ELECTRO-PNEUMATIC INTERLOCKING SYSTEM.

This system is widely used as an apparatus for working switches and signals, probably more so than any other system. It is made by the Union Switch and Signal Company and there are more than 75 plants in use in the United States.

The most elaborate power interlocking plant in the world is the one at the South Station, Boston, and this is equipped with the Westinghouse Electro-Pneumatic System.

In this system the switches and signals are moved by compressed air at 70 lbs. pressure per sq. in. acting on a piston in a cylinder fixed to the signal post or on sleepers near a switch. The air compressor and reservoir, near the cabin, are connected to each switch or signal-cylinder by iron pipes buried in the ground. The compressed air is passed through cooling pipes to precipitate any excess of moisture. Near each cylinder is a small reservoir to provide a sufficient supply of air to insure quick action of the piston at every operation. The valves at each cylinder, admitting air to it, are opened and closed by electro-magnets, controlled from the cabin; the wires, like the air pipes, being laid underground. At each switch is a circuit closer for giving an electric "indication" in the cabin when the movement and locking of a switch are completed. In the machine the levers are only a few inches long and very light, their main function being to open and close electric circuits. (These "levers" appear as cranks on the front of the machine, shown in the perspective view, Fig. 368.) Each lever has a latch by which it is fastened, but this does not have anything to do with the interlocking of one lever with another. Instead of pulling or pushing a lever the operator turns it to the right or to the left, and it performs its function through the medium of a long horizontal shaft, which

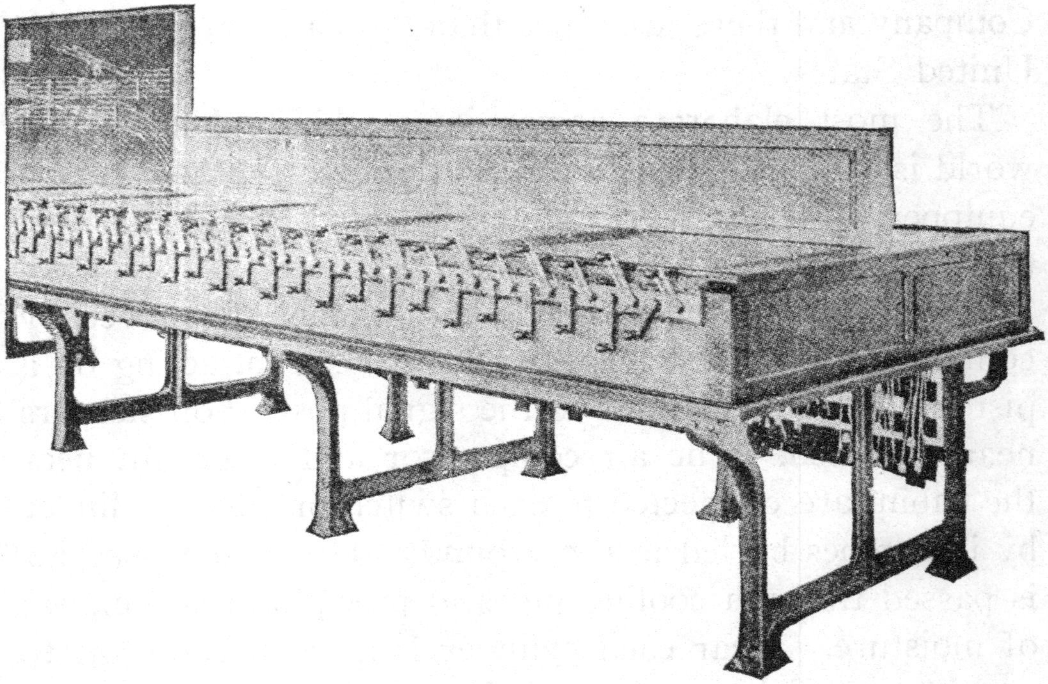

Fig. 368.

is revolved on its axis, through 60 degrees of a circle; and the shaft actuates the interlocking (mechanical) by dogs turned into the paths of sliding bars. The "indication" from a switch or signal actuates a magnet fixed near that shaft (in the machine) which controls that switch or signal; and to insure that the signalman shall not clear a signal before its switch is in the right position, this magnet locks the shaft, after it has been turned far enough to move the switch, and prevents the man from completing the turning of the shaft until the switch has actually finished its stroke, and sent in its indication.

The "indication" having been received, the stroke of the lever is completed, and the prescribed interlocking having thereby been effected, the lever for the signal to give "clear" for the route over this switch may be moved to the clear position.

The signal used in the pneumatic system, a semaphore, is similar in general appearance to the mechanical semaphore. Where the signal is placed on a bridge its post is short, and precisely the same design is used both for interlocking and for block signaling. This signal is the result of years of development and embraces numerous improvements made by the manufacturers since the system was first used.

In the dwarf semaphore, used for signals situated on the ground, between tracks, where there is but little space (and never for high-speed movements), the return of the arm or blade to the stop position, in case of failure of the apparatus, is secured by means of a spring, instead of by the weight which is used in the full size semaphore. The dwarf signal is shown in Fig. 369. In this design the cylinder is movable and the piston stationary, the pressure pushing up the cylinder, against the downward pressure of the spring. This movement puts the arm down; and on the release of the air, when the signal is to be restored to the stop position, the spring forces the cylinder and its rod downward. The piston rod is hollow, thus serving as a port for the admission of air to the cylinder. The lamp is mounted on top of the post, so that the glasses (red for stop and green for go-ahead), are in a different position, as related to the arm, from that of the glasses in the full size semaphore. To provide against damage by contact with passing trains, the casting to hold the blade and the glasses is of malleable iron, and the blade itself is made of flexible rubber. In mechanical signals the use of a spring in the place of a counterweight would be deemed unsafe, as the breakage of a spring might cause a failure to give a stop indication; but in this system this danger does not exist, as, if the spring should fail to restore the signal to the stop

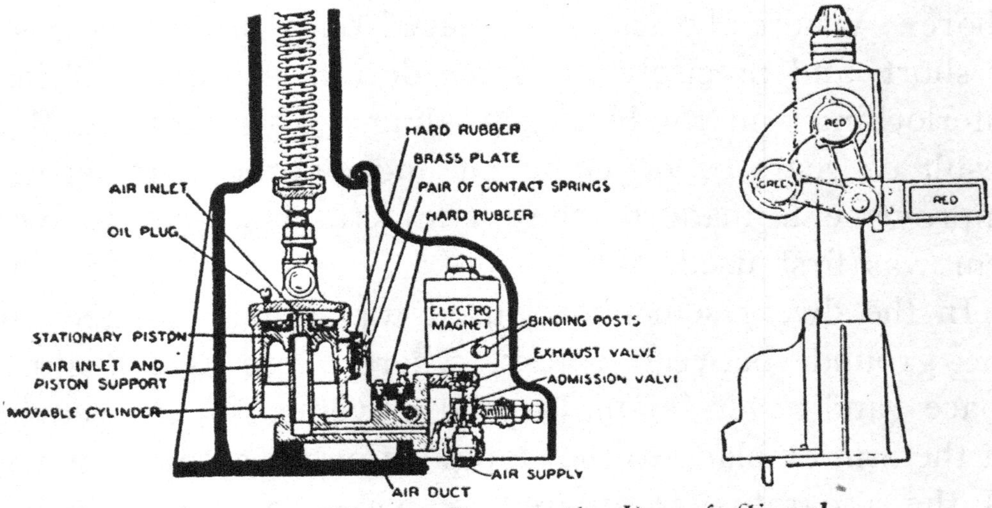

Electro-Pneumatic Dwarf Signal

Fig. 369.

position, the signal lever would be locked, by the electrical appliances, in the go-ahead position, thus compelling attention to the defect in the signal. This safeguard is accomplished by means of insulated metal plates attached to the signal cylinder, which at the proper time close an electric circuit controlling the lever in the cabin.

The electro-pneumatic switch cylinder is shown in Figs. 370 (a and b). The electro-magnets for controlling the admission of air to the cylinder are clearly shown in the engraving. Each cylinder (and its piston) moves not only

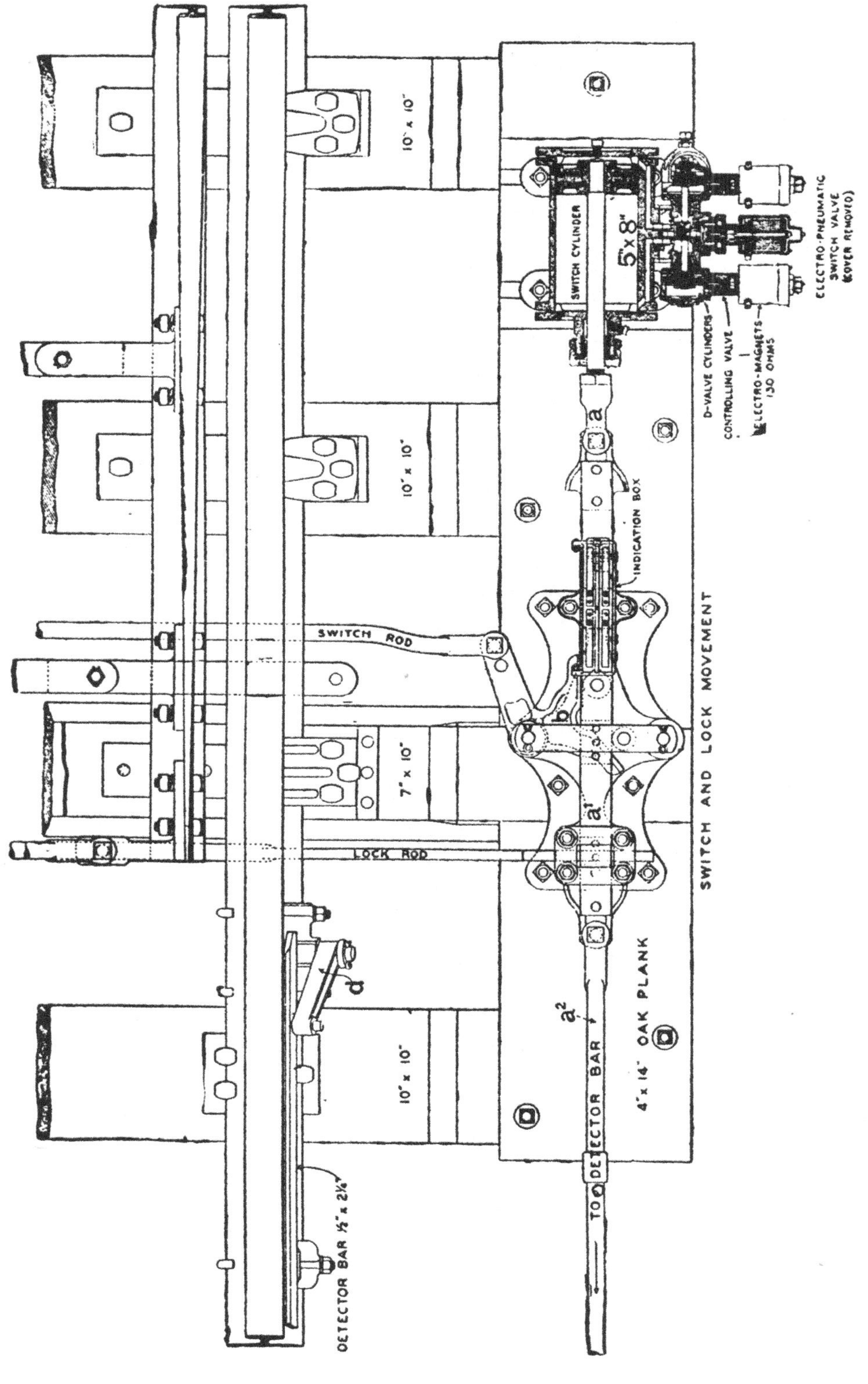
10" x 10"
10" x 10"
7" x 10"
10" x 10"
SWITCH CYLINDER
5" x 8"
ELECTRO-PNEUMATIC SWITCH VALVE (COVER REMOVED)
D-VALVE CYLINDERS
CONTROLLING VALVE
ELECTRO-MAGNETS 130 OHMS
a
INDICATION BOX
SWITCH ROD
LOCK ROD
SWITCH AND LOCK MOVEMENT
a¹
a²
d
4" x 14" OAK PLANK
TO DETECTOR BAR
DETECTOR BAR ½" x 2¼"

Fig. 370a.

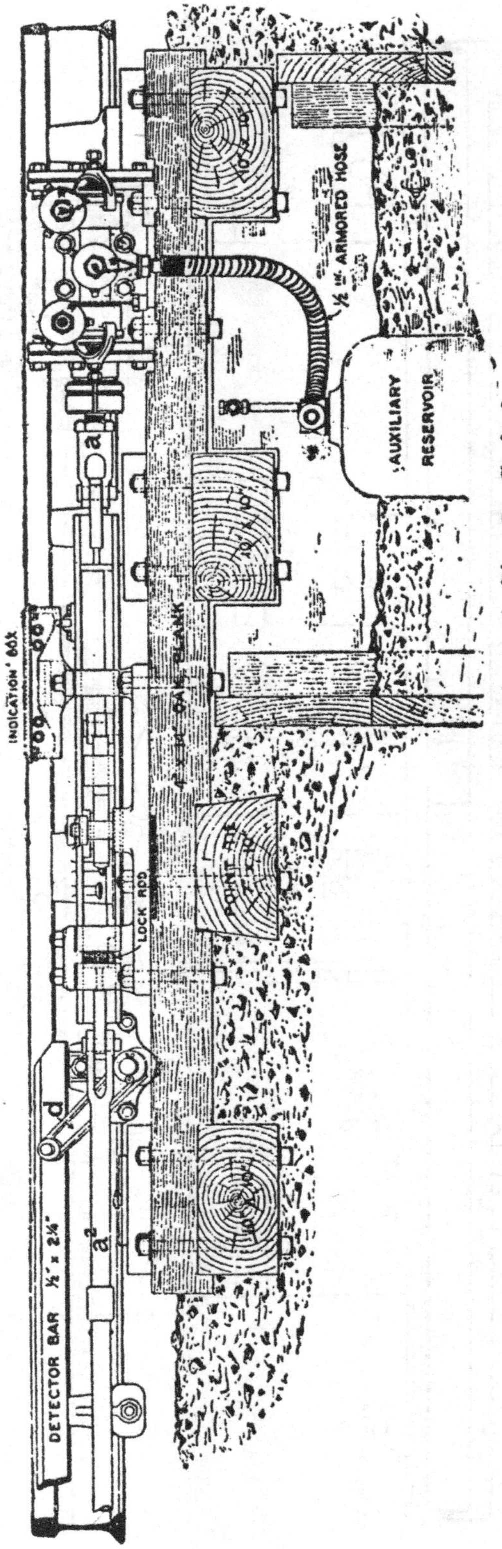

Plan and Elevation of Simple Switch.

Fig. 370b.

the switch but also the detector bar and lock; and two or more switches, near together, are moved by the same cylinder where it is found desirable to do so. This multiplication of functions is provided for by an increase in the size of the cylinder of by a higher air pressure. This concentration of power is, indeed, one of the main advantages possessed by the power apparatus over machines worked by hand; and, moreover, each lever in the electro-pneumatic machine may be made to perform the functions usually performed by two separate mechanical levers, as each one can be moved either to the left or to the right from its normal position.

Portions of the electro-pneumatic interlocking machine are shown in Figs. 371 (a and b) and 372, the former being a section from back to front. Fig. 372 is partly a front elevation and partly a section. The track model, which appears in section in Fig. 371a, is better shown in the perspective view of the machine. It consists of metal rods, representing a plan of the track. Each rod which represents a switch is movable, and is connected to the switch levers in such a way that every movement of the lever, to move a switch in the track, moves the corresponding part of the model. Some models have miniature signals as well as tracks.

The principal parts of the interlocking machine are suitably indicated in the transverse section, Fig. 376. The mechanical interlocking parts supported by the locking bar brackets, are similar in design to the interlocking shown in Figs. 363 and 364. The whole apparatus is enclosed in a wooden case with glass top. The wire connections (not shown in the drawings), are all run to binding posts on the back side of the frame supporting the machine.

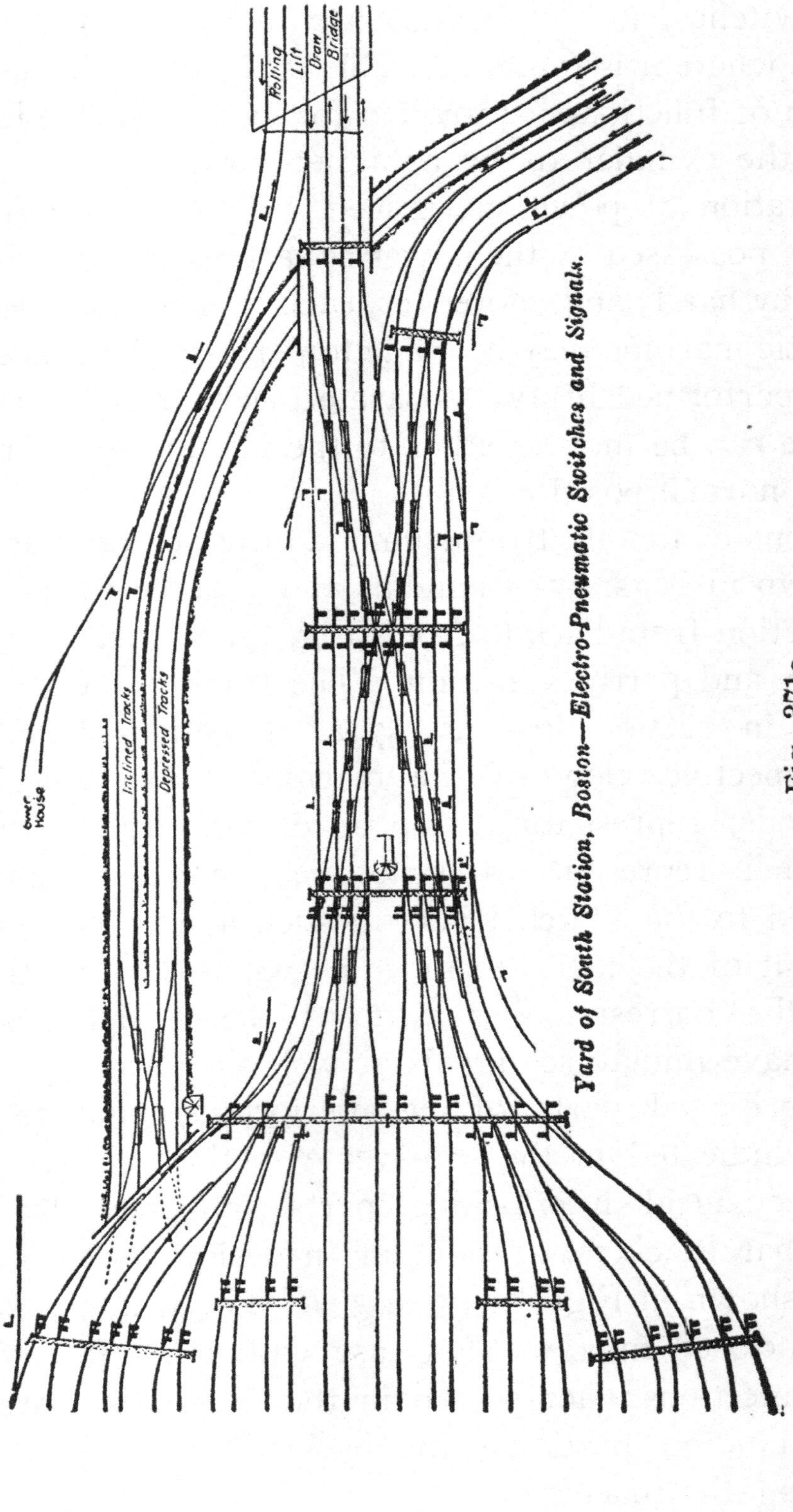

Yard of South Station, Boston—Electro-Pneumatic Switches and Signals.

Fig. 371a.

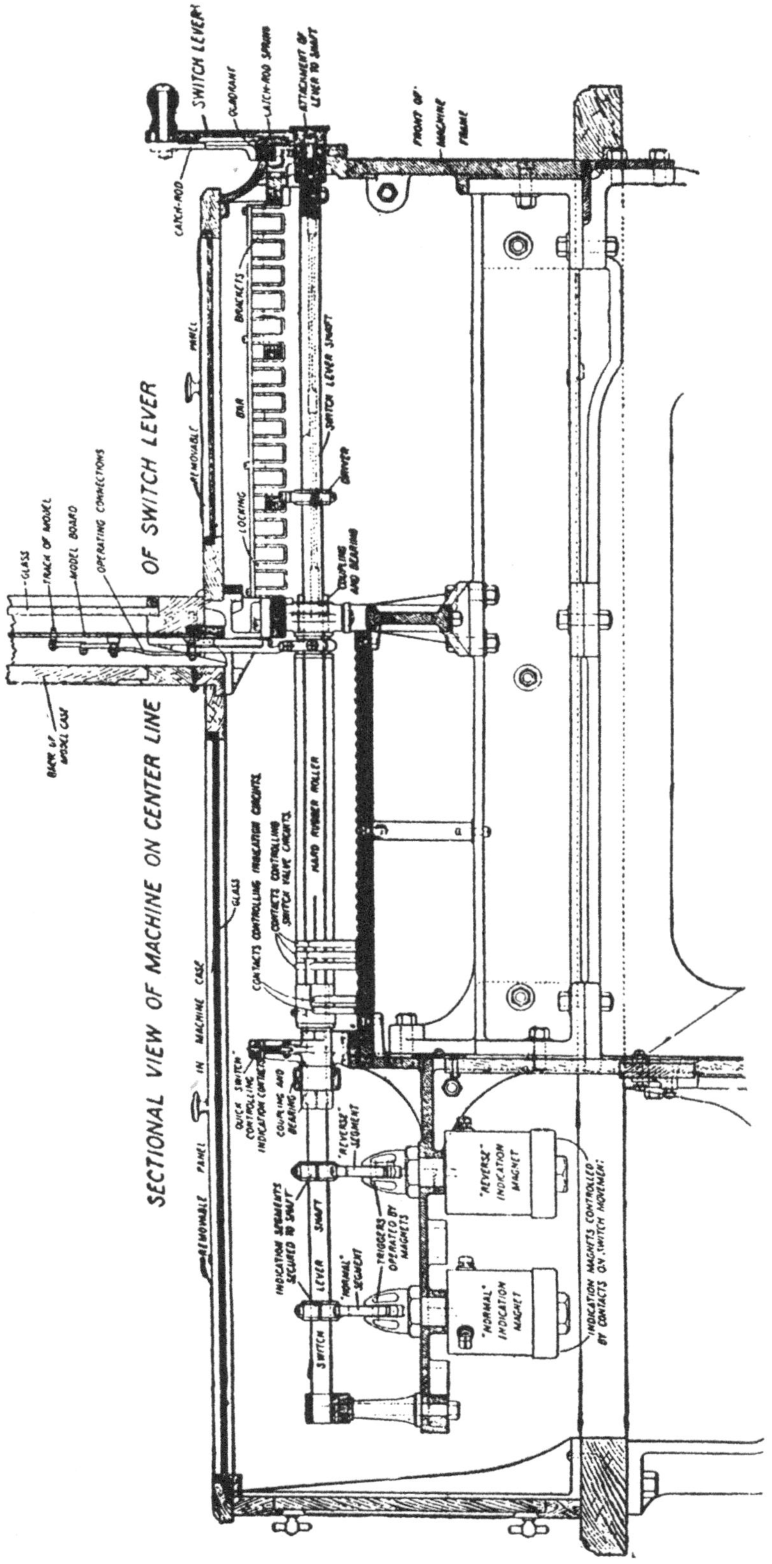
SECTIONAL VIEW OF MACHINE ON CENTER LINE OF SWITCH LEVER
SWITCH LEVER
QUADRANT
LATCH-ROD SPRING
ATTACHMENT OF LEVER TO SHAFT
FRONT OF MACHINE FRAME
CATCH-ROD
BRACKETS
SWITCH LEVER SHAFT
REMOVABLE PANEL
BAR
LOCKING
DRIVER
COUPLING AND BEARING
GLASS
TRACK OF MODEL
MODEL BOARD
OPERATING CONNECTIONS
BACK OF MODEL CASE
HARD FIBRE ROLLER
CONTACTS CONTROLLING INDICATION CIRCUITS
CONTACTS CONTROLLING SWITCH VALVE CIRCUITS
GLASS
QUICK SWITCH CONTROLLING INDICATION CONTACTS
COUPLING AND BEARING
"REVERSE" SEGMENT
REMOVABLE PANEL IN MACHINE CASE
INDICATION SEGMENTS SECURED TO SHAFT
TRIGGERS OPERATED BY MAGNETS
"NORMAL" SEGMENT
SWITCH LEVER SHAFT
"REVERSE" INDICATION MAGNET
"NORMAL" INDICATION MAGNET
INDICATION MAGNETS CONTROLLED BY CONTACTS ON SWITCH MOVEMENT

Fig. 371b.

This sectional view is taken on the center line of a switch lever and shaft. A sectional view of a signal lever would be similar except that there might be no

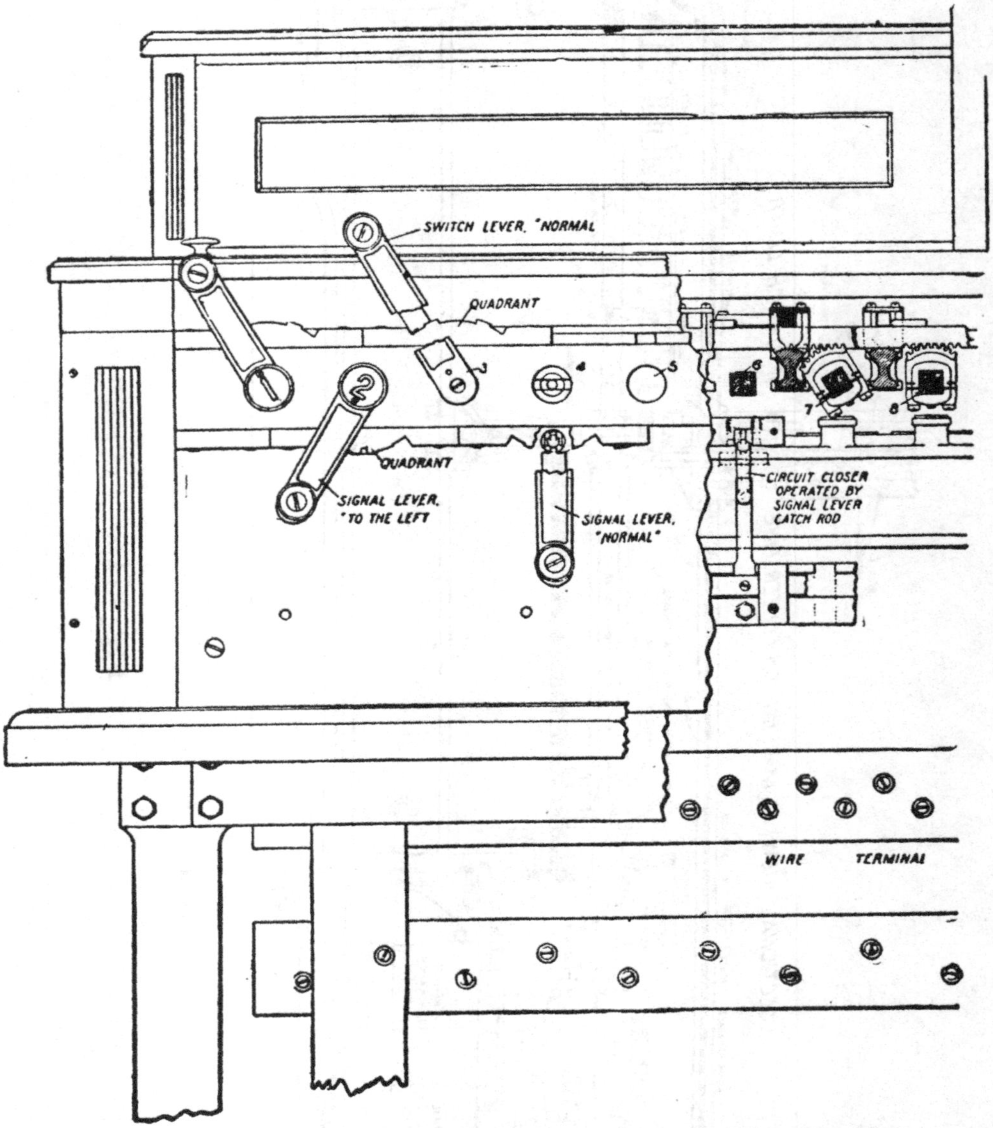

Part of Electro-Pneumatic Interlocking Machine

Fig. 372.

connections to the track model, and there would be one electro-magnet at the back of the machine instead of two. In Fig. 373 is shown a rear view of a small machine. On the first hard rubber roller are seen five electric contact strips. The controlling magnets are below the plate at the right, and the back side of the track model is shown in the upper part.

The process of moving a switch may be briefly described as follows, taking, for example, lever No. 3, Fig. 372. All signals which could give a clear route

Fig. 373.

over the switch in its present position being in the stop position, so that the interlocking will not interfere with the intended movement, the operator grasps the lever and moves it to the right, revolving the shaft. As soon as this revolution has begun, the "driver" on the shaft (Fig. 371) has moved the locking bar a short distance, so as to lock all levers which, if they were to be moved, would permit trains to interfere with the intended movement. This locking effected, the further revolution of the shaft causes a copper strip to close the switch valve circuit, which, by means of the electro-magnet (Fig. 370) admits air to the switch cylinder. The piston movement produced by this air pressure unlocks the switch, moves it to the other position, and then locks it. The locking having been completed, the closing of a circuit in the indication box (Fig. 370) energizes its proper "indication magnet" (Fig. 371), and the operator is then able to finish the stroke of his lever, leaving it inclined to the right. This final movement pushes the locking bar the remainder of its stroke and unlocks the lever or levers by which the operator will clear the signal to permit a train to pass over the switch in its new position.

In Fig. 370 is shown the "switch and lock movement" by which the rod from the cylinder performs the double function of moving the switch and locking it. This device is similar in principle to the locking apparatus used with all kinds of interlocking, where such a combination of functions is desired; and a brief description of it in this place will, therefore, serve as a supplement to the descriptions which have been given of other kinds of interlocking. The principal parts are so lettered as to explain what they are used for. The flexible connection from the auxiliary reservoir to the cylinder is employed to provide against leaks due to unequal pressure or settling on or in the different parts. The rod *a*, and the longitudinal moving part of the "movement," a^1, together with the detector bar connection a^2 are permanently connected, and may be treated as a single rod, which we will call *a*. This rod has a stroke of eight inches; but its action on the switch (by means of crank *b*) does not begin until it has moved about 2 in.; and the switch action is finished in the next 4 in. of the movement; so that the first two inches can be used to unlock the switch and the last two inches to lock it in its new position.

Rod *a* unlocks the switch by withdrawing a lug from the lower notch *y*, which lug is indicated in the elevation, Fig. 370, by dotted lines; and it locks, after moving the switch (and changing the position of the lock rod) by pushing another lug into the upper notch *x*. The vertical pin in a^1, which moves crank *b* is fitted with a loose collar.

The detector bar is provided to prevent the switch from being thrown while an engine or a car is on or near the switch. This bar, about 45 ft. long, with its edge upward, lies alongside one of the rails, and is so supported, on a series of arms ("links"), that, whenever a wheel stands above any part of it, it cannot be lifted. It must be lifted about ¾ in. (and then dropped) whenever the switch is moved, or even unlocked, and, therefore, the presence of a car or an engine prevents any movement of the switch. The lifting is effected by the rod a^2 which is suitably connected. Pushing rod a^2 to the left pushes the detector bar to the right, so that the supporting links *d*, of which there are 10 or 12 for a 45-ft. bar, are pushed over, and left inclined to the right, instead of to the left.

QUESTIONS.

Ques. 29. What can be said in general of the Westinghouse Electro-Pneumatic interlocking system?

Ques. 30. What pressure of air is used in operating this system?

Ques. 31. The air compressor and reservoir being near the cabin, how is the compressed air conveyed to each switch, and signal cylinder?

Ques. 32. How are the valves at these cylinders opened and closed?

Ques. 33. How is the operator in the cabin informed when the movement and locking of a switch is completed?

Ques. 34. What other function does this electric "indication" perform in the cabin?

Ques. 35. After the "indication" has been received, what may be done with that lever?

Ques. 36. What kind of a signal is used in this system?

Ques. 37. In yards where there is limited space, what kind of semaphore is used?

Ques. 38. Describe the dwarf signal used in this system?

Ques. 39. Is the use of a spring in place of a counterweight deemed safe in mechanical signals?

Ques. 40. How is this danger obviated in the Westinghouse Electro-Pneumatic system?

Ques. 41. What are the various functions of the cylinder and piston with which each switch or signal is provided?

Ques. 42. What is one of the main advantages possessed by power apparatus over machines worked by hand?

Ques. 43. What is the function of the detector bar in the Westinghouse Electro-Pneumatic interlocking system?

Ques. 44. Where is it located?

LOW PRESSURE PNEUMATIC INTERLOCKING SYSTEM.

The Standard Railroad Signal Company of Troy, N. Y., manufacture an interlocking apparatus, that is worked entirely by compressed air at a low pressure (15 lb. per sq. in.).

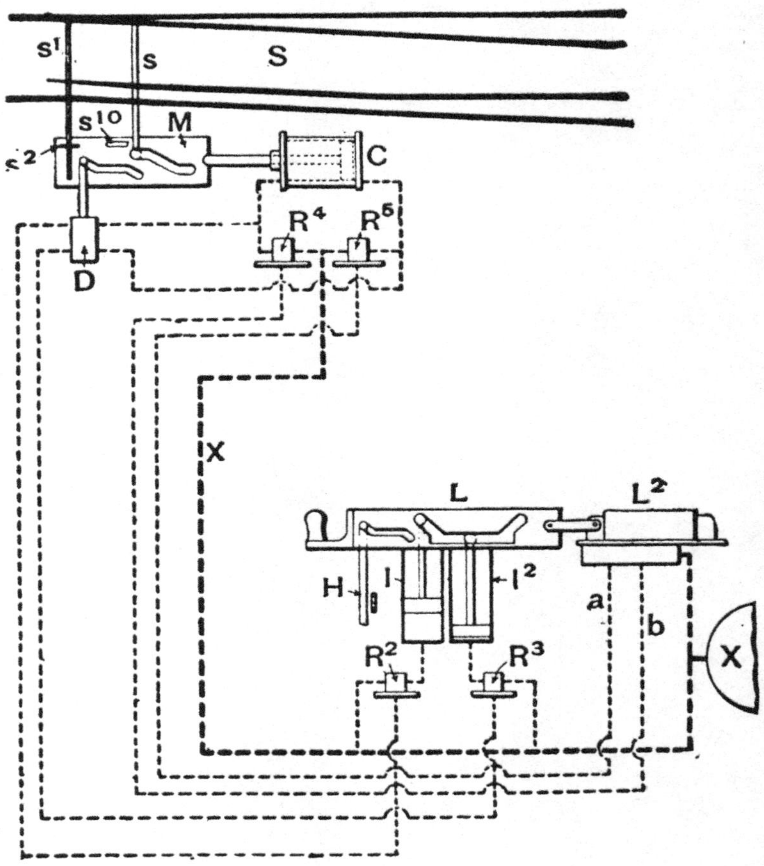

Switch Movement
Fig. 374.

There are no electric features connected with this system. The interlocking is mechanical, the parts being made very small as in the Westinghouse machine.

While the action of compressed air in pipes is not instantaneous like the electric impulse, still it serves the purpose of moving switches quickly enough for all practical purposes.

At ordinary distances the movement of a signal is practically simultaneous with its lever.

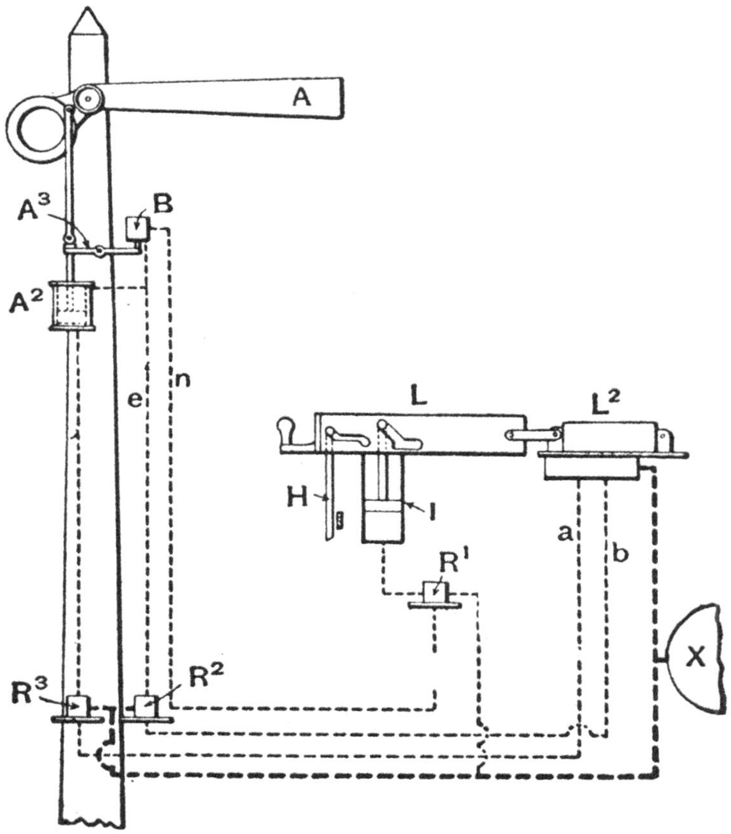

Signal Movement

Fig. 375.

The distinctive features of the low-pressure pneumatic interlocking machine are: (1) A row of slide valves (called "levers"), like that shown in outline at L and L^2, Figs. 374 and 375. As in electro-pneumatic the physical labor imposed on the signalman is inappreciable. (2)

The mechanical interlocking frame, placed vertically, on the front of the machine. The manner of connecting the "lever" with the interlocking is indicated by the position and arrangement of the tappet H in Fig. 374. (3) The indicating cylinders and their relays on each "lever," as shown in Figs. 374 and 375.

The appearance of the machine is shown in Fig. 376, and the way in which the pressure is conveyed and controlled is shown in Figs. 374 and 375. Fig. 374 represents the arrangement of valves and pipes forming the connection between the interlocking machine and a switch cylinder. The principal parts are: S, switch rails; s^1, lock bar; s, switch rod; M, motion plate; C, switch cylinder; D, indicating valve; R^2, R^3, R^4, R^5, controlling valves; L, L^2, operating bar and slide valve; I, I^2, indicator cylinders; H, interlocking tappet; X, air reservoir.

To change the position of the switch the signalman grasps L by the handle and pulls it out. In doing this he admits air (from the main supply X through the valve L^2) through pipe *a* to valve R^5, which opens communication from the supply pipe X to the right-hand end of the cylinder C, pushing the piston to the left. Observing now the slots in L and M, it will be noted that after about one-half of the stroke of L has been completed it is stopped by the piston rod of I^2; but the operation of valve R^5, already accomplished, causes M to move through the whole of its stroke. This stroke of M is uninterrupted, but it performs in succession three functions. The first part of the stroke, say one-third, does not move the switch, but valve D is moved far enough to close the two pipes on its right, while those on its left are open to the atmosphere. At the same time lock bar s^1 has been liberated at s^2. As M moves through the next or middle portion of its stroke, it moves the switch; but it now produces no effect on valve D, because the rod of D is now engaged by the straight portion of its slot in plate M. The switch being set, the third and final part of the stroke of M locks the switch by pushing s^{10} through a hole in s^1; and also (but not until after s^{10} has entered its hole) the plates change valve D so as to connect together the two pipes at its lower end. This conveys pressure from the supply through R^5 and D to valve R^3, which valve then admits air from the supply to I^2, forcing the piston rod upward, and, by means of the diagonal portion of the slot in bar L, forcing this bar to complete its stroke. This return action takes place at ordinary distances in from one to three seconds. In rapid work, this automatic completion of the stroke of the bar saves an appreciable portion of the signalman's time.

By the action of L^2 pipe *a* is now opened to the atmosphere. Valve R^5 is now released from pressure, and R^4 is closed; so that the right-hand pipe to cylinder C and its connection to and through D are open to the atmosphere. All four operating pipes are now at atmospheric pressure.

By the movement of L, tappet H has been moved so as to produce the proper mechanical locking of conflicting levers (in the first part of the stroke of L) and the proper unlocking (in the last part of this stroke) in the same manner and sequence that the same interlocking would be effected in a wholly mechanically interlocking machine.

To move the switch back to its original position, the opposite set of pipes is used. The bar L is pushed to the right; air through *b* actuates R^4, and the return indication to the cabin actuates R^2 and lifts the piston in I.

To work a signal, valves and operating pipes are used of the same general style as those for a switch, but there is only one indicating valve and one indicating cylinder, as it is unnecessary to assure the attendant that a signal is in the go-ahead position. The signal connections are shown in Fig. 375. The principal parts are: A, signal arm; A^2, signal cylinder; A^3, lever to work indicating valve; B, indicating valve R^2 and R^3, diaphragm valves, controlling the admission of air to the top and bottom, respectively, of the signal cylinder; R^1, diaphragm valve controlling admission of air to cylinder I. The signal being in the normal or stop position, the indicating valve B is in a position to maintain a connection between the two pipes attached to it; but the instant the signal arm leaves the horizontal position the valve shuts off this connection.

To change the signal the signalman pulls L to the left, the whole length of its stroke. By this movement L^2, admitting air to pipe *a*, actuates valve R^3, which supplies air to the lower end of cylinder A^2 and pushes up the piston, putting the signal in the inclined or all-clear position. The air impulse is transmitted so quickly that at the average distance—say 500 ft. and less—the movement of the signal is practically simultaneous with the movement of the lever. The signal remains in the inclined position as long as L is pulled to the left. To restore it to the normal or stop position, L is pushed to the right until it is stopped by the piston rod of I (at the end of the horizontal part of the slot in L). With L in this position, pipe *b* is charged and valve R^2 is opened. The passage between pipes *e* and *n* (through B) is now closed, so that the opening of R^2 admits air from the supply to the upper end of A^2. This restores the signal to the horizontal position, and by means of A^3 opens valve B. Air now passes from *e* through B and *n* to R^1, and the latter causes air to enter I and complete the return

Fig. 376.

stroke of L by the action of the piston rod on the diagonal part of the slot. Pipes *b e* and *n* are now at atmospheric pressure, and the parts are in the same position as at the beginning.

In Fig 377 is shown the diaphragm valve, which is called the "relay," its function being similar to that of an electro-magnetic relay in electrical apparatus. This valve is actuated by air at 7 lbs. pressure. This pressure admitted beneath the circular rubber diaphragm 8 in. in diameter, pushes up the cylindrical valve, placed vertically in the upper part of the case, and thereby liberates air at 15 lbs. per square inch to move the piston in the switch or signal cylinder. The movement of the diaphragm is only ¼ in.

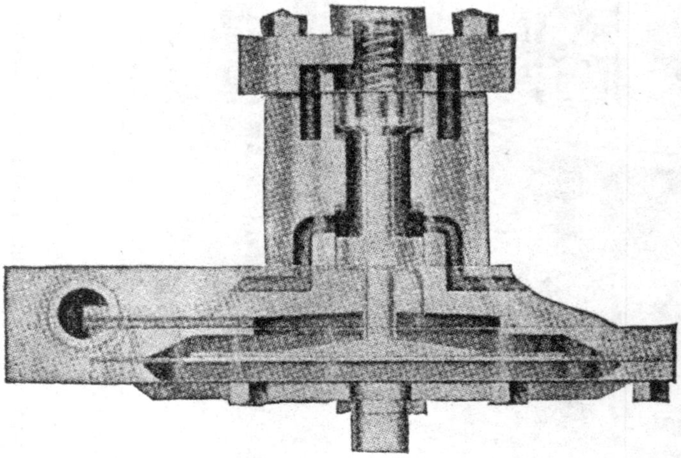

Fig. 377.

The operating and indicating pipes extending to switches and signals are ½ in. in diameter. The supply pipes from the air reservoirs are larger, the size being varied according to the number of switches and signals to be supplied.

The air as it comes from the compressor is run through a cooling frame for the purpose of precipitating moisture; but with low pressure there is practically no trouble from moisture in the pipes.

A low pressure pneumatic machine of 180 levers is used to work the switches and signals in the yard of the Grand Central Station, New York City, a diagram of which is given on the folder. This station is the terminus of the passenger lines of the New York Central & Hudson River and the New York, New Haven & Hartford. This yard, one of the busiest in the country, was for many years signaled with mechanical signals.

The machine in the cabin or tower at 47th street works, besides the switches and signals shown, 28 switches and 22 signals to the left (north) of 48th street, which street forms the left-hand boundary of the diagram. In the drawing the heavy shading at the switches shows the normal position of each. For example, switch No. 115, near one of the baggage lifts, stands normally in position for the straight line, not the diagonal one. Switch No. 92, near the upper left hand corner, stands normally for the straight track; switch No. 140 stands normally for track D. Since this diagram was made, tracks Y and Z have been extended so that they join one another. This yard is the terminus of a four-track railroad, but at the entrance of the yard it is necessary, on account of lack of room, to run all of the trains over two tracks, one inward and one outward. The main inbound track is No. 1, and the outbound is H. Trains run on the left-hand track between this terminal and the junctions a few miles out. The signals and switches marked "fly" are used for inward trains which, before they are run into the station are separated from their engines. A train approaching signals 66, 67, 68 and 70 on track No. 1 receives notice about 600 ft. back, by a two-arm signal, whether or not a "fly" switch is to be made; and, when it is made, two arms, 68 and 70, or 67 and 70, are pulled down at the same time. When a fly switch is made, the engine is run to track No. 9 and the cars are diverted either to the left at switch 71 or to the right at switch 72.

QUESTIONS.

Ques. 45. How is the interlocking apparatus of the Standard Signal Co. of Troy, N. Y., worked?

Ques. 46. What are the distinctive features of this machine?

Ques. 47. What is the maximum pressure of air used in the operation of the machine?

Ques. 48. What is the function of the diaphragm valve or "relay" in connection with this machine?

Ques. 49. What pressure of air is required to actuate the diaphragm valve?

Ques. 50. Describe the action of the diaphragm valve.

Ques. 51. Of what size are the pipes leading to switches and signals?

Ques. 52. What advantage is there in connection with the use of air at a low pressure?

Ques. 53. How are the switches and signals in the yard of the Grand Central Station in New York operated?

Ques. 54. How was this yard formerly signalled?

THE JOHNSON INTERLOCKING SYSTEM.

In the Johnson interlocking machine the locking bars are arranged in a vertical, instead of a horizontal position. This arrangement economizes room to some extent, but the weight of the vertical bars or tappets is added to the load to be moved by the man in the cabin, and it is not so easy to provide against the wear of parts.

Fig. 378 shows a side view of one of the levers, and an end sectional view of a machine, while Fig. 379 will serve to illustrate the interlocking feature. Fig. 380 shows a view of an interlocking machine of the same general type as the Johnson, but differing some in detail.

The lever shown in Fig. 378 has connections for a signal. For a switch the rod would be connected to it at *H*. The bar *K* is for use where a signal is connected by two wires. The lever is held in each of its two positions by the latch rod V^1, which engages with notches in the segment *B*. When the signalman, preparatory to clearing a signal, grasps the lever at its upper end he moves V^1 upward; and in so doing actuates the interlocking, through the tappet *N*, attached at *T*. In lifting the tappet he locks all conflicting levers which need to be locked to make it safe to move this one. In pulling over the lever the rocker *R* is also pulled; but the slot in it is radial to the center on which the lever turns, so that, during the stroke, *N* remains motionless. On the completion of the stroke and the dropping of V^1, *N* is raised still further, and this unlocks such levers as should

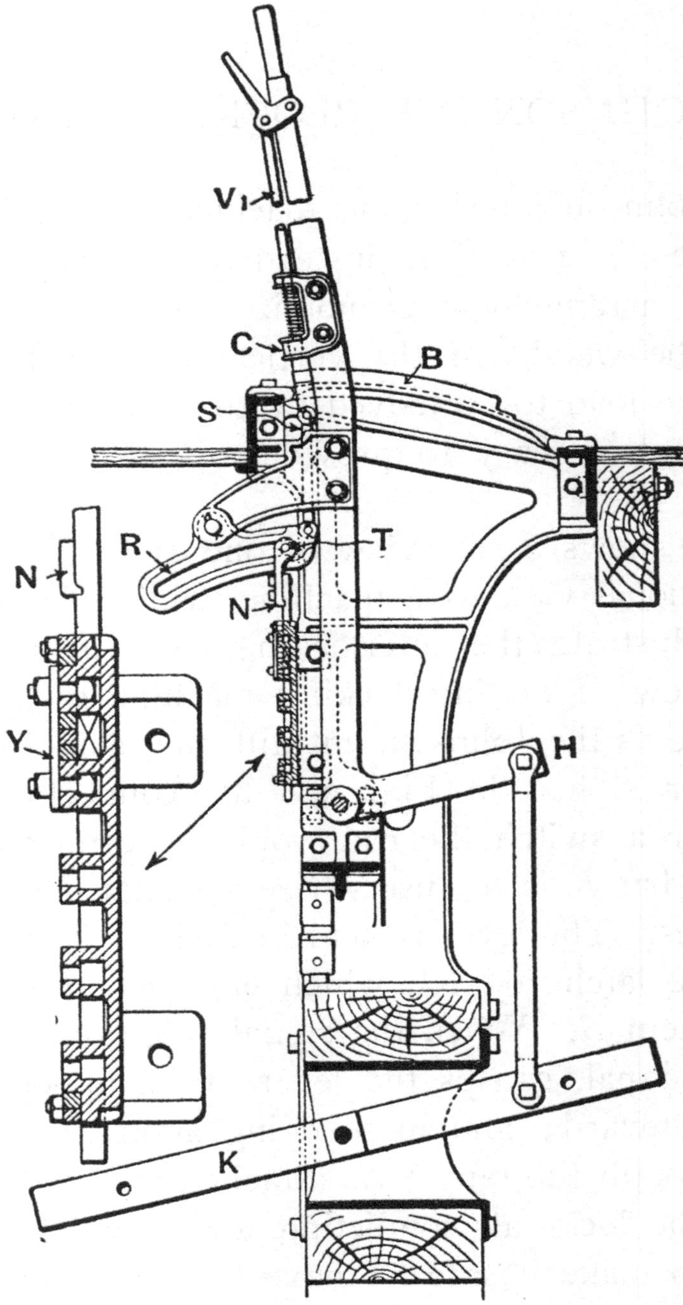

Johnson Interlocking Machine

Fig. 378.

be unlocked after this lever is pulled ("cleared" or "reversed"). It will be seen that whenever the tappet *N* of any lever is locked in the position shown in the engraving, it is impossible to raise V^1, and therefore impossible to move that lever. The action of tappet *N* and the way in which it is locked may be understood by

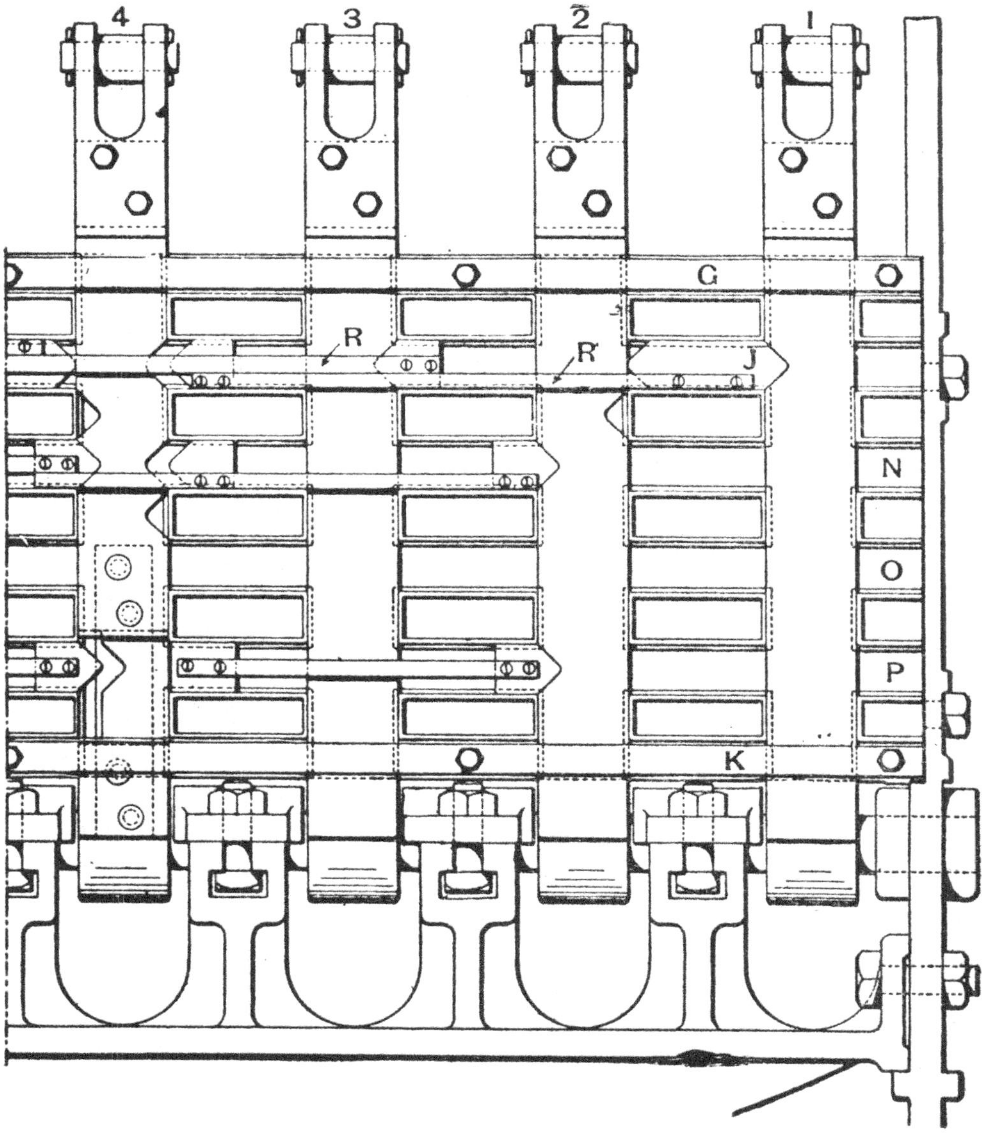

Fig. 380.

reference to Fig. 379. A tappet, say No. 3, slides vertically in a planed recess in the locking plate, being held in place by horizontal strips *G* and *K*. Transverse grooves *N, O, P,* carry dogs, such as *J*. Two dogs may

Fig. 380.

be connected together by bars, R. The dogs are held in place by straps, *Y*.

Locking is effected by sliding the dogs horizontally; for example, dog *J* has been pushed into the notch in tappet 1, holding it in the normal position. If tappet 2 were raised, a distance equal to the width of a dog, its notch would come opposite dog *J*; and then the lifting of 1 would lock 2 by pushing *J* to the left. By means of rod *R*, which connects *J* with a dog which will engage with tappet 4, the lifting of 1 also locks 4. If 4 were already up it would be impossible to lift 1.

QUESTIONS.

Ques. 55. How are the locking bars arranged in the Johnson Interlocking System?

Ques. 56. Is there any advantage in this arrangement?

Ques. 57. Mention some of its disadvantages.

Ques. 58. When the signal man takes hold of and grips a lever preparatory to clearing a signal, how are all conflicting levers affected?

Ques. 59. Does this make it safe for him to move the lever he has hold of?

Ques. 60. After the signal man has pulled the lever to the completion of its stroke, and relinquished his grip upon the latch rod what occurs?

Ques. 61. When the tappet of any lever is locked in position, can it be unlocked and again moved independent of the other levers?

THE BLOCK SYSTEM ON SINGLE TRACK ROADS.

Several prominent single track roads in this country are using the block system of running trains, among which may be mentioned the Erie, the Atchison, Topeka and Santa Fe, and the Wabash.

On the Erie lines west of Salamanca the block system has been in use since 1891. According to the rules then prescribed and which have been continued in use to the present time and are now in force from Salamanca to Chicago, 574 miles, the block signaling is under the immediate active control of the train despatchers, the length of road to be managed by each despatcher having been made short enough to enable him to order signals for all trains at all stations without delay.

This is certainly a good feature, and should commend itself to officials in charge of the running of trains.

On the busiest portion of the road a despatcher controls about 40 miles. The signal is kept normally in the stop position and, except in case of failure of the telegraph wire, is not moved from its position except by order of the despatcher. Ordinarily the despatcher gives each station the proper instruction before the train arrives, but in the absence of an order the operator notifies the despatcher as soon as a train is known to be approaching. The operator says "No. —— coming," and the despatcher replies "red," "green" or "white," according as he desires to have the train stopped or sent forward permissively or sent forward with a clear block. A train having been stopped, it must not proceed without a clearance card, or until the signal is shown in the caution or the safety position. Where a train is to use a side track beyond the station, the conductor remains at the telegraph office until the signal to proceed is shown and then repeats it to the engineman by white or green flag or lantern.

As some side tracks are not conveniently located, as related to the telegraph office, there is a rule requiring the conductor to go personally, on arrival, to the telegraph office, when necessary, in order to enable the operator to promptly give a clear signal to other trains after the main track is clear. Where an operator has to give a go-ahead signal for two or more trains to go out of a side track which is beyond his signal, he is required to put the signal back to the stop position after each train and then clear it for the next following train only on the order of the train despatcher.

All conductors and enginemen are required to follow the ordinary time-table and train rules as though no block system existed.

On this road the permissive or caution signal is given by showing a semaphore arm inclined upward at an angle of 60 degrees. To provide for this signal, Mr. Mozier, the Superintendent of Transportation, designed the semaphore shown in the engraving, Fig. 381. The arm A is shown in the normal or stop position, in which it is held by a counterweight B, which is attached to the arm A by chains as shown. The signal is pulled to the safety or all-clear position by pressing downward on the lever C in the cabin, and to the caution or permissive position by moving this lever upward. If either or both of the connections from the office should break, the counterweight would pull the arm to the "stop" position if it were not already in that position. When the semaphore arm is pulled down, the lamp which in the drawing is hidden behind the red lens, is uncovered

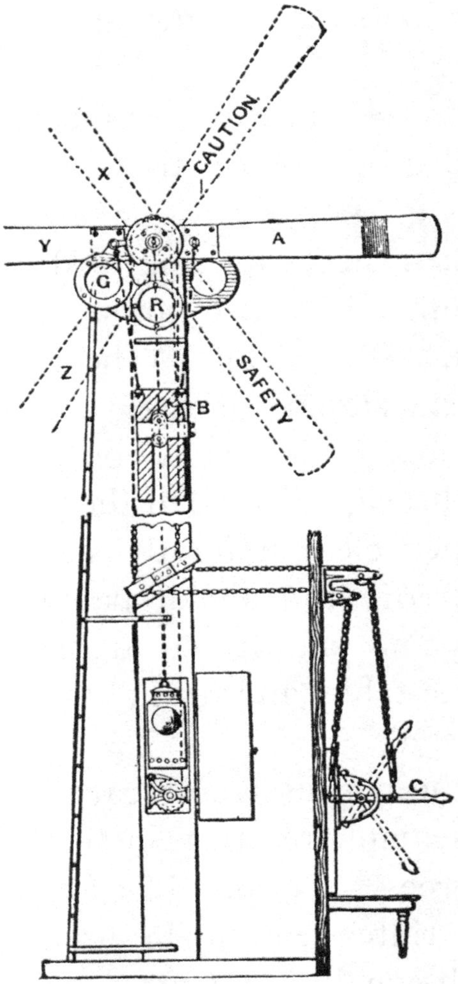

Fig. 381.

and shows white or all-clear. When the arm is moved upward, the green lens G is moved to a position in front of the lamp. The letters X Y Z indicate the positions of an arm on the other side of the post, which is used for signaling trains moving in the opposite direction. In the

Mozier signal a hollow post is used, and the lamp, which is suspended in position by a chain running over a pulley, is lowered within the post for cleaning and inspection.

The signal levers in the office are fitted with Mozier's patent shunt for cutting out the keys of the telegraph instruments. By this arrangement the signal may be fastened in the all-clear position, without danger of accident from forgetfulness, as the lever when in that position closes the telegraph circuits around the telegraph keys, while not cutting out the relays, so that it is impossible for the operator to accept another train, or to communicate at all by telegraph, until he has put his signal back to the stop position.

Officers of the Erie say that they are well satisfied with the plan of having all block-signal operations performed by direct orders of the despatcher. The chief advantage claimed is that the station man does not need to give so much time and attention to the telegraph, and therefore can better attend to his other duties. At many small stations the agent performs all his duties, including telegraphing, without an assistant, whereas if he had to be on hand at the telegraph instrument to give prompt attention to the needs of trains both at his own station and at the stations east and west of him he could not do so. More despatchers are required than on other roads, but not nearly so many station men. The despatchers say that they like to control the block signals in this way. They have a more satisfactory feeling of safety in handling the trains, and the prevention of blockades at terminals and junctions is made easier.

The despatchers on this line have for several years taken advantage of the block system to make easier the work of sending telegraphic train orders. According to the usual practice, a despatcher's telegraphic order, directing a train to run ahead of or later than its regular time, or changing the place at which it shall meet a train running in the opposite direction (on single track), must be repeated back to the despatcher by the operator who receives it, must be read aloud to the operator by the conductor or engineman, or both, and must be signed by one or both. The despatcher must not finally approve the order until these signatures have been affixed. These safeguards are adopted to make sure that no collision shall occur from a misreading of the order. But with the block system in force, collisions are provided against at all times by the block-signaling; not only when it has been necessary to issue telegraphic orders, but for every regular movement as well. This being so, despatchers' orders may be delivered without signature (and even without repetition); for a mistake either in writing or reading could do no worse harm than to cause delay (by giving preference to a wrong train). Taking advantage of this, the Erie despatchers issue many of their orders on "Form 19," requiring no signatures; and operators are able to deliver to both enginemen and conductors without requiring the train to be stopped. In the case of a heavy train, and especially on an ascending grade, this often obviates annoying and expensive delays. By attaching an order to a light hoop of rattan, wood or metal, it can be delivered to a train running at full freight-train speed, and even at a higher speed, the hoop being held at the end of a stick in such a manner that the engineer can safely catch it on his arm in passing.

The Atchison, Topeka and Santa Fe Co. like the Erie uses a signal by which the permissive or caution signal is indicated by inclining the arm of the semaphore upward. This company has introduced the use of the block system over several hundred miles of road, including some lines on which the number of trains operated is comparatively small.

Fig. 382 shows a design of signal for this purpose, called "Style D three phase," made by the Union Switch and Signal Company. The spectacle casting attached to

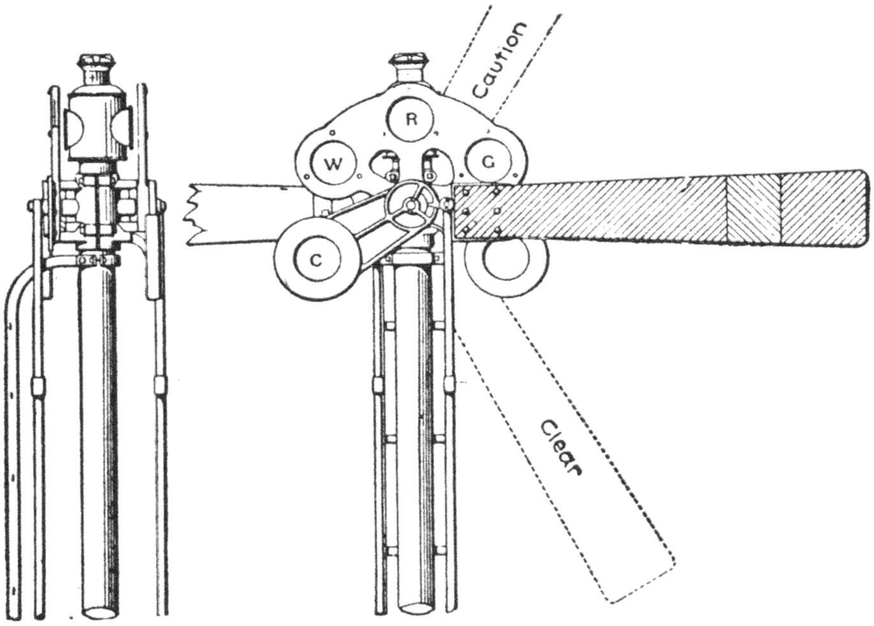

Three-Phase Semaphore—Atchison, Topeka & Santa Fe Railway

Fig. 382.

the semaphore blade has three openings, one being filled with red glass, one with green and the third with no glass. The casting is made in this form so as to permit the substitution of a color in the place of "white" for the all-clear night indication when that shall be found desirable. The counterweight, C, is an iron casting which acts as a counterweight and is put on for no other purpose. When the arm has been pulled down to the all-clear position this counterweight acts, as in the case of the ordinary semaphore, to restore the signal to the horizontal position in the event of the breakage of the connecting rod; but when the signal is moved upward to the caution position (by pushing the rod upward) the counterweight becomes disengaged from the blade and remains in the position shown in the drawing. It rests against a stop fixed to the post. In case the connections break while the arm is inclined upward, it falls to the horizontal or stop position by its own weight.

As will be seen from the drawing, this signal is mounted on a cylindrical iron post, and both the eastbound and westbound signals are fixed to the same post. A single lamp answers for both.

Where the three-position signals have not been introduced, the Atchison uses a card for permissive block signaling.

The Atchison road also uses the Hall automatic block signal system to some extent, and on one section the train staff is used.

AUTOMATIC BLOCK SIGNALS ON SINGLE TRACK.

The first important use of automatic signals on a single track road was on the Cincinnati, New Orleans & Texas Pacific. Three kinds of signals are there used, the clock work, the enclosed disk, made by the Hall Company, and a similar signal made by the Union Switch & Signal Company. On the line of this road from Cincinnati to Chattanooga, there are 27 tunnels, and the signals were at first introduced in isolated sections for the protection of the line through these tunnels, but additions have been made from year to year so that now about 230 miles of the line is signaled. There are no distant signals, the overlap being used in all cases. The block sections are from one mile to 3.5 miles long. Trains in each direction are governed by the signals on the right hand side of the track as seen by the engineman. The overlap is in most cases about half a mile long, and its end is indicated by a white post, fixed at the side of the road, bearing the number of the signal controlled by that section. The beginning of each block section is also indicated by a post. This post is about 7 ft. high, and it bears at the top the words, "Electric Signal Block," the signal itself being set 210 ft. within the block section, so that the engineman may see it move from the clear to the stop position.

The drawing, Fig. 383, shows a sketch of three consecutive block sections with the signals, wires and batteries as they are arranged to protect trains from both rear and butting collisions. The letters, A, B, C, etc.,

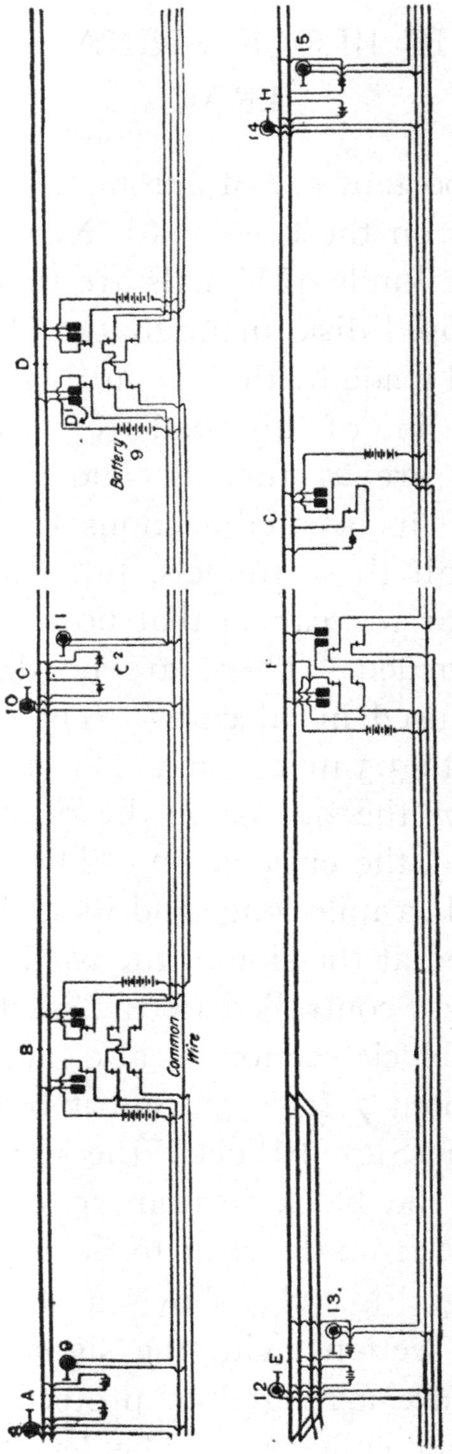

Fig. 383.

indicate the ends of the track circuits, not the block sections. The block sections are from A to C, from C to E, and from E to H. These letters are not spaced to scale, and parts of the rail lines and wires have been broken out to save space. The odd-numbered signals govern southbound trains and the even numbered signals northbound trains. The successive positions of a series of signals as they are operated by a southbound train are shown in the following table, in which C means clear signal and R means red signal.

Signal Nos.:	7	8	9	10	11	12	13	14	15	16
A to B..........	R	C	R	R	C	C	C	C	C	C
B to C..........	C	C	R	R	C	R	C	C	C	C
C to D..........	C	C	R	C	R	R	C	C	C	C
D to E..........	C	C	C	C	R	R	C	R	C	C
E to F..........	C	C	C	C	R	C	R	R	C	C
F to G..........	C	C	C	C	C	C	R	R	C	C
G to H..........	C	C	C	C	C	C	R	R	C	R
H—	C	C	C	C	C	C	R	C	R	R

For example, a train in section C-D holds signals 9, 11 and 12 in the stop position. So far as the action of the signals is concerned, it is to be remembered that these three signals would stand in the stop position for a train in this section without regard to which way it is moving. The way in which a train protects itself may also be clearly understood by following the connections to one signal, say No. 9. The presence of the train in section C-D opens the relay D^1 by short circuiting practically all of the current from battery C^2. The dropping of the armature of D^1 opens the three contact points, and it will be seen that the uppermost one of these opens a wire running, on poles, to signal 9. The same wire, it will be seen, runs through two relay points at B so that signal 9 may be similarly demagnetized by trains running or standing in sections A-B and B-C.

To illustrate the head end protection, consider a train going from E to D and beyond. When the first pair of wheels passes D, signal No. 9 is set at danger in addition to No. 11, which has been previously set.

In the table it will be seen that when a train enters section A-B it sets No. 9, Nos. 7 and 10 having been previously set; reaching section B-C, signal No. 7 is cleared and No. 12 is set; reaching section C-D, No. 10 is cleared and No. 11 is set, and so on from section to section.

The rules for the operation of automatic signals on a single track railroad are the same as those for a double-track road, except in the important particular that in case of failure of any signal a flagman must be sent forward to guard against a butting collision. The rule of the C., N. O. & T. P. on this point is as follows:

"If the signal is red before the train enters the block or is white and does not change to red before the train reaches the signal, it indicates danger and the train must stop; send a flagman in advance immediately; wait five minutes after the flagman has started and then proceed under full control, not exceeding six miles an hour, until the obstruction is reached or the train passes out of the block."

The flagman or rear brakeman of every train is required to observe the signals which control movements of trains running in the direction opposite to that of his own train, and to promptly report if such signals fail to clear when he passes out of the sections of track which they control.

To avoid unnecessary delays, each passenger train carries, in the baggage car, a light three-wheel handcar with which a brakeman can quickly run ahead of the train to the next block signal in advance.

QUESTIONS.

Ques. 62. Is the block system used to any great extent on single track roads?

Ques. 63. What officials control the block signalling on the Erie lines?

Ques. 64. How many miles of road does a dispatcher ordinarily control?

Ques. 65. What is the rule in regard to the progress of a train after it has been stopped?

Ques. 66. When a train is to use a side track beyond a station, what are the regulations?

Ques. 67. If an operator has to give a signal to two or more trains to go out of a side track, what rule is he to be governed by?

Ques. 68. What rules are conductors and engineers to follow in the running of trains?

Ques. 69. How are the caution and the safety signals given on this road?

Ques. 70. If either, or both the connections from the office should break, what takes place with the Mozier semaphore?

Ques. 71. What kind of a device are the signal levers in the office fitted with in order to prevent accident through forgetfulness?

Ques. 72. Has the plan of having the train dispatcher control the block signalling proven satisfactory?

Ques. 73. What is the chief advantage claimed for the system by the officers of the Erie?

Ques. 74. What do the dispatchers say in regard to it?

Ques. 75. How are the train dispatchers' orders delivered to the engineers when the train is in motion?

Ques. 76. What kind of signals are used on the Atchison, Topeka & Santa Fe?

Ques. 77. In case the connections break while the arm of the signal is in the caution position, what happens?

Ques. 78. If the semaphore arm is standing at "all clear," and the connections should break, what position will the arm assume?

Ques. 79. What other system of automatic signalling is used on the Atchison, Topeka and Santa Fe?

Ques. 80. On what single track road were automatic block signals first used to any extent?

Ques. 81. Why was the system first introduced on that road?

Ques 82. How long are the block sections in this tunnel?

AUTOMATIC BLOCK SIGNALS.

Automatic signals, actuated by the passage of the wheels of the cars and engines along the track, were tried in this country almost or quite as early as manual block signaling; and as the wages of competent men have been and are much higher in this country than in Europe the need of automatic apparatus was from the first felt more strongly here than in England. Consequently the automatic signals of the present day are the result of American development. In Europe they have not even been tried, except in a very few places. Thomas S. Hall, of Connecticut, began his experiments in 1866, and in 1871 had 16 miles of road signaled on the Eastern Railroad (now a part of the Boston & Maine). Rousseau's apparatus was in use at Philadelphia at the time of the Centennial Exhibition in 1876. These early inventors used track instruments or treadles which, pressed down by the wheels of a passing train, closed or opened the electric circuit in a wire carried on poles from one end of a block section to the other. Hall and his successors developed this method to a high degree of perfection; but wire-circuit signals, except the 16-mile section on the Boston & Maine, have been superseded by the rail-circuit, in which the rails of the track take the place of the wires; so that the oldest form of automatic signal which is still in general use is the clock-work apparatus of the Union Switch & Signal Company. The first important permanent installation of this was on the Fitchburg Railroad (now a part of the Boston & Maine), from Boston to

Waltham, Mass., in 1879-80, though Frank L. Pope had made experiments on the Boston & Lowell eight years earlier, and at Philadelphia in 1876 had worked a signal through a track submerged in water. Robinson's experiments were also important.

THE UNION CLOCK WORK SIGNAL.

This signal is a disk, turned on a vertical spindle, by clock work moved by a weight. The arrangement of

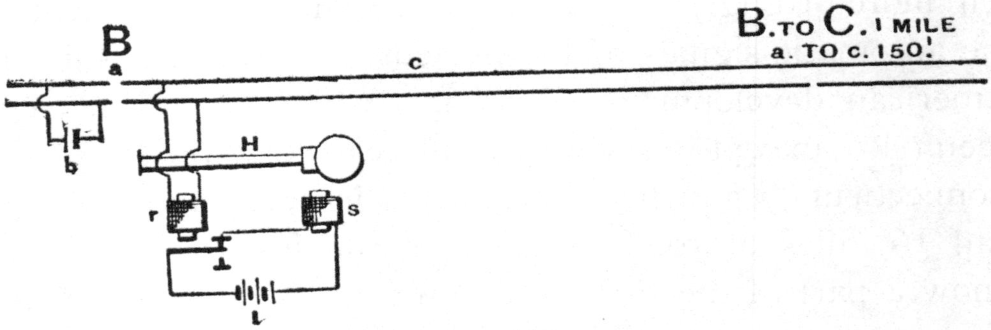

Fig. 384.

the signals and their relation to the track is shown in Figs. 384 and 385, in which the parallel lines represent the track. H and HC represent the signals.

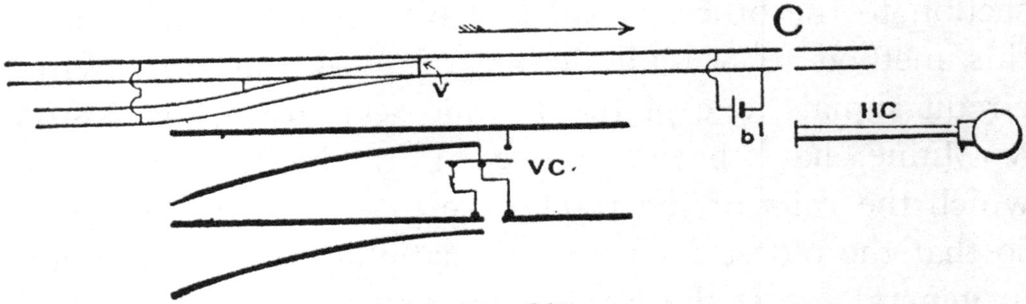

Fig. 385.

The block section extends from B to C. The rail circuit battery for this section is at b^1. When there is no train in the section this holds relay *r* closed, which in turn, closes the circuit through the signal magnet *s,* which holds the signal in the all-clear position. The presence of a train in the section, or any metallic connection between the left hand rail of the track and the right hand, makes a shorter path for the current and withdraws practically all of it from relay *r,* allowing its armature to drop, demagnetizing *s* and thus turning the signal to the stop position. The rail circuit, connected by wires, is carried through the rails of the side track at *v* for a short distance, so that a train entering the side track is protected, the same as when on the main track, until all of its cars are fully clear of the main line. The switch, or any number of switches in a block section, must be connected with the signal so that when set for the side track the signal will indicate "stop" to approaching trains. The rail circuit may be carried through a spring connection contained in a box at the switch, so that when the switch rails are moved the circuit will be broken on one side, and the signal turned to "stop." This arrangement is shown (enlarged at *vc*) in Fig. 384. Another method is to run the circuit from the local battery *L* direct to the switch, as shown in Fig. 385. With this arrangement the signal is controlled by the switch, even if the track circuit should be deranged; but if the switch is a considerable distance from the signal and the wires are run on poles there is increased liability to disturbance by lightning.

On some roads the signal, *H,* is set about 150 ft. within the block section, say opposite *c,* Fig 384, so that the engineman as he enters the section at *a* can see the signal turn from all-clear to "stop" and thus be assured that it is in working order and that his own train is protected. The battery, *b,* in Fig. 384, furnishes the current for the section A B. A battery for a track circuit must be of low intensity on account of the poor insulation of the rails from the ground. It usually consists of two jars of the gravity type, kept in a well far enough below the surface of the ground to be free from the effects of frost. At each rail joint the two adjoining rail ends are connected by a bond wire to insure a good electrical connection, at the ends of the sections as at *a* insulating material is put inside the splice bars and between the rails and around the bolts.

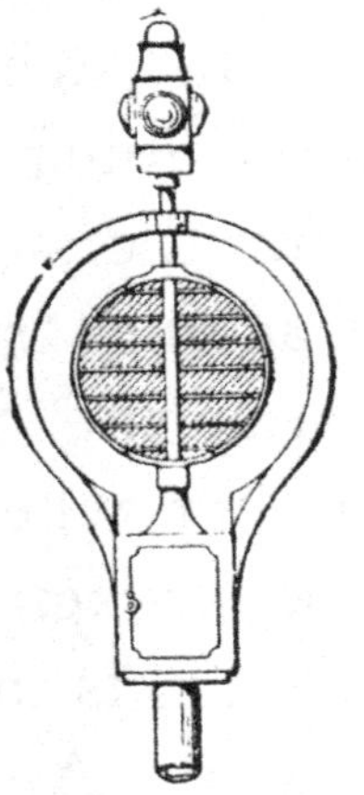

Fig. 386.

The appearance of the signal is shown in Fig. 386. The clock work is contained in the box immediately beneath the disk or target, and it is worked by a weight which hangs within the hollow iron post, 16 ft. high, which supports the signal. The target is fixed to the vertical axis supporting it, and when it is set so that its face is visible to the engineman it indicates "stop." For the all-clear indication it stands with its edge toward the train, or, as is the case on some roads, another target of a different shape, fixed at right angles to this, indicates all-clear. The stop target is usually painted red and the all-clear target, when used, is painted white. Where a signal of this kind is used as a permissive signal, green is used in some cases instead of red. The lamp fixed to the top of the spindle has, in two of its sides, opposite to one another, red lenses; and in the other two sides white or green lenses, according to which of these colors is used to give the all-clear indication.

The vertical axis of the signal is connected by miter gears with a horizontal shaft having a ratchet wheel and a winding drum which are turned by the weight, as just explained. The horizontal shaft carries on one end a cross-head which has in its arms 4 pins so placed as to engage with a detent on the end of a horizontal bar pivoted near its middle, and having a slight motion up and down, at right angles to the axis of the winding drum. The horizontal bar is attached at its rear end to the armature of an electro-magnet and can engage one or another of the pins in the cross-head, according as it is in one or the other of its positions. The place of the pins is so chosen with reference to the detent on the horizontal bar, that when the armature at its rear end is attracted, and the bar is at its highest point, the disk will be held in the position indicating "all clear," but if the current ceases in the electro-magnet, its armature falls away and the detent releases the clock work, which then turns one-quarter of a revolution, carrying the signal to the "stop" position. It is arrested by the detent engaging with the next pin on the cross-head. When the electro magnet is again energized the clock work makes another quarter revolution and brings the signal again to "all-clear." The current through the electro-magnet in the signal is controlled by the track circuit relay, as shown in the diagram, and this last is open or closed according as the block is occupied or unoccupied by a train. A broken rail, by interrupting the current through the relay, acts on the signal the same as a train.

Since the movement of the signal is dependent upon the clock work this must be wound at frequent intervals. That this may not be neglected and by chance leave the signal standing all-clear when the weight is run down, the current which operates the signal is made to pass through a pair of springs which are held together when the signal is wound up, but which are separated when the weight has nearly reached the bottom; so that just before the clock work is run completely down the circuit through the signal is opened and the signal left standing at "stop." Track circuit batteries must ordinarily be renewed about once in two months; the signal batteries will run longer.

In consequence of the imperfect insulation, rail circuits do not work well for a length of track greater than about one mile, and the most common length is shorter than this. Where it is deemed unnecessary to have the block sections so short as this a number of successive rail-circuit sections are connected one to another, so that the signals may be placed two or three miles apart. When this is done the circuit of one track section is opened and closed by a relay worked by the adjoining section. For example, if the point B, Fig. 384, were an intermediate point in a block section, the circuit made by battery *b* would be arranged to open and close by the action of relay *r*. This can be repeated as many times as may be found necessary.

The Overlap.—The arrangement of distant signals in automatic systems, and the use of automatic signals on single track, will be taken up in a later chapter. Most of the early automatic block signals had no distants, and to avoid the necessity of requiring trains to slacken speed at the signals (as in case of fog) the "overlap" is used. The arrangement of overlap circuits is shown in Fig. 387. The block section *B C* is of the usual length, as in Fig. 384, but the rail circuit is divided into two parts, one from *B* to *a* and the other from *a* to the outgoing end. The signal magnet, *s*, is controlled by the relays (R^2 and R^3) of both these track sections, and of course will thus stand in the stop position as long as the train is on the rails of either section; and in addition to this the relay, R^2, of the overlap section, controls relay R^1, so that, until a train

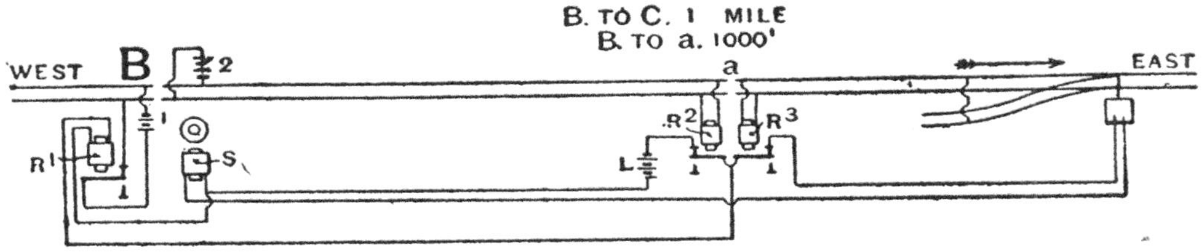

Rail Circuit With Overlap

Fig. 387.

passes *a*, the signal at *A* (the beginning of section *A B*, the next section to the west) remains in the stop position, thus keeping trains from approaching *B* until the preceding train has passed beyond *a*. The switchbox *x* contains a circuit breaker, controlled by the switch rails, so that whenever these are set for the side track the signal at *B* will indicate stop.

No new clock work signals have been put up since 1894, as other kinds have been found more desirable. The enclosed disk signal, first extensively used by Hall, was early in the field, but until about 1893 it was used only with wire-circuits. The track-circuit was from the first in great favor because, aside from its other merits, it usually detected broken rails, and because its efficiency was never impaired or interfered with by the accidental breaking apart of a train. The electro-pneumatic system was brought out soon after the clock-work, but its progress was slower because of its greater cost. The most extensive installations of clock-work signals are those of the Boston & Albany, the Old Colony (New York, New Haven & Hartford) and the Boston & Maine. On each of these roads large numbers of these signals have been in use 10 or 15 years and more. The regulations for their use are quite simple. The essential clauses of the Boston & Albany regulations are as follows:

When the number has been placed on an automatic signal, it denotes that it has been put in operation, and should be observed under the rules.

Enginemen should notice that an automatic signal changes to danger when a train approaches within 200 ft. of an all-clear signal.

When a train finds an automatic signal at danger (that is, showing a red target or light), or if an all-clear signal fails to change to danger, its speed will be reduced to three (3) miles per hour, and [it will] proceed only as the way is known to be clear; but the signal must be reported as if causing a stop. If the number of an automatic signal is covered or erased, enginemen will proceed with caution through the section to which that signal applies.

A train need not stop at a signal which is not lighted; but it must be reported, unless the number is covered or erased.

Enginemen must report all stops on the proper blanks.

It will be observed that the Boston & Albany sets the signals 200 ft. in advance of the entrance of a block section. This is to enable each engineman to know positively that the signal is in working order and has moved to the stop position, so as to protect his train. This 200-ft. arrangement is not universal with automatic signals. It is objectionable, at least theoretically, to have enginemen run past a signal which is in, or is moving toward, the stop position. Again, a signal might be moving to the stop position, just a second or two before the engine wheels enter the section, by reason of the misplacement of a switch; and the engineman would think that the movement was caused by his own engine. On the other hand, if the signal is arranged to move to the stop position after the engine has passed it, there is no check on its operation. A signal might be out of order and stand continually in the all-clear position and the engineman never know it. To have a brakeman or flagman on the rear end of the train watch the signals and record their movements is attended with some difficulty and has never been systematically undertaken. On railroads where the signals move to the stop position behind the engine, the officers assert that the design and construction of the signals and the efficiency of the care and inspection warrant the assumption that dangerous failures—failures of the signals to go to "stop" when a train enters the section—will not occur. The "normal-danger" arrangement, hereafter to be described, obviates some of the objections to setting the signal behind the engine. It was formerly customary to require trains to be brought to a stop on encountering a signal showing red, or not in working order; but with suitable discipline the present Boston & Albany rule is found satisfactory. The same rule prevails on the Central of New Jersey, where automatic block signals of another kind are used. On the Lehigh Valley and the Pennsylvania the rule requires a train to stop one minute, and then proceed, expecting to find an obstruction. These three roads, unlike the Boston & Albany, have distant as well as home signals in their automatic systems.

Enginemen's reports of stops, are, of course, promptly examined, and if found to indicate any defect or derangement of signals are at once attended to by an inspector.

Trouble is sometimes experienced by using hand cars on tracks connected with clock-work signals, because the hand car will not always make a thoroughly good connection between the rails, and the signal operates to show danger or safety according as the circuit is repeatedly completed or broken by the hand car, and the clock work is entirely run down; it may happen therefore that the signal requires to be wound several times a day in order to keep it in service. This can be prevented, of course, by using hand cars with wooden wheel centers, but then the signal would never be operated by hand cars, and if one should be left on the track the signal would not give warning to trains.

The efficiency and reliability of automatic signals depend on constant and intelligent inspection and care. The number of failures of clock-work signals is ordinarily but a very small fraction of 1 per cent of the number of operations, but nevertheless is probably larger than from other forms of automatic signals. A variety of causes may make the signal show "stop" when no train is in the block. A broken rail, a broken wire above or below ground, a bad connection or a broken spring in the switch box, failure of the battery or of the relays, or the breaking of a battery jar; also, since the rails require to be bonded with a wire past the joints, a failure may be caused by a broken wire, or by rust in the holes in the rails; cinders on the track may also cause so great a leakage of current as to make the signals inoperative. Breakage of battery jars has been a great source of trouble, but this can be almost entirely done away with by using first-class material for these jars.

The worst failures of an automatic signal are, of course, cases where the signal should show "stop" but is actually standing clear. These are very rare and the record is constantly becoming better in this respect. The principal causes are lightning (in the summer), sleet storms (in the winter), fixed magnetism in the relay or the signal magnets, or the sticking or binding of some part of the clock-work. Better mechanical work will eliminate most of the failures of the clock-work; such as remain being due to causes inherent in the principle, and only to be gotten rid of by abandoning the clock-work altogether. A heavy rain in winter turning quickly to snow or sleet will sometimes stick all the clock-work signals in the position in which they happen to be, and they will not work again until set free by breaking away the ice or by moderation of the temperature.

QUESTIONS.

Ques. 83. In what manner are automatic block signals actuated?

Ques. 84. What is the oldest form of automatic block signal which is still in general use?

Ques. 85. What is the Union clock work signal?

Ques. 86. How does the pressure of a train in a block section affect the signal?

Ques. 87. How are the rails connected at the joints to insure good electrical connection?

Ques. 88. Where is the clock work located?

Ques. 89. Describe the target or signal used in this system?

Ques. 90. What kind of a lamp is used at night?

Ques. 91. Describe the working of the device that moves the signal.

Ques. 92. Since the movement of the signal depends upon the clock work which must be wound frequently, what precautions are taken to prevent the attendant neglecting to wind the clock?

Ques. 93. How often must track circuit batteries be renewed?

Ques. 94. What is the longest distance over which rail circuits will work satisfactorily?

Ques. 95. Have any new clock work signals been installed in recent years?

Ques. 96. Upon what roads are the most extensive installations of clock work signals still in use?

Ques. 97. Mention the principal regulations regarding their use.

Ques. 98. In some instances the signals are located 200 feet in advance of the block section. Is this a desirable arrangement?

Ques. 99. What are some of the objections to this method?

Ques. 100. What is the rule on the Lehigh Valley and the Pennsylvania roads in this respect?

Ques. 101. What serious trouble results from the use of hand cars on tracks connected with clock work signals?

Ques. 102. How may this trouble be prevented?

Ques. 103. Upon what does the efficiency and reliability of automatic signals depend?

Ques. 104. Mention some of the causes that would make a signal show "stop" when no train is in the block.

Ques. 105. What is the worst failure that is likely to happen to an automatic signal?

Ques. 106. Do such failures happen very often?

Ques. 107. What are the principal causes of such failures?

THE ENCLOSED DISK SIGNAL.

The distinctive merit of this signal is that the moving parts are protected from wind, rain, snow and ice, and therefore can be made light enough to be moved by an electro-magnet of moderate size. Thus the use of a more powerful force, such as the weight and clock-work, is un-

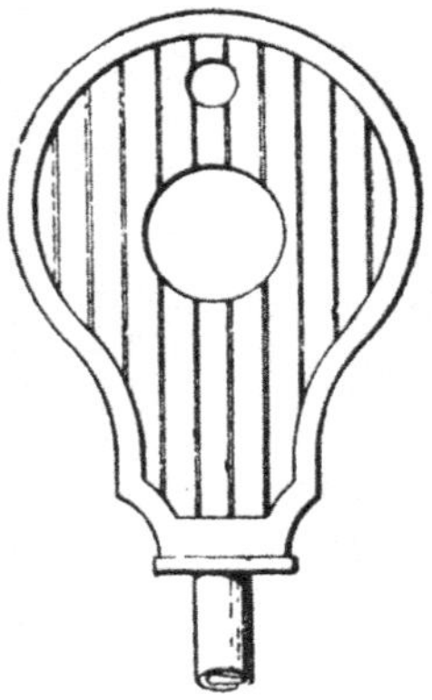

—*Case for Enclosed Disk Signal*

Fig. 388.

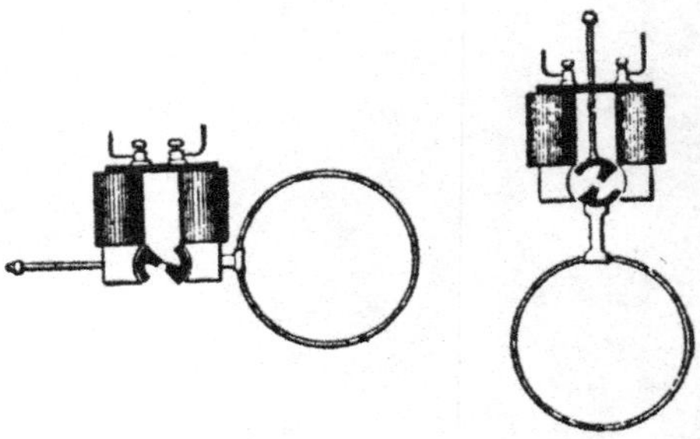

Hall Automatic Signal Magnet

Fig. 389.

necessary. The signals of this type made by the Hall Signal Company, are constructed on the same general principle as those used by that company at first with wire circuits. The arrangement of circuits shown in connection with the clock-work signals (Figs. 384 and 385) is applicable to all automatic signals. The outward appearance of a wooden case for an enclosed signal is shown in Fig. 388. The case is supported on a post of suitable height, or on a bridge, is usually painted a dark color. The disk, about 18 in. in diameter, made of silk or other light fabric, stretched on a metal ring, or of aluminum, very thin, is fixed to the armature of an electro-magnet in the way shown in Fig. 389. When the block-section is clear and the signal magnet is consequently energized, the armature holds the disk up, in the position shown at the left of Fig. 389; and the approaching engineman, looking at the glass-covered opening, sees the back of the interior case, which is painted white; this is the all-clear signal. On the withdrawal of the electric current the disk drops by gravity to a position before the window (the armature and rod turning on the axis as shown), thus displaying a red, or stop, signal. For a distant signal a green disk is used.

In the older type of signal the night signal is a lamp fixed at the back of the disk, outside the case, showing uncolored for "all-clear," and illuminating the cloth disk for the opposite indication.

In the latter design the lamp is placed back of the smaller opening near the top of the case, and the upper end of the rod which carries the cloth disk (see Fig. 389) has fixed to it a small transparent disk, colored to correspond to the color of the day disk.

The Hall enclosed disk shown in Fig. 388 is usually mounted on the top of a post or pillar, as before stated.

When home and distant signals are used, the cases are arranged as shown in the cut, which represents a signal on the Philadelphia & Reading Railroad in the Schuylkill valley. The upper signal has a red disk, and is the home signal, while the lower one has a green disk, being the distant for the block section next beyond. This style is used also on the Lehigh Valley road. On the Illinois Central for a few miles out of Chicago, some parts of the line have four, six and eight main tracks.

The arrangement of Hall disk signals on a bridge on this line is shown in Fig. 391. The latest design of the Hall Signal Company's signal instrument, described in connection with Fig. 389, is shown in perspective on this page. The lower disk, the daylight signal, is made of cloth (red for a home, and usually green for a distant signal); while the upper disk, the night signal, is made of thin glass, of the same color as the lower one.

Fig. 390.

The enclosed disk signal as described in connection with Fig. 388 is now made both by the Hall Signal Company and the Union Switch & Signal Company. In this design the case is of metal (pressed steel). It is made very narrow, leaving room enough only for the disk, and the electro-magnet is contained in a box outside the main case. The joints of the case are caulked, making it airtight, so that the disk cannot be affected by moisture. The box containing the magnet is also tight, but can be unlocked by the repairman. The front and rear glasses are of heavy plate fixed in brass rings. The lamp is supported on hinges, so that it can be turned around to be cleaned and filled. The whole case and the parts supported by it are very light, important parts being made of aluminum, and it may be revolved on its

Fig. 391.

bearings. With this arrangement the repairman can, while standing on the ladder, turn it around for cleaning or inspection, and if a signal is put out of service the case can be turned with its edge toward approaching trains.

Where it is desired to use green for the all-clear-color indication in the daytime the inside of the back of the case can be painted that color, so that when the disk is withdrawn the approaching engineman will see a green disk.

Enclosed signals may become obscured by damp snow sticking to the outside of the glass, and sometimes the rays of the sun, or light reflected from the sky, will make it impossible to see the color of the signal promptly on approaching it. These difficulties may cause delay to trains, but they cannot be called elements of danger, as a signal imperfectly displayed must, according to a rule which is universal, be regarded as a stop signal. The officers of the railroads using these signals do not seem to look upon these delays as very serious.

QUESTIONS.

Ques. 108. What is the distinctive merit of the enclosed disk signal?

Ques. 109. Of what kind of material is the disk made?

Ques. 110. How are the night signals arranged in this system?

Ques. 111. How are the enclosed disks moved from one position to another?

Ques. 112. Mention some of the elements of danger attending the use of this type of signal.

THE ELECTRIC SEMAPHORE.

The acknowledged inefficiencies of enclosed signals, and the fact that disks are not as distinctive to the eye as semaphore signals are, have inspired many inventors to experiment with automatic apparatus of various kinds for working full size semaphores.

The result is that on a number of roads the electric semaphore is in use, and the results are satisfactory.

The principal advantage possessed by this type of signal is, that each semaphore post is equipped with its own independent motor, and this motor is powerful enough to operate the semaphore in all kinds of weather, regardless of wind, snow or sleet.

The semaphore is moved from the top to the all-clear position by an electric motor of one-sixth horse-power, which is supported on a bracket attached to the side of the post below the semaphore arm.

When released, the arm is returned to the stop position by gravity.

By an arrangement of electro-magnets and wires, presently to be described, the signals are arranged to stand normally in the stop position. The semaphores are practically identical in design with the ordinary standard semaphore, except that ball bearings are introduced to reduce the friction. Each block section has a distant as well as a home signal, and each switch on the main track is provided with a vibrating bell which sounds, giving warning not to move the switch, as soon as a train enters the section in the rear of the one in which the switch is situated. Whenever a block section is clear (no bell ringing) a switch may be opened to connect a siding with the main track; and as soon as the main track is broken by the movement of a switch the bell begins to ring, thus apprising the attendant that the signal at the entrance of the block section has been set to stop approaching trains.

To work a signal motor, from 10 to 16 cells of Edison-Lalande battery are used. The battery is placed in a box at the foot of the signal post. When the motor is

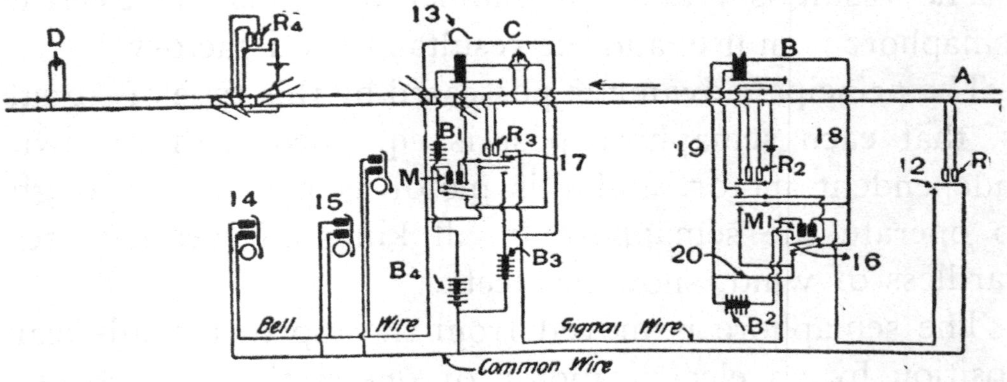

Electric Circuits for "Normal-Danger" Automatic Block Signals

Fig. 392.

run it winds up a cable, ¼ in. in diameter, which lifts the counterweight attached to the signal arm and thus causes the arm to assume the inclined or all-clear position. The motor has an automatic brake, consisting of a resistance coil which retards the revolutions of a soft iron circular disk fixed to the armature shaft.

The arrangement of the two signals, home and distant, controlling the entrance of trains to a block section, is shown in Fig. 392. The several rail circuits extend from A to B, B to C and C to D. The block section which we are now considering extends from C to D, say one mile. From B to C is about 1,800 ft., and from A to

B the distance is about 1,200 ft. R_1, R_2, R_3, R_4 are track relays; M and M_1 are relays directly controlling the motor circuit; B'_1 and B_2 are the batteries operating the motors, B_3 is the battery for operating the bells, and B_4 is the battery for operating the relays M and M_1. All the track relays are normally closed and the relays M and M_1 are normally open. The entrance of a train on the section A B shunts the track battery from track Relay R_1, closing the path for battery B_4 through 12, thus energizing relays M and M_1. The energizing of relay M closes the front contacts on that relay, thus completing the circuit for battery B_1 through this contact and the home signal motor, thus pulling this signal to the clear position. A similar result follows with relay M_1, battery B_2 and the distant signal.

When the home signal is in the clear position a circuit is closed through the circuit closer 13 (attached to the signal), battery B_3 and the bell magnets 14, 15, causing the bells to sound at these switches Any or all switches in the section of track controlled by this signal would be connected in the same way.

When the train enters the section B C, Relay R_2 is demagnetized, and the distant signal arm resumes the horizontal position, this being effected by the breaking of the circuit for battery B_2 through the opening of the front contact of this relay. The train entering section C D demagnetizes relay R_3, and the home signal resumes the stop position by reason of the breaking of the circuit of battery B_1 through the opening of the circuit through relay M, at 17. It will be noticed that the home signal is in the stop position and the circuit closer open; but the bells will continue to sound as long as the train is in the section C D, because a circuit is completed through the back contact point of R_3, battery B_3 and the bell magnet coils. When the train passes out of section C D all apparatus is restored to the normal position.

If, at the time a train passes "A", the block "CD" is occupied by a preceding train or car, or a switch is open, the demagnetization of "R" by the action of the wheels has no effect on relays "M" and "M_1", because the circuit of battery "B_4" is held open at "R_3" by the preceding train (or car), and as a consequence the signals remain in the horizontal position, and stop the train at "C."

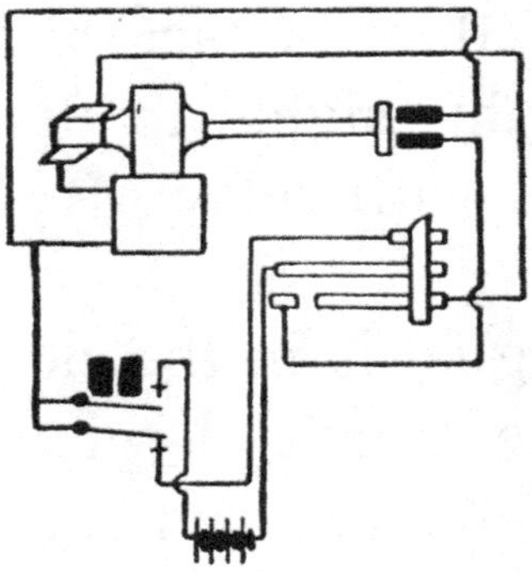

Fig. 393.

The lower armature of relay "M_1" when down, that is, when in the position that it takes when a train passes it, and the signal at "B" goes to the horizontal position, closes a circuit through wires 18, 19 and 20.

This circuit, which has no battery, runs through the motor, and the office of the lower armature of relay M_1 is to close a circuit through the motor, outside of the motor battery, and thus aid in retarding the motion of the signal arm as it approaches the end of its stroke. The relation of relay M_1 to the motor is shown in Fig. 393. Relay M has connections to its motor similar to those of relay M_1.

If, while a train is between B and C, a following train

should pass A, this second train, by holding M_1 closed, would prevent the closing of the no-battery circuit described in the preceding paragraph; to avoid this the dropping of the third armature of R_2, which is effected by the presence of the first train in the section B C, is made to close the no-battery circuit around M_1.

At stations and at all places where there are switches, an advance or starting signal is provided, so that a train having to stop at the station can be moved past the home signal while the block section to the next station in advance has not been cleared. With this arrangement the relative positions of the signals are as shown in Fig 394, in which *b* is the home and *s* is the starting signal.

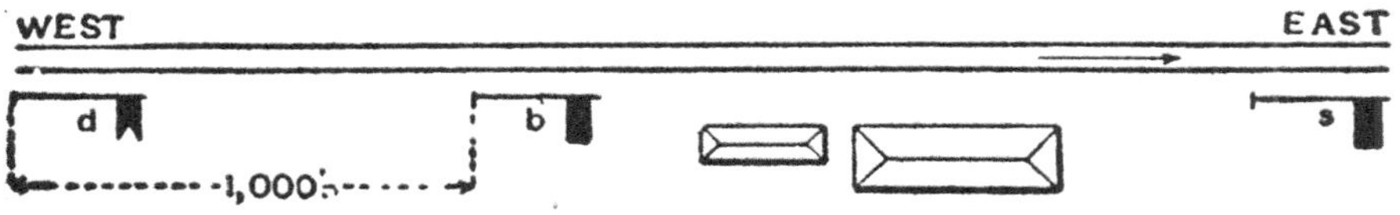

Fig. 394.

At the station between *b* and *s*, there is usually a cross over track, or a switch leading to a side track.

The normal position of each semaphore arm is horizontal, therefore a train approaching the distant signal will upon entering the preliminary section AB, Fig. 392, clear the distant home, and starting signals provided the track beyond the home signal is not occupied, but if the line beyond the starting signal is occupied, the system is so arranged that only the home signal would be cleared.

In later installations of this system, the distant signal B instead of being cleared by the same current that clears the home signal, and simultaneously therewith, is made to depend for its movement on the previous clearing of the home signal.

This is accomplished by either running an additional wire from the home signal, or else by the use of a polarized relay at B, to be worked by the track circuit BC.

The polarity of this track circuit is changed by the movement of the home signal.

ELECTRO-PNEUMATIC AUTOMATIC BLOCK SIGNALS.

Automatic block signals operated by the Westinghouse electro-pneumatic system are in use on some of the most prominent railroads in this country, as for instance, the Southern Pacific, the Chicago & Northwestern, and the Chicago, Burlington & Quincy. A portion of the New York Central, and the Pennsylvania lines are also equipped with this system.

The interlocking machine, and the pneumatic apparatus for moving switches, are the important features of the Westinghouse system. Block signals are worked in this way chiefly on very busy lines, where interlocking switches and signals for switches are numerous.

In such a situation a single air compressor serves for the yard work, and also for several miles of block signals.

The essential parts of the system are, first, the battery, rail circuit and track relay and connections; second, the signal, which is a full size semaphore; third, a cylinder, with piston, for moving the signal arm by compressed air, and an electro-magnet for controlling the admission of air to the cylinder; fourth, an air compressor (with reservoir and with cooling pipes to precipitate any excess of moisture) and a pipe laid along the side of the railroad, above or below the surface of the ground, to convey air under pressure to all the signals. There is a small auxiliary reservoir at the foot of each signal post. On four-track railroads the signals are arranged on bracket posts or on bridges. On the Pennsylvania, bridges are used in most cases, and the cut, Fig. 395, shows an iron-post signal designed to go on a bridge. In this signal each arm is arranged for giving two color-indications at night, having two spectacles. On the home signal one is

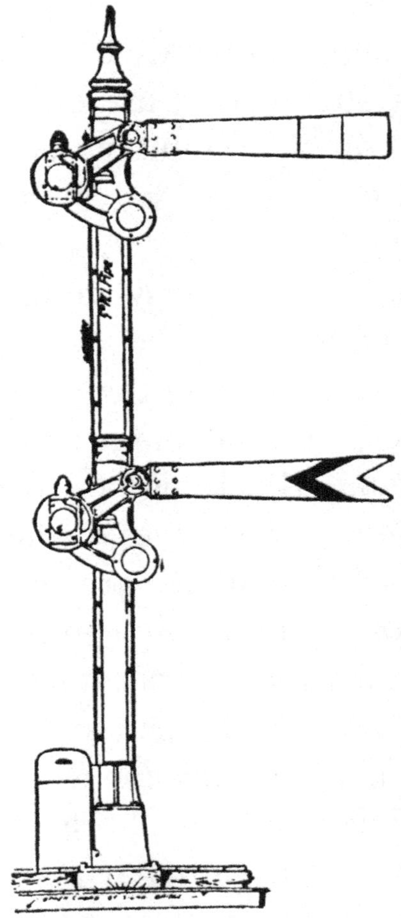

Two-Arm Signal, for Bridge; Iron Post

Fig. 395.

fitted with red glass to indicate "stop" and the other with green to indicate "all-clear." The distant signal shows a green light for "proceed" and a yellow one when the arm is horizontal (indicating "be prepared to stop at the home signal"). In Fig. 396 the lamps (and counter-weights) are not shown. The arms have each one glass, the go-ahead indication being given by showing no colored glass in front of the light. This is the Pennsylvania Railroad practice. Fig. 400 is an engraving from a photograph of a signal bridge on the Pittsburgh division of the Pennsylvania Railroad, where five tracks are signalled. The observer is looking west.

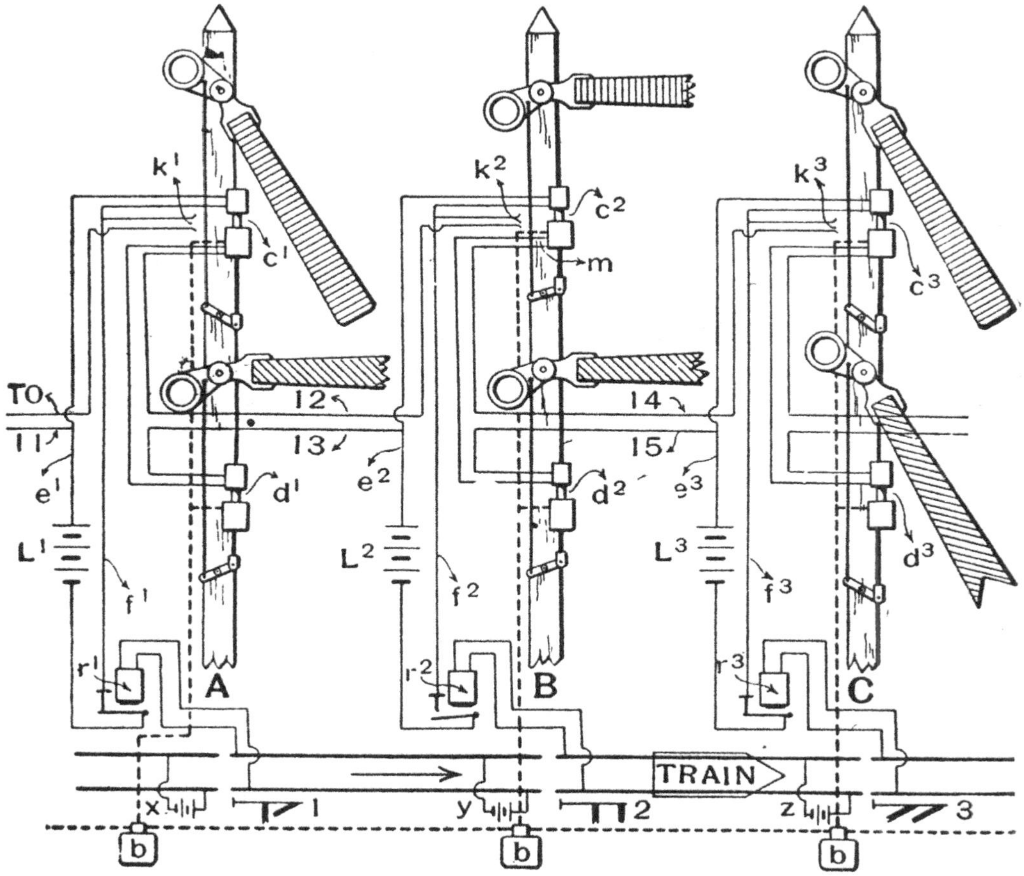

Electric Circuits for Electro-Pneumatic Block Signals

Fig. 396.

Where electro-pneumatic signals are used the block sections are usually short—not over 4,000 ft. long—and the distant signal for each section is placed on the same post with the home signal of the next section in the rear, as shown in the diagram Fig. 396. This diagram shows the condition of the signals, thus arranged, for three block sections as follows, A to B, B to C, and C to D.

There is a train in the middle section, but the other two are clear. The location of the signal posts A, B, C, is more appropriately indicated at 1, 2 and 3.

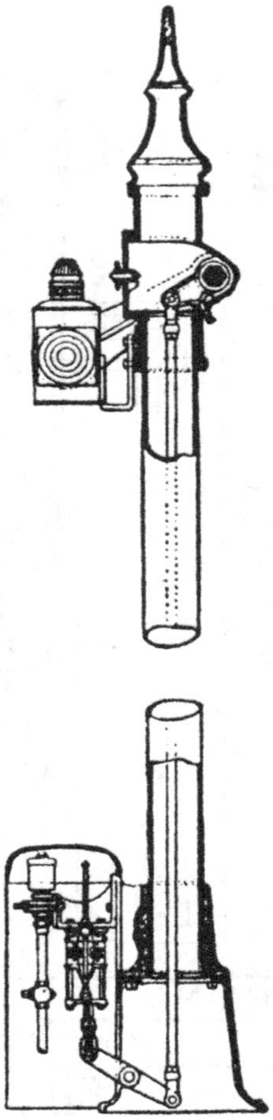

Sectional View Electro-Pneumatic Signal

Fig. 397.

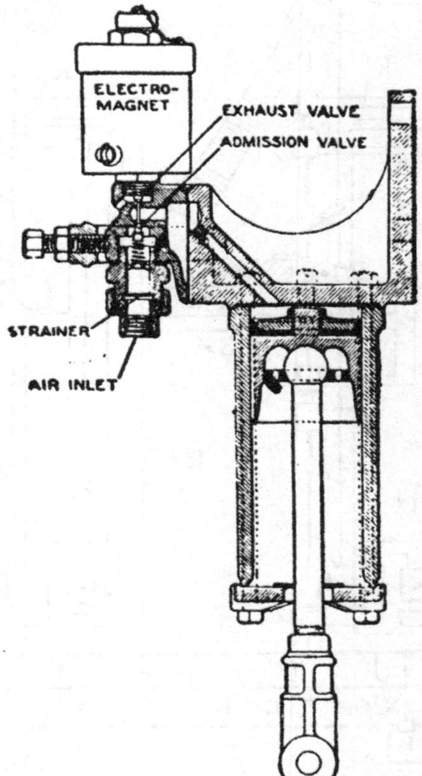

—Electro-Pneumatic Signal Cylinder

Fig. 398.

The home signal arms at the tops of these posts are moved to all-clear by the pistons in the cylinders c^1, c^2 and c^3, and the distant signals are similarly moved through cylinders d^1, d^2 and d^3. Air pressure is supplied to the cylinders through pipes from the compressor, indicated by heavy broken lines, and this supply is controlled by valves moved by electro-magnets, worked through wires e^1 and f^1, e^2 and f^2, etc. These circuits, energized by local batteries L^1, L^2, L^3, are opened and closed, as in all track circuit systems, by the relays r^1, r^2, r^3.

The action of the signals as affected by the passage of a train may be traced by beginning with section *CD*. This section being now free of trains, relay r^3 closes the circuit through the valve in cylinder c^3 and admits air, putting the home signal in the all-clear position. When the signal assumes this position it closes, at k^3, the circuit through wires 14 and 15 to cylinder d^2, controlling the lower arm at *B*, which is the distant signal for section *CD*. This signal, however, in the present case, remains in the horizontal position, because its circuit is held open at *m*, by the home signal at *B* being in the "stop" position. This break at *m* is provided so that an engineman approaching *B* may not be confused by finding a stop signal against him while the lower arm appears to give him permission to go on. A home signal, when it goes to horizontal, always causes the distant signal on the same post to do the same.

The presence of a train in section *B C* causes relay r^2 to remain open, and this leaves the home signal at *B* in the stop position. The circuit from battery L^2 being open, cylinder d^1 remains unchanged and the distant signal at *A* remains horizontal, corresponding to the position of the home signal at B.

The home signal at A indicates all-clear, in response to its track circuit relay r^1, the track section here being clear, the same as at C.

In the latest installations of this system of signals, the

wires leading from the home signals to the distants 10, 11, 12, etc., are omitted, the distant signal being controlled by a polarized relay, actuated by the track circuit.

The track circuit relay most commonly used with these signals has its moving parts enclosed in a circular dust-proof box, with glass sides, the coils, with their cores, being placed in a vertical position on top of the box. A

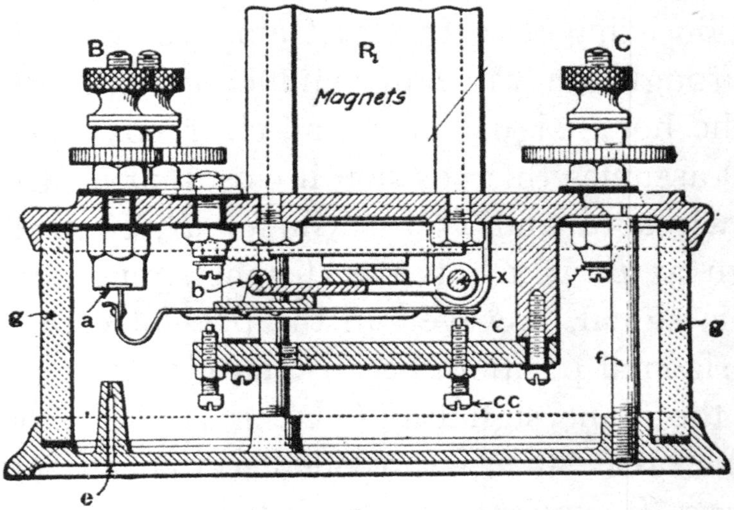

Relay for Automatic Block Signals

Fig. 399.

sectional view of this relay is shown in Fig. 399. The distinctive feature of this relay is the device for preventing the signal from remaining in the all-clear position in case the contact points at *a*, should become fused by lightning and remain stuck together. In such case *a* becomes the fulcrum and when the coils are de-energized, the lever *ac* drops at its right hand end, this being made possible by the introduction of the pivot at *b*. Connection is thus made at *c* and the local or signal current flows from the battery through *ac* and, by a wire connecting the screw *cc* with one of the binding posts at *C*, back to the battery, thus weakening the current through *x* to the signal magnet sufficiently to allow the armature of the latter to drop away, and thus cause the signal to move to "stop," the same as though the circuit had been broken. In normal working the main armature lever, pivoted at *x*, moves up and down, carrying the bar *ac* with it the same as though the two were rigidly fastened together. The current flows from the battery through binding post *B*, contacts *a*, the armature, and thence, by a wire attached at *x*, to the signal magnet and back to the battery. If the points at *a* become fused, leaving the signal circuit closed, the demagnetization of the coils *R*, by the passage of a train, will not open the signal circuit, and the vital function of the relay is interfered with; but the withdrawal of the current from *R* allows the right hand end of bar *ac* to drop by gravity, as before described, and withdraw nearly all of the current from the signal magnet by making a shorter return to the battery.

The circular box which encloses the working mechanism has top and bottom plates of brass; and sides, *gg*, made of a ring of plate glass. By using the small hole at *e* the armature can be moved with a long needle. One of the screws, *f*, by which the case is held together, can be covered at the top with a wax seal, thus enabling the chief inspector to keep a record of all occasions on which the box is opened. The delicate parts being enclosed in the air-tight box, this relay can be used in battery wells or other damp places.

The armature pivots supporting the bar *x* have glass bearings. The screws supporting this bar at its ends are hollow; and the bearing, which is a short piece of cylindrical-shaped glass, is inserted in the end of the screw. This glass is about ⅝ in. long; and to avoid failure in case of fracture it is encased in a hard rubber thimble which is screwed on to the brass screw. The binding post connections of this relay are so arranged that the inspector can take out one of the coils without disturbing the other; and each coil can be tested separately from the outside.

The upper contact points of track relays can be either carbon or platinum. Carbon is used, where its conductivity is sufficient, to avoid fusing by lightning. The U-shape of the end of the armature, combined with its lightness, makes it sufficiently elastic so that, when it is drawn up, the contact point is made to slide on the carbon or platinum surface, which makes the connection self-cleaning. In two-point relays the contacts are balanced so that the pressure is equal on both, even if the adjustment be not the same.

The arrangement for working automatic distant signals without a line wire is quite simple. Each distant signal being on the same post with a home signal, the track circuit governing the home signal is made to govern the distant also, by means of a polarized relay; or, rather a combined simple and polarized. The relay opens and closes the local circuit for the home signal in the usual manner, without regard to the direction of the current, by means of an armature at the ends of the cores; but it closes the circuit for the distant only when the current flows in a given direction, an armature to be moved by the reversal of the direction of the current being added between the coils. This reversal is effected, of course, by the movement of the home signal at the next post in advance—the second block section in advance—which block section is the one for which this distant signal gives indications. Referring to Fig. 396, relay r^2, for example, closes the circuit through c^2 in the usual manner, but when the home signal at C moves from the stop to the clear position, it changes the polarity of the track circuit from battery z, by a pole changer fixed to the signal post, and suitably connected to the track circuit, and thus causes the polarized relay, which takes the place of r^2 to close a circuit through d^3.

Fig. 400.

When the polarity of the track circuit is changed there is a brief period of time during which there is no current flowing through the circuit. To prevent this from opening the local signal circuit the relay is so wound and adjusted as not to respond instantly to the withdrawal of the current; before the armature falls away the current is restored.

The signal shown in Fig. 395, which is the latest form, has a hollow iron post, and the rod from the cylinder to the arm is within the post. This arrangement is shown in the sectional view, Fig. 397. The cylinder, fixed at the foot of the post and covered by an iron box, is shown on a larger scale in Fig. 398. The weight of the rod

Fig. 401.

within the post, with that of the spectacle casting, serves to throw the signal to "stop" when the air pressure is withdrawn, so that no other counterweight is needed.

Switches can, of course, be treated the same in this as in other automatic systems. Where mechanical semaphores are found, having been erected before the introduction of the block signals, as is often the case at large yards, circuit breakers can be fitted to the signal levers in the cabin, so that whenever the signalman prepares to move a mechanical signal to the "stop" position he will first cause the arm of the automatic home signal to assume the "stop" position. Any mechanical signal can be made to automatically indicate "stop" immediately after the passage of every train, by means of the electric slot already described.

On the Pennsylvania road mechanical signals are usually changed, when the automatic signals are introduced, to "semi-automatic." Cylinders are added to move them by air-pressure, and the signalman can clear them only while the main track is clear; though he can open the wire circuit and throw the signal to "stop" at any time. Those mechanical signals which govern movements off from the main tracks are left in the same condition as they were in before.

QUESTIONS.

Ques. 113. What are the essential parts of the electro-pneumatic automatic block signal system?

Ques. 114. Is this system much in use?

Ques. 115. On the Pennsylvania road what does "red" indicate?

Ques. 116. What does the green signal indicate?

Ques. 117. How long are the block sections, where the electro-pneumatic system is used?

Ques. 118. How are the distant signals controlled electrically, in the latest installations of this system?

Ques. 119. What is the distinctive feature of the relay in this system?

Ques. 120. What facilities are provided for the inspection of these relays?

Ques. 121. Describe the arrangement for working automatic distant signals without a line wire.

Ques. 122. How does the direction in which the electric current flows affect the working of the relay?

Ques. 123. How may a mechanical signal be made to automatically indicate "stop" after the passage of every train?

THE GRAFTON SYSTEM.

The Pittsburg, Fort Wayne & Chicago Railroad Company is using on some portions of their system an arrangement of automatic block signals devised by Mr. W. M. Grafton, signal engineer of the road.

In this system, which was first described in the Railroad Gazette, of Jan. 18, 1901, the distant signal is dispensed with, and by using a single semaphore arm in three different positions, it is made to serve the purpose of both a home and a distant signal.

Three-Position Automatic Block Signals

Fig. 402.

The arrangement of the signals for three consecutive sections is shown in Fig. 402. The signal at the entrance of a section which is occupied by a train, stands in the horizontal position, as in the usual practice; but the signal at the entrance of the next section in the rear, which under ordinary practice would have two arms, one indicating all-clear for that section, and the other "caution" for the succeeding or occupied section, stands in the diagonal position. This indicates that the section immediately in front (B C) is clear, but that the signal at the entrance of the next section (C D) may be expected to be found in the stop position. The signal in the vertical or all-clear position as at A indicates that two sections in advance, A B and B C, are clear. In speaking of these signals, the terms clear, caution and stop are used for the three positions shown at A, B and C respectively.

Each signal is controlled by the track section immediately in front of it by means of the track circuit in the usual manner; but the control of signal B, for instance, by a train in section C D, is effected by means of a wire on poles from C to B. The signals are worked by electric motors of one-twelfth h.p., one at each signal.

Fig. 403.

Current for these motors is supplied by a generator in a central station, and in addition to this, there is a storage battery at each signal bridge.

On that portion of the road, which is four-track, all of the signals are on bridges spanning the tracks. One of these bridges is shown in Fig. 403.

At all places where there are switches in the line, automatic signals are placed close enough to the interlocking

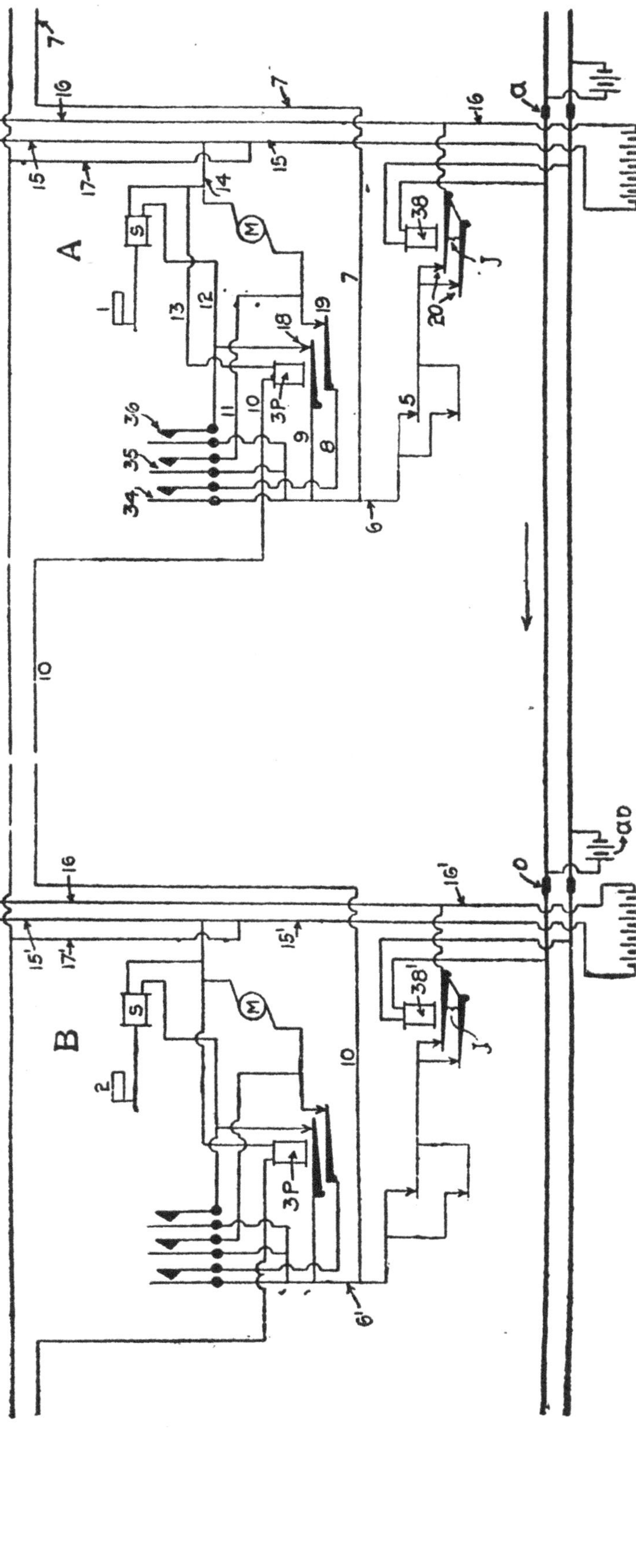

Fig. 404.

to serve as convenient starting signals for trains which have passed through the interlocking.

The motor moves the signal arm by means of an apparatus analogous to the electric slot, and it is called the

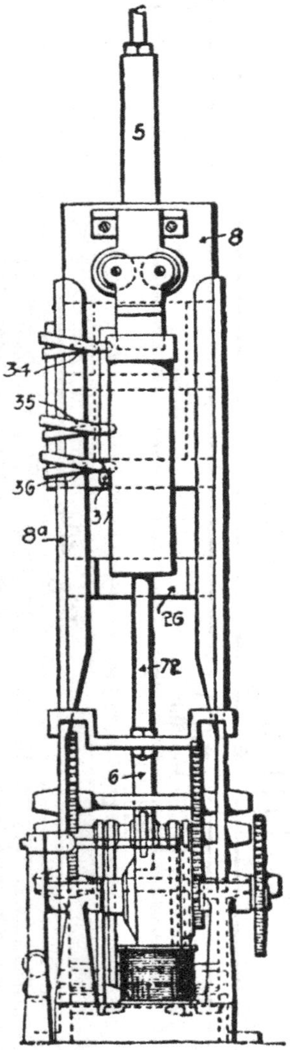

Fig. 405.

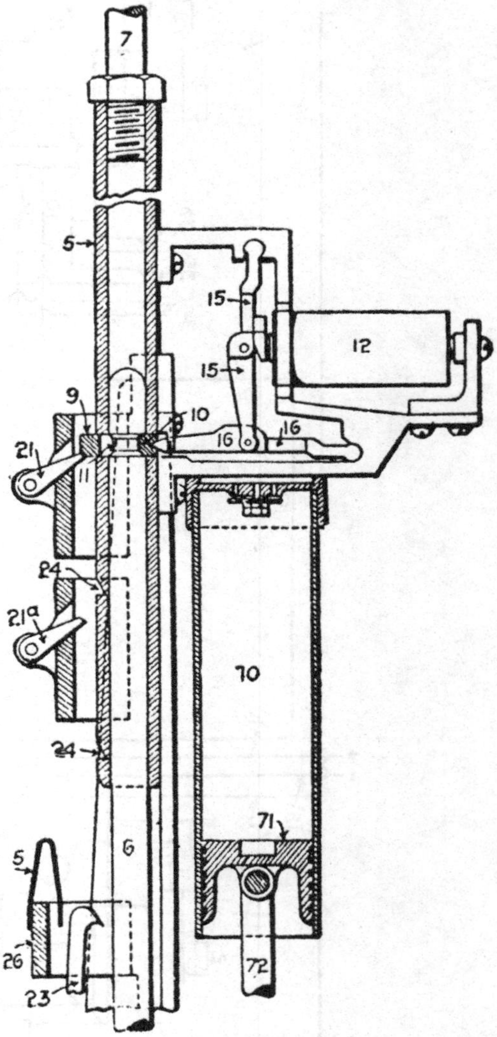

Fig. 406.

slot instrument. The track relay, by which trains control the signal, is connected as shown at 38, station *A*, Fig. 404, S being the "slot" mechanism and M the motor. When the block section A B is clear 38 is energized, and holds up its armature. This energizes the slot magnet 12 in Fig. 406, S in Fig. 404, and locks the signal rod 5, Fig. 405, holding the signal in the all-clear position.

The entrance of a train de-energizes 38 and 12, and thus unlocks the signal.

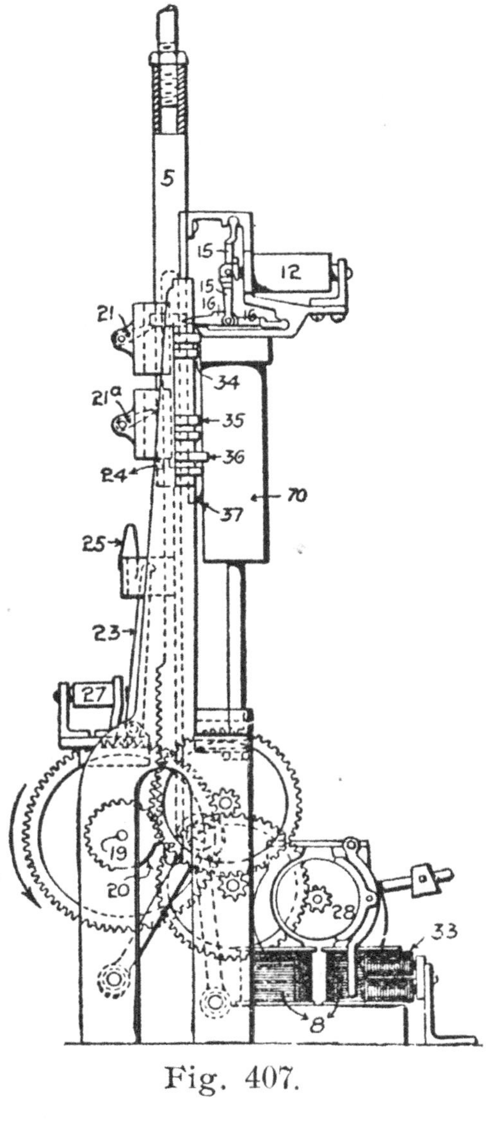

Fig. 407.

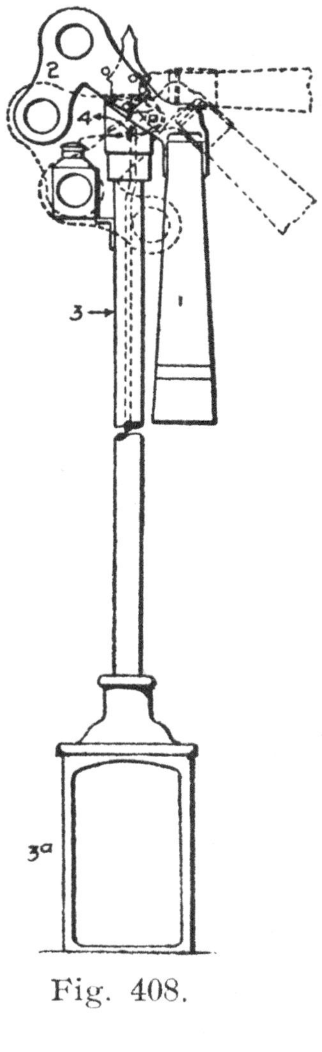

Fig. 408.

Before examining the action of this signal system, the function of the different parts of the operating mechanism should be thoroughly understood.

The motor and "slot" are practically one machine, contained in an iron box at the foot of the signal post.

Fig. 404 is a diagram showing the simplest arrangement of circuits for the signals for two adjoining block sections.

Figs. 405, 406 and 407 show the operating mechanism of the signal. Fig. 408 is the signal, and Fig. 409 shows a detail.

As shown in Fig. 406, the connection from the signal to its operating mechanism is formed by a sleeve 5 and a rod 6, fitting within the sleeve. The sleeve forms a guide for the upper portion of the rod 6. The sleeve is provided with wings 8, Fig. 405, which serve as guides sliding in grooves formed in the uprights 8a of the frame of

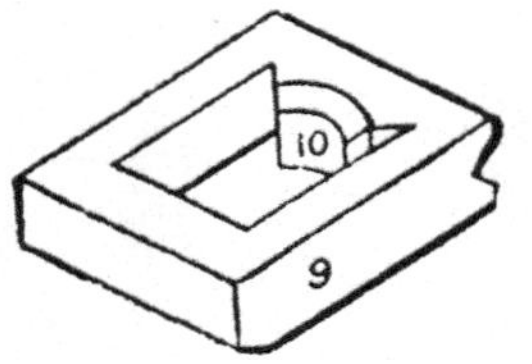

Fig. 409.

the machine. The sleeve and rod are locked together by locking block 9, which has a locking projection 10 (see Fig. 409), to engage with notch 11, in the rod 6. This rod moves freely through the block when the latter is not held in position by lever 16 controlled by electro-magnet 12, which is supported on sleeve 5. Links 15, 15, 16, 16, form toggle joints by which magnet 12, when energized, presses block 9 to the left; while, on the other hand, whenever 12 is de-energized the weight of sleeve 5, with that of the casting on the signal arm is sufficient to force block 9 to the right, and thus unlock rod 6 and permit it to slide freely in the sleeve. The reader should remember that to move the signal arm down, the rod 6 and sleeve 5 have to be moved up.

The signal is shifted to caution and clear positions by the electric motor, shown at the bottom of Figs. 405 and 407.

The armature shaft is connected by a train of gear wheels to shaft 19, carrying pinion 20.

The teeth of this pinion engage a series of teeth on rod 6, and the rotation of the pinion thus lifts the rod, and shifts the signal to the caution, and clear positions.

In order to permit rod 6 to move down, after the signal has gone to "stop," without reversal of the motor and its train of gearing, some of the teeth on the pinion 20 are removed; when the untoothed portion of the pinion comes opposite the rack on rod 6 the latter is free to drop.

In order to relieve the train of gearing and motor from the weight of the signal when in clear and caution positions the pawls 21 21a are mounted so as to project into the path of block 9 when it is in locking position. These pawls are so located vertically on the frame that when the signal has been moved down to caution or clear position the projecting end of block 9 will have moved just above one or the other of them, which will drop under the projecting end of the block, and thereby prevent any downward movement of the sleeve (= upward movement of the signal arm) so long as the block is held locked. Block 9 and the pawls 21 21a are so beveled that the weight of the signal and sleeve will have a tendency to move the block to unlocking position. Arm 23 is a lock for holding the signal in stop and caution positions so long as track conditions require such positions. It is mounted on the frame of the machine in such position that the hook on its end will engage notches 24 on the sleeve when the signal is in stop or caution position. The hook is shifted to engage these notches by spring 25, which is held in a position in which the sleeve will be free to move by a magnet 27, Fig. 407, controlled by train movements.

The drum shown at 28, Fig. 407, is a brake to prevent over-running of the signal-shifting mechanism when the circuit of the motor is broken, but this is not used in the machines on the Fort Wayne road. When the brake is used its shoe is held away from the drum by means of electro-magnets 33, which are included in the circuit of the motor.

In order to prevent shocks to the mechanism when the signal is shifted, a cylinder 70, Fig. 407, constituting a dash-pot, is provided. Stem 72 of the piston is secured to the frame of the machine.

The motor circuits and the circuits for energizing the slot-magnet 12 are controlled by the movements of the signal, and to this end circuit-breaking mechanisms 34, 35 and 36, Figs. 404, 405 and 407, are so arranged as to be opened and closed in a certain pre-determined order. These lie in the path of a projection or rib 37, Fig. 407, carried by the sleeve 5. Circuit-breaker 34 and 35 are opened by the movement of the sleeve while 36 is closed thereby. These circuit-breakers form a part of the motor circuits, one of which consists, starting from the battery *p*, Fig. 404, of the wire 16, armature of relay 38, wire 6, circuit-breaker 35, wire 11, motor M; thence by wires 14 and 15 back to the battery. This is called the caution circuit, as when closed it moves the signal arm from the stop to the caution position. The other motor circuit, for moving the signal-arm from caution to clear consists of battery *p*, wire 16, armature of 38, wire 6, circuit-breaker 34, wire 8, contact points 19 of relay 3P, motor M, wires 14 and 15 to the battery. Circuit-breaker 35 is closed when the signal is in the stop position. Circuit-breaker 34 of the clearing circuit is so located as to be closed as long as the signal is in either the stop or the caution position, but it will be opened by rib 37 as soon as the signal reaches the clear position. Both of these motor-circuits are controlled by the track-relay 38, so that as long as the track-relay is de-energized the motor-circuits will remain open. The "catch-magnet" 27 is also de-energized, permitting spring 25 to press 23 against 5, so as to engage with one or the other of the notches 24. The signal-clearing circuit is also controlled by the relay 38′ at station B, the circuit of which consists, starting from battery p' at B, of wire 16′, contact points of 38′; wire 6′, wire 10, relay 3P at A, wires 13, 14, 15 and 17 to the common wire and thence by 17′ and 15′ back to the battery p'. As the circuit for clearing the signal at A is controlled by track-relay 38′ it cannot be cleared until relay 38′ becomes energized, and this cannot occur as long as a train is on section B C.

As the circuits for the motor are the same at each station, it will be understood that the signal at B will go to stop as soon as a train enters upon its section.

When rod 6 is raised as in Fig. 407, the teeth on the rod are engaged by only a few of the teeth of the mutilated pinion 20, and those adjacent to the mutilated portion. As soon as the slot-magnet is de-energized the sleeve 5 drops along the rod 6, carrying with it the slot mechanism. When the motor is energized the pinion 20 will be rotated in the direction of the arrow and its teeth disengage from the rack of the rod 6, which will then drop sufficiently far to bring the notch therein in line with locking-block 9. As the slot-magnet is energized simultaneously with the closing of the motor-cir-

Signal Tower.

cuit, block 9 will lock the sleeve and rod together as soon as the notch comes into line with the block. By the continued rotation of pinion 20 the rod 6, and with it the sleeve and signal, will be shifted to caution position, where the circuit-breaker 35 will be opened, opening the motor-circuit and de-energizing magnet 27; catch 23 will then engage with the lower one of the notches 24, thereby locking sleeve 5 so that the signal cannot be shifted to clear; but will not interfere with its further movement to stop position. When the signal moves to stop, in the manner described, catch 23 will engage the upper one of the notches 24 and hold the signal in that position, the circuit of magnet 27 being broken.

Fig. 404 shows the arrangement of the electrical apparatus and wires at two adjoining signals or stations, A and B. The two signals 1 and 2 are shown in the all-clear position, as this section (A B), also sections B C and C D, are clear of trains.

We may now follow the movement of a train from A to B, Fig. 404. The entrance of a train at A to section A B (or the presence of a pair of wheels at any point on the track between *a* and *b*) provides a short circuit for the current from battery *a b* and de-energizes track relay 38. This breaks all circuits leading to signal 1 at A through wires 6, 8, 9, 11, 12, 13 and 14. Slot magnet S at A is de-energized, sleeve 5, Fig. 406, drops and the signal goes to the stop position. It is held in that position by arm 23 (engaging in upper notch 24) until the train passes out of the section at B.

When the train passes B relay 38 is again energized and closes the motor circuit, (battery *p* wire 16, contacts 38, wire 6, contacts 35, wire 11, motor M, wires 14 and 15). The slot circuit is also closed, (battery *p* wire 16, contacts 38, wire 6, cut out contacts 36, wire 12, slot magnet S, wires 14 and 15). S and M being energized, and 27, Fig. 407, being also energized, the signal is shifted from "stop" to caution. When it reaches the cau-

Interior of Signal Tower.

tion positions it separates the points at 35, and then opens the motor circuit. The signal is held in this position until the train passes out of block section B C. The sleeve 5 is held down by hook 23 engaging in lower notch 24, and it is held up by pawl 21a dropping under block 9.

When section B C is cleared, by the train passing C, relay 38′ at B is energized; and this energizes relay 3 P at station A by the completion of a circuit from battery *p*′ by wire 16′, points of 38′, wire 6′, wire 10, relay 3 P, wires 13, 14, 15, 17 and back as before described. Contacts of relay 3 P now being closed, a new path is provided for the slot circuit from battery *p* by wire 16, contacts 20 of track relay 38, wire 6, wire 9, contact 18, wires 12 and 14, and wire 15 back to battery. The motor circuit is also closed from battery *p* by wire 16, track relay contacts 20, wire 6, cutout 34, wire 8, 3 P relay contact 19, motor M, wires 14 and 15 back to battery. The motor and magnet 27 being energized the signal is shifted from the caution to the clear position. When it reaches that position the motor cutout 34 is opened and the signal is held there by pawl 21, block 9 being pressed to the left by action of magnet 12. In the movement from caution to clear, cutout 36 is opened and the slot circuit passes through relay contacts 18 and wires as last described.

In case a switch in section B C should be opened, or any break should occur in that track circuit, magnet 12 (S, signal 1, station A) would be de-energized (through 38′) and signal 1 would go to stop position; but 12 will be immediately re-energized, as, although 3 P at A is opened, by the opening of 38′, a path for the motor circuit is provided by circuit breaker 36, which is closed by the movement of the signal from the clear to the stop position.

The track relays used are two-point or double-contact relays and were made to carry two separate circuits. A wire, termed a "jumper," is soldered across, connecting together the two armature levers and the relay is used as a single contact relay. This was done to increase the carrying capacity if the current should become excessive.

At 5, Fig. 404, is shown the location of a cutout in the main feed wire, which is operated by the movement of the signal mechanism, and which is designed to automatically open the circuits leading to the signal and prevent waste of battery, if wires 9 and 12 should become broken or disconnected.

Fig. 410 shows the arrangement of signals and electrical connections at a junction where the automatic circuits must be under the control of an attendant. The switches and switch locks and the signals (like 2 and 12) which govern movements that are wholly within the yard, are worked by levers in a mechanical frame in the usual way; signals *y* and *z*, which are for movements wholly outside the interlocking limits are entirely automatic; and the signals (1 and 14) which control high-speed movements through the yard are worked by electric motors and controlled by the track circuit, but are also controlled by the signalman, who can set them in the stop position at any time by opening the electric circuit which holds them in the clear position.

The lower arm of signal 14 is fixed in the stop position, and always shows a red light at night. According to a general rule in force on the Pennsylvania Lines west of Pittsburgh this "dummy" arm is always supplied in

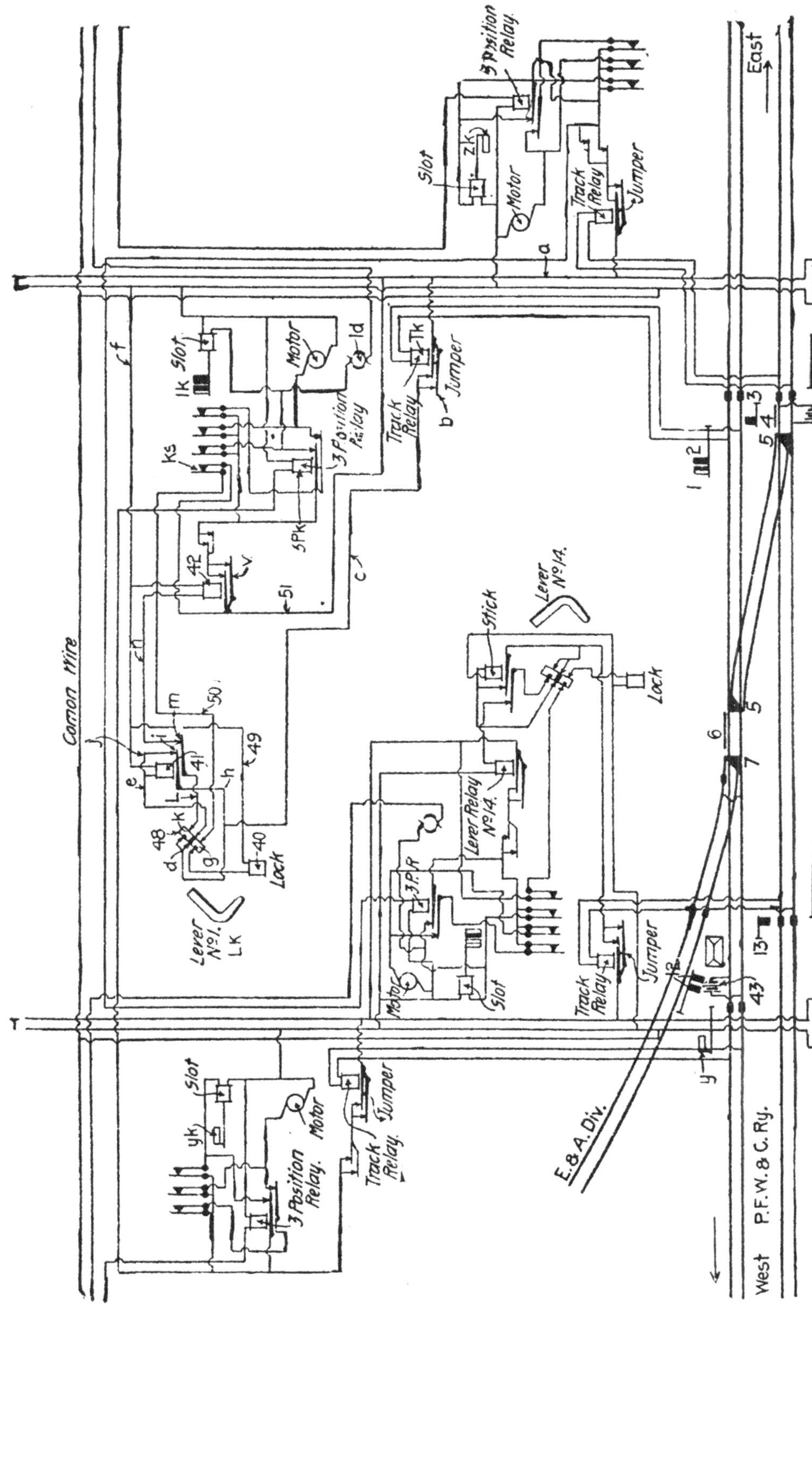
Comon Wire
Slot
Motor
3 Position Relay
Track Relay
Jumper
Lever No. 1. LK
Lock
Stick
Lever No. 14.
Lever Relay No. 14.
East
West
P.F.W. & C. Ry.
E. & A. Div.

Fig. 410.

situations like that here shown, in order to make the practice uniform with that where there are diverging tracks, as at signals 1 and 2.

Signal *z* is shown, with its electrical connections, at *zk* and signal *y* at *yk;* these are controlled by the track circuits the same as the signals in Fig. 404.

Signals 1 and 2 are shown at 1 *k*. Signal 1 is controlled by relay *Tk,* the same as by relay 38 in Fig. 404; and relay 3 *Pk* performs the same function for the track circuit beyond signal *y* as does the similar relay in Fig. 404 for section B C.

The circuit breaker at 1 *d* is open whenever signal 1 is in the stop position, thus insuring that the distant indication for signal 1 shall always be "caution" when 1 indicates stop.

Lever No. 1 for signal No. 1, is shown at *Lk*. The signalman is free to reverse the lever whenever it is not locked home (it is locked home mechanically whenever switch 5 or switch 7 is not set for the main track, and whenever signal 12 or signal 13 is cleared); but reversing lever No. 1 does not clear its signal unless the track is clear from signal 1 to signal *y;* for the track circuit of this section (battery 43, relay *Tk*) controls the lever by circuits presently to be described.

At 40 is shown the electric lock for lever No. 1. This lock when de-energized holds lever No. 1 immovable (if the lever is in the normal or "stop" position) and it is energized (unlocking the lever) only when its circuit is closed at *g* and at *ks*.

At 48 is shown a rotary circuit controller which is operated by the movement of the latch lever on lever No. 1. The "stick relay," 41, is to compel the signalman to return the lever to the normal position before clearing the signal a second time. The "lever relay," 42, is energized only when lever No. 1 is reversed. In these semi-automatic signals this relay controls the same circuits that the track relay does in the entirely automatic.

The arrangement of circuits is as follows: From battery by wire *a,* track relay contacts *b,* wire *c,* circuit controller 48 at contact *d,* wire *e,* stick relay 41, wire *f* back to battery. When 48 is shifted by lifting the latch lever, contact *g* is closed and *d* is opened, the stick relay circuit remaining complete through wire *h,* stick relay contact *i,* relay 41, wire *f* to battery. Contact *k* of the circuit controller 48 has also been closed, completing the lever relay circuit through *a, b, c, k,* wire *L,* contact *m,* wire *n,* lever relay 42, wire *f,* back to battery.

When relay 42 is closed by the action of the signalman on the latch the signal circuits will be completed through its contacts, the same as they are through the track relay in the entirely automatic, and the signal is shifted by the motor from "stop" to go ahead.

When a train enters the block section between signal 1 and signal *y,* track relay *Tk* is opened, the circuits through stick relay 41 and lever relay 42 are broken and the signal is shifted to stop. Stick relay 41 will remain open, as no current can pass through its coils until lever 1 is returned to its normal position completing the circuit for 41 through *d.*

Thus the putting of the signal to the stop position is done in a manner similar to that described in Fig. 404; but the movement of it the other way is controlled by the signalman, through 48 and 42. Closing 42 has no effect on the signal unless the track is clear, as the presence of a train on the track between 1 and *y* will hold the motor circuit open at *Tk,* and a train between *y* and the next block signal will hold it open at 3 *Pk,* the same as in Fig. 404. When, by lifting the latch preparatory to pulling the lever, the signalman closes 42, he closes the motor circuit at *v;* but if the track is clear only to *y* the signal will be moved only to the caution position; while if the section beyond *y* is clear the signal will be moved to the clear position.

When lever lock 40 is de-energized and its armature down, it locks lever No. 1 immovable, provided the lever is in its normal or "stop" position. When the signalman lifts the latch preparatory to pulling the lever he closes the circuit of 40 at *g* and thus causes 40 to unlock, provided its circuit is also closed at *ks* (wires 50, 51 battery, wires *f* and 49); and it is closed at *ks* if the signal is in the stop position.

On the approach of a train from the east the signalman, if the westbound track is clear as far as *y,* proceeds to pull lever 1 to put the arm of signal 1 in position to admit the train. He must first release the lever by lifting its latch; this moves 48, closing *g* and opening *d;* (41 is not affected, as it has two connections from 48); lifting the latch also closes *k,* closing 42; and this last starts the motor to move the signal arm downward.

The slot machine circuit closers shown on this plan are the same as those shown in Fig 404, except that there is a fourth one, *ks,* its function being to close the lock circuit when the signal is at "stop."

The control of lever No. 14 is arranged in the same way as that of No. 1.

QUESTIONS.

Ques. 124. In what respect does the Grafton system differ from most other systems?

Ques. 125. How is each signal controlled and operated?

Ques. 126. What is the power of these motors?

Ques. 127. How is current supplied to them?

Ques. 128. How are the signals supported on the four track portion of the road?

Ques. 129. What are the functions of the automatic signals in the Grafton system?

Ques. 130. Where is the motor located in this system?

THE TAYLOR SIGNAL CO.'S SYSTEM.

The Taylor Signal Co.'s system of electric switch and signal apparatus consists of interlocking and signal machines moved and controlled entirely by electric power.

The switch movement protected by a weatherproof box is fixed to a sleeper close to the switch and signal motors similarly protected are placed upon the signal posts.

Electric circuits passing through the rails are used in the place of detector bars.

Current from a 60-volt storage battery is the power for all of the functions in this interlocking device, except the rail circuits, which are worked by gravity batteries in the ordinary way.

Current from the storage battery is used only while switches or signals are being moved, therefore the consumption is very small.

A gasoline motor operating a dynamo is used to charge the accumulators.

As in other interlocking systems there is an indicator acting on the lever in the cabin to indicate to the signalman that the stroke of the track rails is completed, and permit him to finish the stroke of the lever.

The switch motor is shown in Fig. 411. Through the gearing shown the motor M turning 20 revolutions in about 1½ or 2 seconds, revolves wheel 1 one revolution to make one revolution of the switch.

The switch is connected by rod 2, pivoted at a to cam movement 3 revolving on pin b.

Crank-pin *c* on wheel 1 besides moving the switch by means of the cam, moves the lock directly by rod 4, connected to the lock through 5 and 6. To the end of

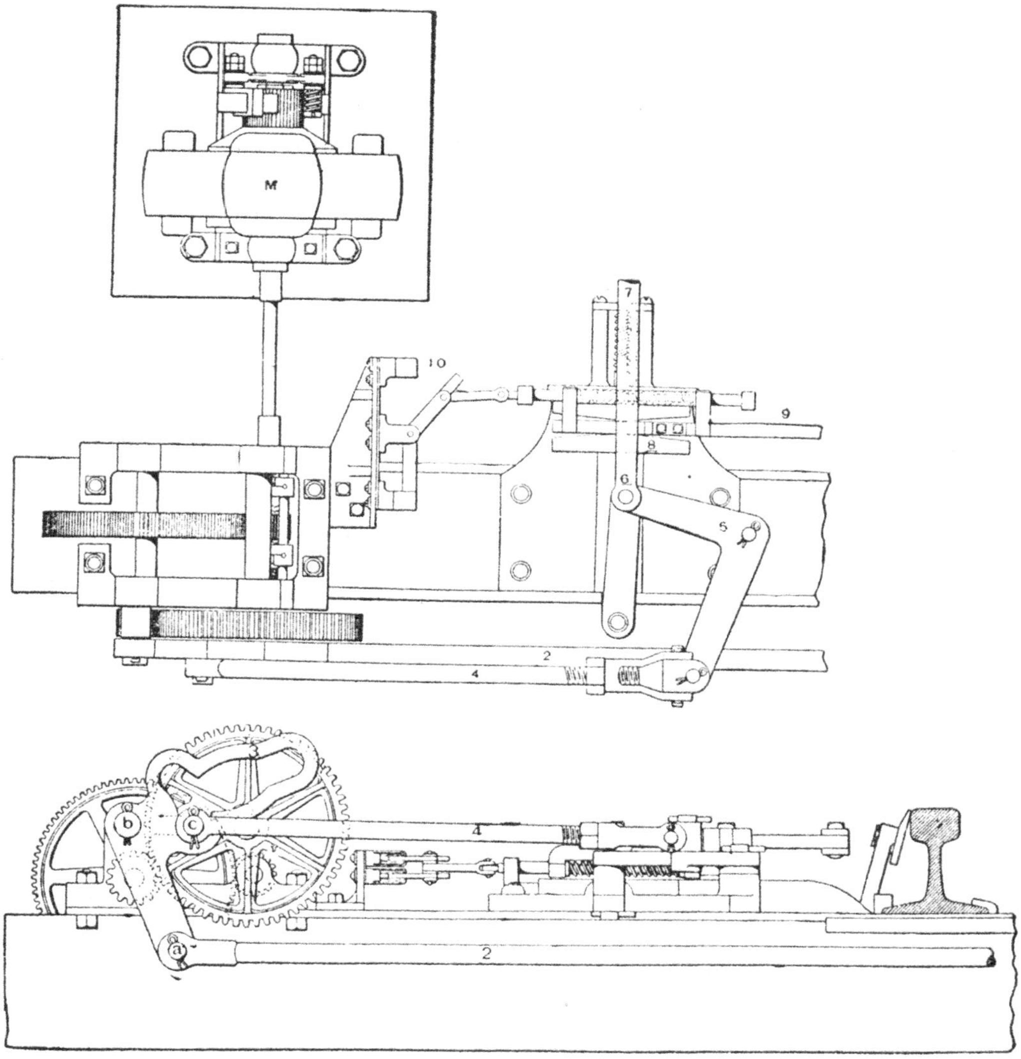

Fig. 411.

6 (7) may be attached, if desired, a crank to work a mechanical detector bar. In the elevation a detector bar is shown. The lock bolt 6 is withdrawn from lock rod 9 before rod 2 begins its stroke, and it is reinserted

after 2 has come to rest. The switch being locked, the final movement of the lock bolt causes the reversal of electric switch 10 which opens the power circuit and closes the indication circuit. The power for moving this circuit closer or switch is furnished by the spiral spring encircling its rod. The spring is compressed by the movement of rod 9 and, at the proper moment, by the final movement of lock bolt 6, is released so as to close 10.

When a switch movement is completed and the motor circuit is broken the motor is at the same moment converted into a generator. This is done by closing the indicator circuit (at 10) and this current, generated by the motor, lasting only a fraction of a second, is what gives the indication.

The movement of the switch back to its original position is accomplished by turning wheel 1 in the opposite direction, the movements of the cam being the same as before, but in reverse direction. The polarity of the motor was changed (by the preceding movement) at 10. If a motor should run too long it is automatically thrown out of gear. For single switches or derailing switches a 1-h.p. motor is used, but the power required is usually only 7 amperes at 60 volts.

A signal motor is 1-6 h.p. It puts its signal in the all clear position by winding up a chain which lifts the weighted end of the balance lever, pushing up the signal rod. This winding up requires about 2 amperes.

When a signal has been put in the clear position a pole changer operates as at 10 on the switch machine, closing a circuit through a brake magnet, to hold the signal down.

The motor is at the same time de-energized. To return the signal to the stop position the brake magnet

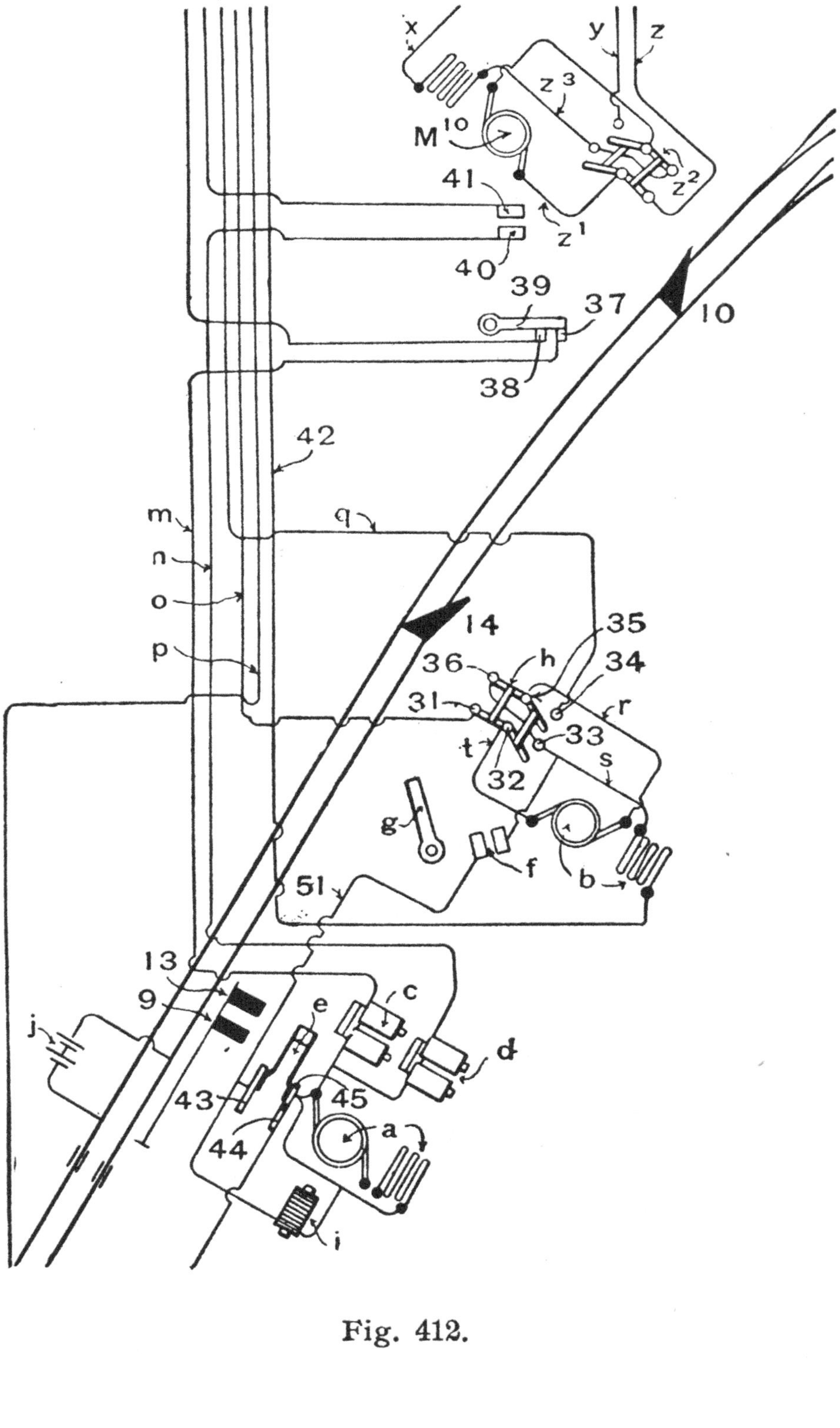
x
y
z
z^3
M^{10}
z^2
41
z^1
40
39
37
10
38
42
m
n
o
p
q
14
35
36
h
34
31
r
t
33
s
32
g
f
b
51
13
9
c
e
j
d
43
45
44
a
i

Fig. 412.

is de-energized and the counter weight causes the semaphore blade to take the horizontal position.

The brake magnet requires only .1 ampere.

In Fig. 412—(A) is a motor for two signals, and **b** is a motor for a derailing switch (14) on a track approaching a grade crossing of another railroad.

The signal is fitted with two arms for the reason that there is a diverging track (10) just before reaching the crossing.

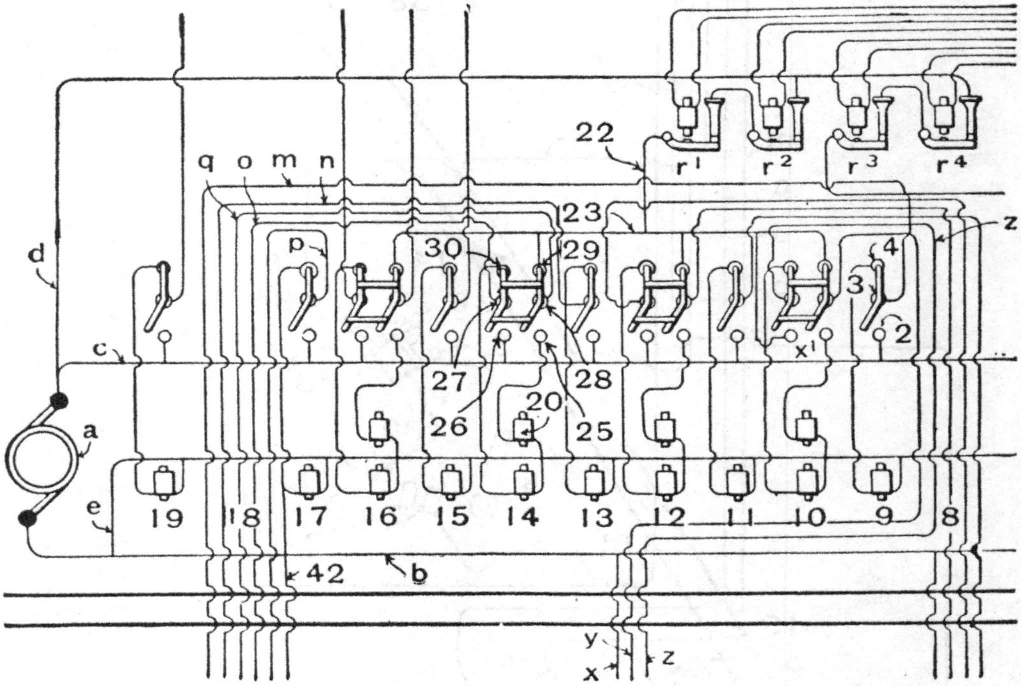

Fig. 413.

The wires m, n, o, p, q, x, y, z and 42 lead to the machine in the cabin, which is described in connection with Fig. 415.

Magnets c and d are fixed in the signal post, and one or the other depending upon which circuit is closed, causes a hook to engage a rod for working the proper signal.

Referring again to Fig. 412—(e), is the circuit closer for opening the motor circuit, and closing that to the electric brake, or vice versa.

At *f* are shown contact points inserted in the main wire, as an additional safeguard to insure that a high

Fig. 414.

speed signal shall never be given with a facing-point switch or a derail open. When the switch or derail is closed the points *f* are closed by *g*. The pole changer is shown at *h; i* is the brake magnet and *j* is the track circuit battery.

In Fig. 413—(a) is the generator (or storage battery) for supplying current to the principal circuits; *b, c, d* and *e* are principal leads out; *m, n, o, p* and *q* correspond with the same letters in Fig. 412; r^1, r^2, r^3, r^4 are track relays.

A view of an interlocking machine is shown in Fig. 414. In this cut "levers" (bars) with handles extending upward are for switches, and those with downward handles are for signals. A switch lever, when moved, opens two circuits and closes two; the signal lever opens one and closes one.

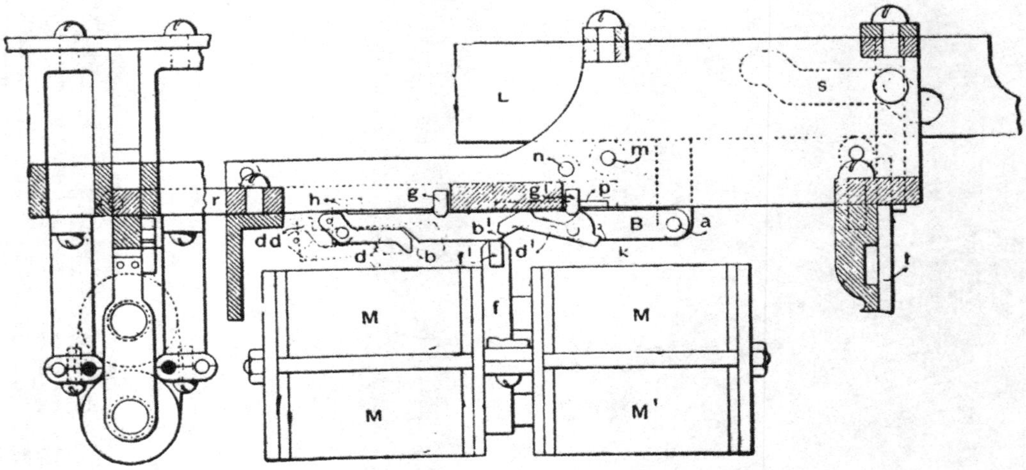

Fig. 415.

The vertical tappets which do the interlocking are actuated by the movements of a slot in the lever. (See *s*, Fig. 415.) Fig. 415 is an outline drawing of a "lever" and its connections (with handle broken off).

The "indication" acts by means of the solenoids *M* and M^1, Fig. 415. When a switch (having completed its stroke) energizes one or the other of the pairs of solenoids *M* or M^1, the movement of the armature *f* controls the bar *B*, supported by *L* and pivoted at *a*. The armature *f* is rigidly fastened to the cores of the magnet.

In Fig. 415 the lever *L* has been pushed in about three-fourths of its stroke. The first part of the stroke, by the action of the upper left hand portion of slot *s* has depressed tappet *t* far enough to lock all conflicting levers. During the intermediate part of the stroke, while the pin of the tappet is in the horizontal position of slot "s", the lever "L", through bars "B" and "r", has changed the circuit closer, or controller from the reverse to the normal position.

The terms reverse and normal are used in the sense common in interlocking practice.

The controller being normal it sends a current to the switch motor changing the track rails to normal. This being done the "indication" sent back from the switch, as before described, energizes magnet M^1.

This attracts armature "f", and permits bar "B" to drop out of notch "h", by which it moves bar "r". The operator can then push "L" the remainder of its stroke forcing tappet "t" farther down, and thus unlock such levers as it is proper to unlock after the movement of this switch.

In case the "indication" should not come, as would be the case if the switch should fail to move "home", and "L" is pulled back towards the position from which it started, the pawl *d* engages the lug f^1 on the side of armature *f* and pushes the armature to the right; *d* is then lifted over f^1 by striking against another lug, *g*, extending downward from the frame. The pushing of *f* to the right prevents *B* from dropping without being released by an indication current. If there is a failure to indicate on this side also, and the operator makes a second attempt to move the switch by again pushing in *L* the dog d^1 will push *f* to the left. Bar *B* can never drop except after *M* or M^1 has pulled *f* out from under the point b^1 or *b*. It is possible to move the lever inward to the full extent of its stroke after having moved it partly outward, but it is impossible to move it through the last part of its stroke in either direction without getting a release by means of the "indication."

The dotted lines at *dd* indicate the position of bar *B* when it is disengaged from *r*. Rollers *m* and *n* support or guide the sliding bars. The function of the recess at *p* is to limit the motion of sliding bar *r*.

The movement of the switch in the opposite direction is effected in the same manner as here described, but by pulling instead of pushing lever *L*.

The dwarf signal Fig. 416, is worked by a pair of solenoids *S*, S^1. These, being energized by a current from the cabin, lift the connecting bar *B*, force up the signal rod *R* and clear the signal. A spring in the column, *C*, returns the arm to the horizontal position when the current is withdrawn. In this signal the indication current is closed by the contact points shown between the solenoids. Just before the signal reaches the clear position the low-resistance winding of the solenoid is automatically put in series with the high resistance, so that the signal is held up at all clear by ¼ ampere, although a current of 6 amperes was used to move it.

The arrangement of the connections between a lever and its switch or signal can be seen by reference to Figs. 412 and 413. Take switch number 10. On closing the contact points at "x", Fig. 413 (lever number 10), current from the supply "a", flows through wire "d", forward stops, and armatures of relays "r^2" and "r^1" (in series) contacts of circuit controller "x^1", wire "z", pole changer "x^2" Fig. 412, motor of switch number 10, wires "x" and "b", back to generator *a*. Motor M^{10}, being actuated, moves the switch. The movement of the switch changes the connection at z^2, so that the next movement of lever 10 will send a current through *y* instead of *z*.

Fig. 416.

The control of this lever by the rail-circuit is effected through track relay r^2, Fig. 413, which is actuated by a current from battery *j*, Fig. 412, which flows through the track from signal 9-13 to the crossing beyond switch 10. The armature of r^2, when dropped, in consequence of the presence of a train on the track between *j* and the crossing, opens the circuit through which the signalman works switch No. 10.

Both the "normal" and "reversing" circuits of switch No. 10 are taken through the contacts of the relays r^1 and r^2 because a train may pass over this switch in either of its positions. Only the "normal" circuit of a derailing switch is taken through the relay contacts because a train is not to run over a derail when "normal" and it may be desirable to "reverse" it with a train on the track section.

The connections to switch 14 and signal 9-13 may be traced in Figs. 412 and 413. To reverse switch No. 14, or in other words to set for the main track, lever 14 is reversed (pulled out) which puts the arms 27 and 28, Fig. 413, in contact respectively with blocks 26 and 25. This closes a circuit from generator *a* through wire *c,* block 26, arm 27, wire *o,* contact block 31 (Fig. 412) arm 32, armature of motor *b,* arm 35, block 36, wire *s,* field coils of motor *b,* wire 42, back to generator *a.* When the switch is completely closed and locked the pole changer *h* is thrown over so that arms 32 and 35 are taken away from 31 and 36 and put into contact with the blocks 33 and 34 respectively. This closes a circuit from the switch motor to indication magnet 20 (see Fig. 413) through wire "r", arm 35, block 34, wire "q", arm 28, block 25, magnet 20, wires "e" and 42 to field coils of motor "b".

This energizes magnet 20 and releases latch "B" (see Fig. 415) so that lever 14 can be pulled to the end of its stroke.

This final movement of lever 14 releases lever 9. Reversal of lever 9 puts the arm of the electric controlling switch 3 (Fig. 413) into contact with block 2.

This closes a circuit from generator "a", through wire "c", block 2, arm 3, wire *m*, contacts 38, 39, 37, controlled by the rails of switch No. 10; it continues through wire *m*, magnet *c*, signal motor *a* and signal circuit controller *e*, wire 51, controller *f g*, operated by switch No. 14, wire *s*, field coils of switch motor *b*, wire 42, back to the generator *a*.

When the signal arm has moved to all-clear the circuit controller *e* is moved so that it no longer connects 43 and 45; but the above described circuit is shunted around this break by the brake magnet *i*. This holds the signal in the all-clear position as long as lever 9 remains reversed. The movement of the circuit controller *e* when the signal goes to clear, connects contacts 43 and 44, thus closing the circuit to the distant signal. This circuit thus depends on the home signal being clear.

When lever 9 is put back into its normal position arm 3 is put into contact with block 4. Then as the signal returns to its normal position the last part of the movement of the blade replaces circuit controller *e* so that it connects 43 and 45. This closes the circuit containing the signal motor *a* and the indication magnet 9 in the cabin. The signal motor armature is rotated by the fall of the blade, and the current generated by this rotation flows through wire *m*, arm 3, block 4, magnet 9, wires *e* and 42, field *b*, (Fig. 412), wire *s*, points *f* and 43, circuit controller *e*, field coils of the signal motor *a* and back to the armature. This current serves the purpose of retarding the fall of the counter-weight of the signal and also of energizing the indication magnet 9 (Fig. 413) which releases the latch and permits lever 9 to be moved through the final portion of its stroke.

QUESTIONS.

Ques. 131. Of what does the Taylor Co.'s signal system consist?

Ques. 132. How is the switch movement protected, and where is it located?

Ques. 133. Where are the signal motors located?

Ques. 134. What takes the place of detector bars, in the system?

Ques. 135. From whence is the current supplied for operating the device?

Ques. 136. How is the signalman informed when the stroke of the switch rails is completed?

Ques. 137. When a switch movement is completed, and the motor circuit is broken, what occurs to the motor?

Ques. 138. How is this done?

Ques. 139. If a motor runs too long what happens to it?

Ques. 140. What is the power of the switch motors?

Ques. 141. What is the power of a signal motor?

Ques. 142. How does it put the signal in the "all clear" position?

Ques. 143. How much current does this winding up require?

Ques. 144. When a signal has been put in the clear position, what then occurs?

Ques. 145. How is the signal returned to the stop position?

STANDARD SEMAPHORES.

Fig. 417.—The Chicago & Northwestern R'y Standard Semaphores. These semaphores are singular in the colors adopted for night signals, and in the fact that the home signals on this road are always operated by pipe; the distant signals, like those on almost all other roads, are operated by wire.

The home signal in the danger position exhibits a red light, while in the clear position a green light is shown.

The distant signal carries a peculiarly shaped lamp, which has on its face two openings; the one on the left is filled with a green glass, and is never obscured, while the opening on the right is protected with a white lens, which, however, is never visible. The upper hole in the distant signal semaphore casting carries a red glass, the lower a metallic shield; therefore, when a distant signal is in the danger position two lights are visible, a green and a red, but when the signal is in the clear position, only the green light is visible, owing to the presence of the metallic shield in front of the white lens.

The practical result of this method of lighting is an entire avoidance of the danger due to the breaking of the colored glasses ordinarily carried in semaphore castings, for in this system a white light is not construed as a signal at all.

This method was conceived by Mr. E. C. Carter, Principal Assistant Engineer of the Chicago & Northwestern R'y.

Fig. 417.

Fig. 418.—Station Block Signals. The figure on the left illustrates a standard form of block signal as it is arranged when placed in front of a telegraph office or block signal station. The two blades control trains run-

Fig. 418.

ning in opposite directions and by the use of a semaphore casting, one lamp only is required. This signal may be operated either by pipe or wire, although when located near the cabin, wire is much the cheaper arrangement, and is quite strong enough.

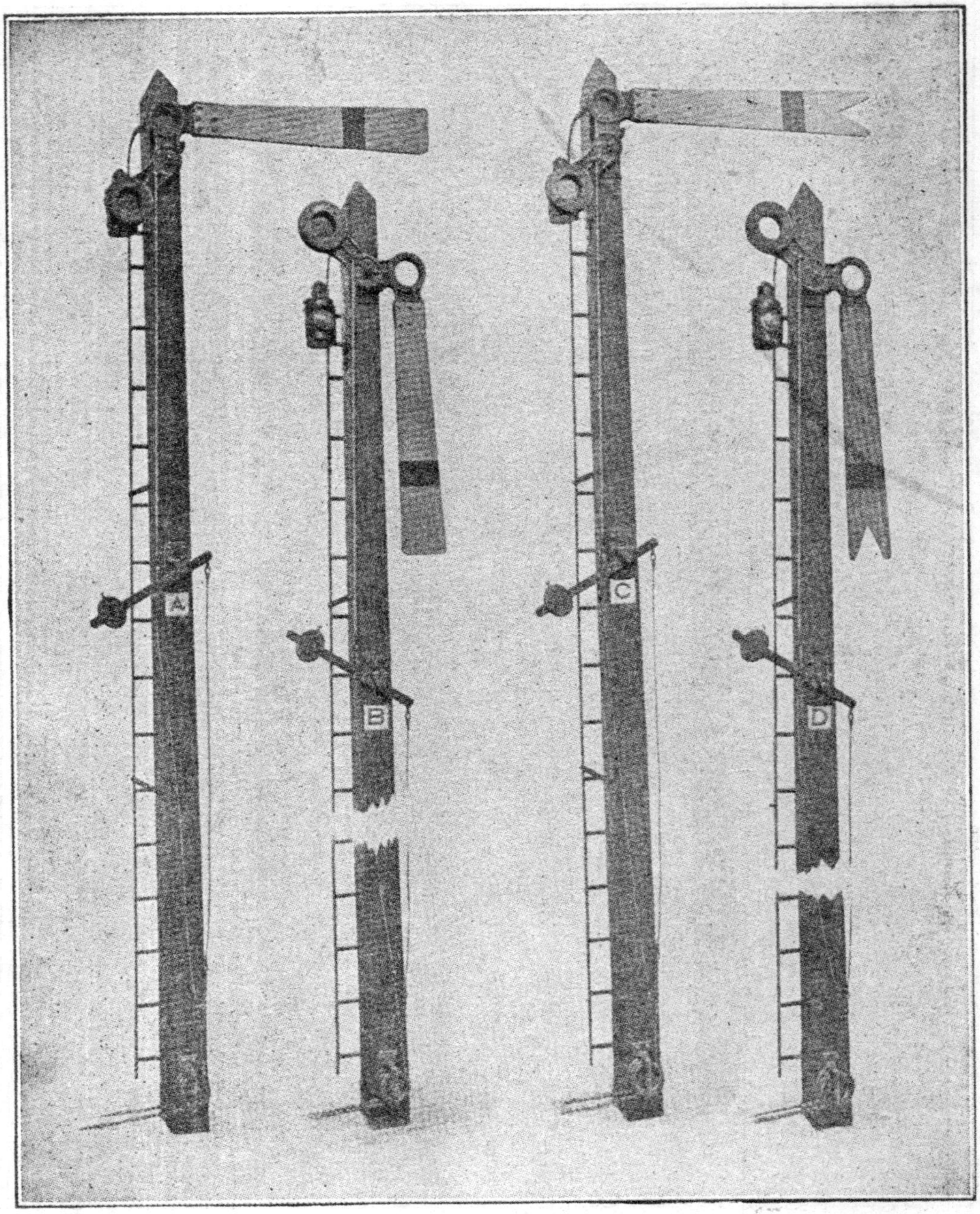

Fig. 419.

On the right is illustrated the block signal in use on the Pennsylvania Lines West of Pittsburg. The blade pointing to the right occupies the cautionary position, which is made possible by the great angle (90°), which separates the danger (horizontal) and clear (vertical) positions. One lamp only is required with each of these double signals, since it is placed on top of the post and shines in both directions through the ring attached to the top of the semaphore casting.

Fig. 419.—Standard Semaphores of the Pennsylvania Lines West of Pittsburg. The main objects of this system, which was designed by Mr. W. McC. Crafton, Signal Engineer of the above-named road, are: first, to

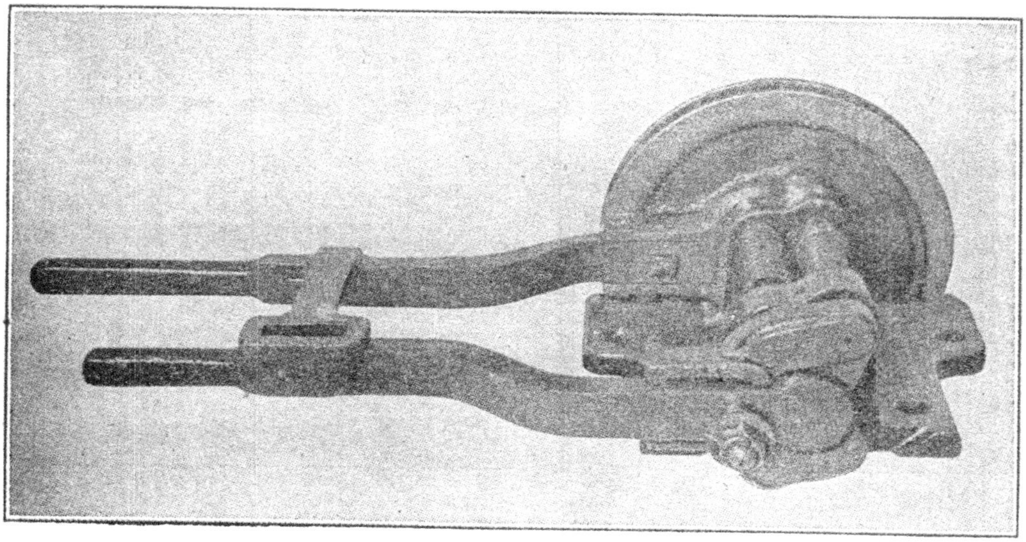

Fig. 420.

secure a more consistent signal than the ordinary one, by painting the blades a negative color and depending for their identification solely on form, thus doing away with the contradiction incident to passing a clear signal that is painted red; second, to make a greater distinction between a danger and a clear signal, by increasing the arc through which the blade must travel.

The ordinary night signals, red for danger, green for caution, and white for safety, are retained.

Fig. 420.—Double Ground Lever. This device is designed for throwing a switch or cross-over and distant signals at outlying points on the main line of a railroad, and is of particular value on curves or at other places where the switch lights cannot for any reason be seen.

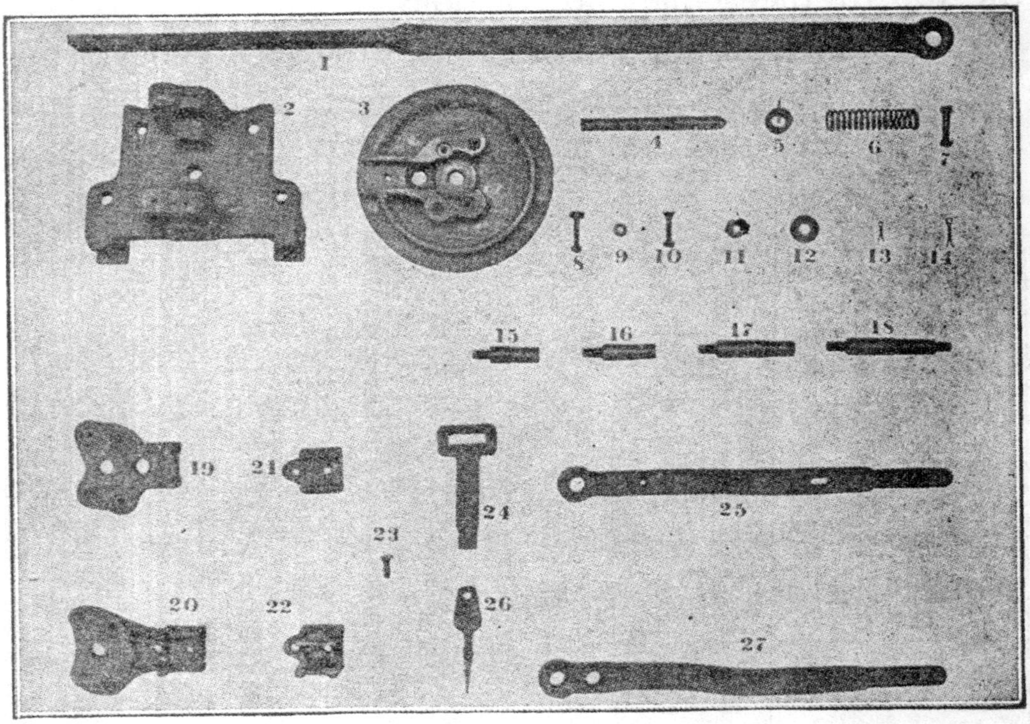

Fig. 421.

It is positively interlocked in such a manner that the signals cannot be cleared until the switch is right for the main track, and, conversely, so that the switch cannot be moved from the main track route until the signals have been thrown to danger.

Fig. 421.—Parts of Double Ground Lever.

1. Lock Bar.
2. Base.
3. Chain Wheel.
4. Locking Pin.
5. Collar and Dowel Pin, for bearing of No. 6.
6. Spring, for No. 4.

7. Bolt and Nut, ½"x2½", for fastening No. 25 to No. 3.

8. Bolt and Nut, cheese head, for filling lock holes in Nos. 3, 19 and 20.

9. Washer, ⅜", for Nos. 8 and 10.

10. Bolt and Nut, countersunk head, for filling lock holes in Nos. 3, 19 and 20.

11. Hex Nut.

12. Washer ¾".

13. Split Cotter, 5-32"x1¼".

For left-hand ends of Nos. 15, 16, 17 and 18.

14. Split Cotter, 7-32"x2", for right-hand end of No. 18.

15. Stud, for switch rod connection on No. 27; this also helps to join Nos. 19 and 27.

16. Stud, used only when a pipe connection is desired with No. 3.

17. Stud, for joining Nos. 2, 19 (or 20), and No. 27.

18. Stud, for joining Nos. 2, 3 and 25.

19. Switch Lever Casting, this is fastened to No. 2, and is used in connection with No. 27.

20. Switch Lever Casting, this is fastened to No. 2, and is used in connection with a separate ground lever, in which case No. 27 is not provided.

21. Clip, used with No. 20, when the separate ground lever is of flat iron.

22. Clip, used with No. 20, when the separate ground lever is of round iron.

23. Rivet, for fastening No. 21 or No. 22 to No. 20.

24. Cross Arm, for No. 25.

25. Signal Lever.

26. Staple.

27. Switch Lever.

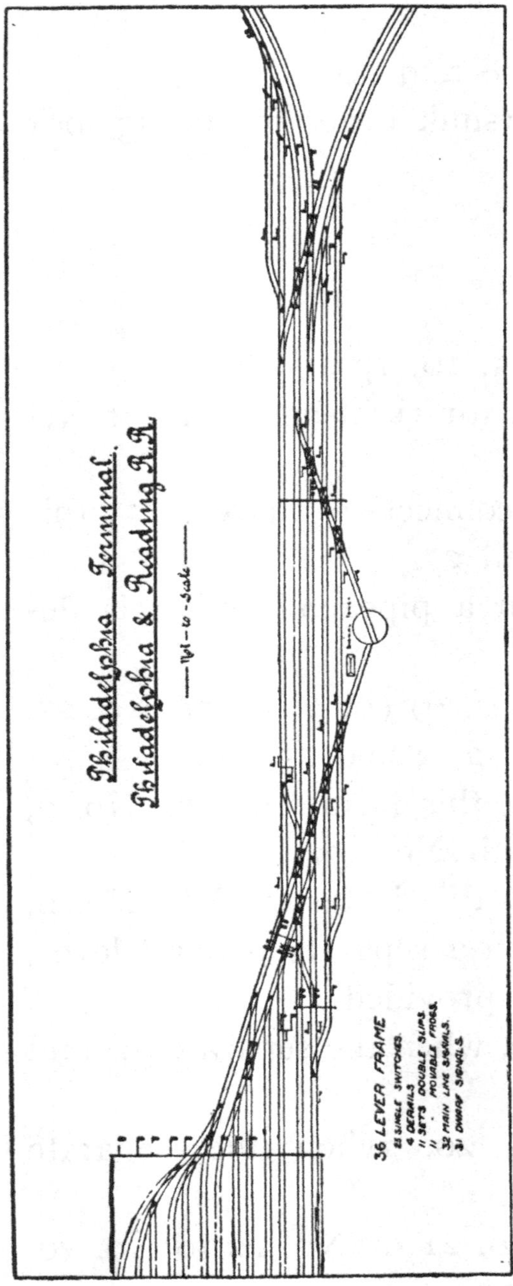
Philadelphia Terminal.
Philadelphia & Reading R.R.
Not-to-Scale
36 LEVER FRAME

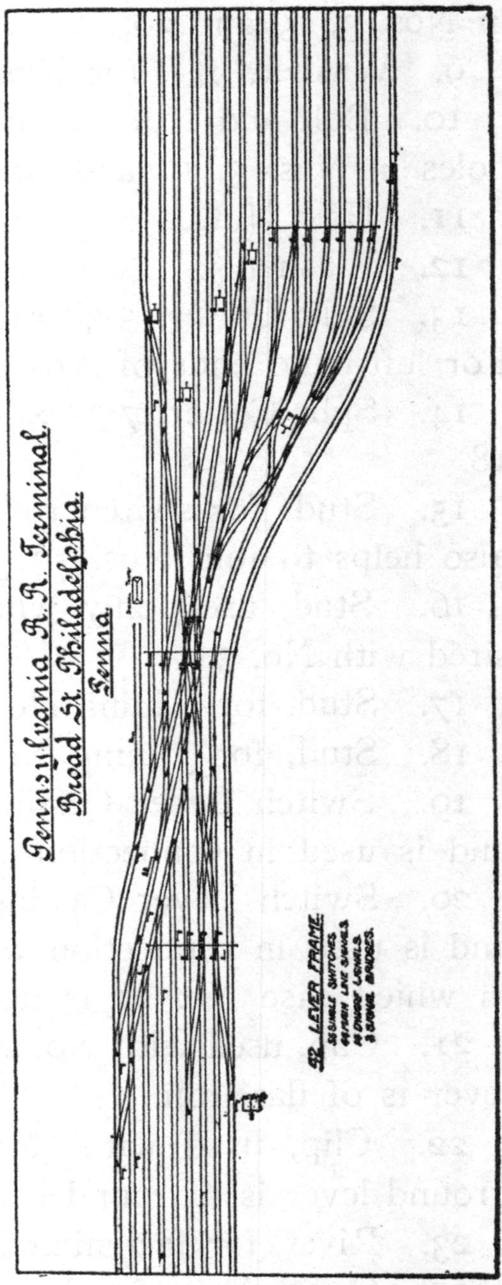
Pennsylvania R.R. Terminal
Broad St. Philadelphia.
Penna.
92 LEVER FRAME

Fig. 422.—Triple Ground Lever. The Triple Ground Lever is an interlocking machine with the Stevens locking attached to it. It is of great value where one or two switches and one or two signals must be operated, and where it is not desired to maintain a signal man. It is used at the end of double track and out-lying crossovers. Since the end of double track is frequently moved, the interlocking device, through the simplicity of its construction, can easily be moved with it.

Fig. 422.

Fig. 423.—Parts of Triple Ground Lever.

1. Base.
2. Top.
3. Shaft, 1¼" diam.
4. Set Screw, ⅜"x¾", for shaft.
5. Locking Bar, ⅜"x1¼".

6 and 7. Two forms of Locking Dog.

8. Rivet, ¼"x1", for locking dog.
9. Pin, ½"x3", for fastening lever.
10. Chain,
11. Rivet,

for attaching pin No. 9.

12. Bolt, ⅝″x1½″, for fastening top.
13. Lever.
14. Rack.
15. Pinion.
16. Bolt, ½″x1¼″, for pinion.

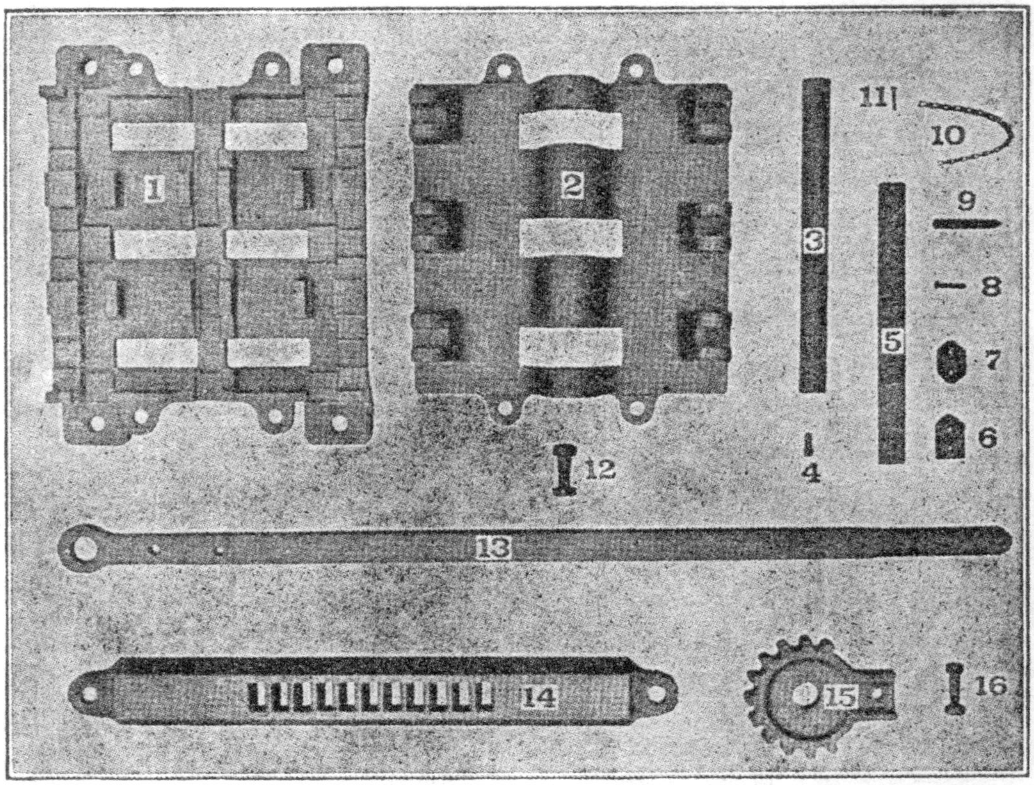

Fig. 423.

Fig. 424.—Multiple Bridge Coupler. The Multiple Bridge Coupler accomplishes the movement of several signals and switches on shore by a machine located on a revolving drawbridge. The figure on the left shows the position of the coupler when the drawbridge is closed and one of the levers in the machine is reversed. The figure on the right shows the device uncoupled, the shore connections and bridge connections locked, and the drawbridge ready to swing. They may be made

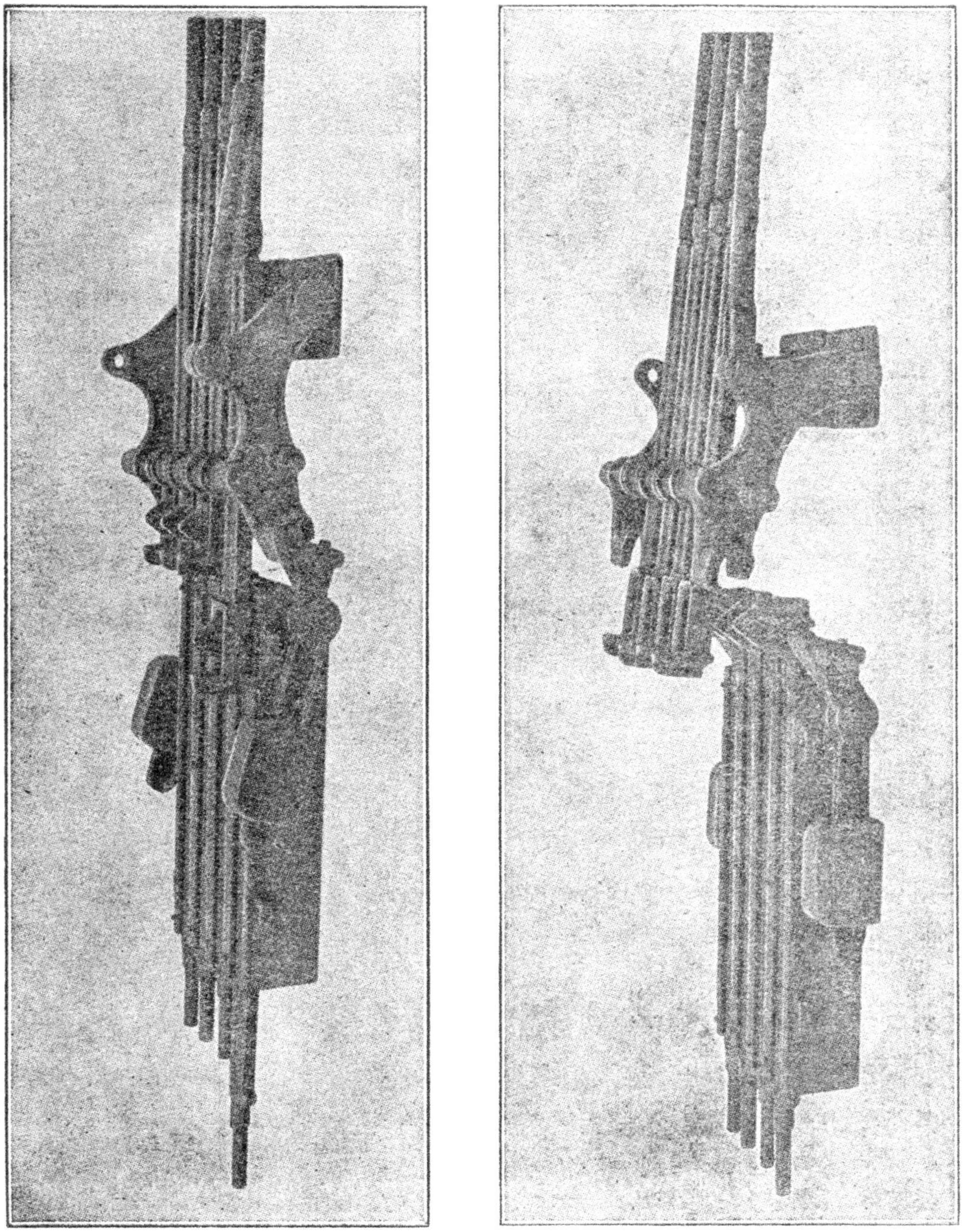

Fig. 424.

of considerable width, since the effort required to disengage the coupler is very slight. In large plants it

Fig. 425. View of tracks and signals at the Jersey City Terminal of the Pennsylvania Railroad.

Particular attention is called to the location of the starting signals, each of which is placed directly over the track that it governs.

is sometimes necessary to use several couplers, owing to the fact that the distance between the trusses and the rails is occasionally quite limited.

This is a device, the action of which is absolutely reliable under all of the conditions of traffic or temperature.

OTHER BOOKS FROM CGR PUBLISHING AT CGRPUBLISHING.COM

1915 San Francisco World's Fair in Color: The Grandeur of the Panama-Pacific...

San Francisco 1915 World's Fair: The Panama-Pacific International Expo.

1904 St. Louis World's Fair: The Louisiana Purchase Exposition in Photographs

Chicago 1933 World's Fair: A Century of Progress in Photographs

19th Century New York: A Dramatic Collection of Images

The American Railway: The Trains, Railroads, and People Who Ran the Rails

The Aeroplane Speaks: Illustrated Historical Guide to Airplanes

The World's Fair of 1893 Ultra Massive Photographic Adventure Vol. 1

The World's Fair of 1893 Ultra Massive Photographic Adventure Vol. 2

The World's Fair of 1893 Ultra Massive Photographic Adventure Vol. 3

Henry Ford: My Life and Work - Enlarged Special Edition

History of the Crusades Volumes 1 & 2: Gustave Doré Restored Editions

Ethel the Cyborg Ninja Book 1

The Complete Ford Model T Guide: Enlarged Illustrated Special Edition

How To Draw Digital by Mark Bussler

Best of Gustave Doré Volume 1: Illustrations from History's Most Versatile...

OTHER BOOKS FROM CGR PUBLISHING AT CGRPUBLISHING.COM

Ultra Massive Video Game Console Guide Volume 1

When Trolleys Ruled the Earth: A Photographic History of Streetcars...

Complete Works of Edgar Allan Poe Volumes 1 - 10

Classic Cars and Automobile Engineering: Volumes 1-5

Antique Cars and Motor Vehicles: Illustrated Guide to Operation...

Chicago's White City Cookbook

The Clock Book: A Detailed Illustrated Collection of Classic Clocks

The Complete Book of Birds: Illustrated Enlarged Special Edition

1901 Buffalo World's Fair: The Pan-American Exposition in Photographs

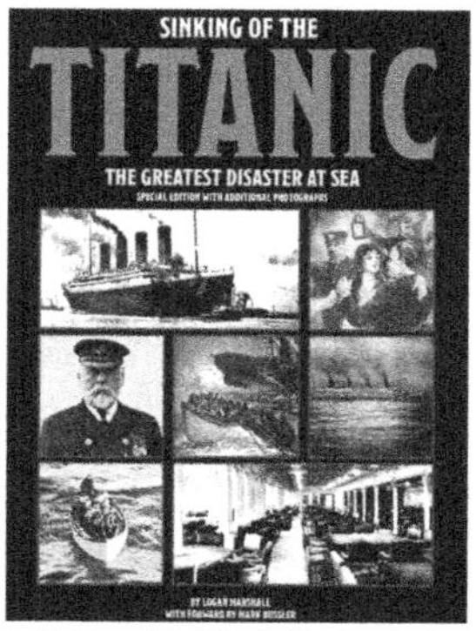

Sinking of the Titanic: The Greatest Disaster at Sea

Gustave Doré's London: A Pilgrimage: Retro Restored Special Edition

Milton's Paradise Lost: Gustave Doré Retro Restored Edition

The Art of World War 1

The Kaiser's Memoirs: Illustrated Enlarged Special Edition

History in the Age of Vikings Series Volumes 1-3

The Complete Butterfly Book: Enlarged Illustrated Special Edition

Made in the USA
Monee, IL
05 February 2026

43455720R00249